物理类专业系列教材

物理学专业英语简明教程

主编 仲海洋 王轶卓 田 莹

参编 付 姚 李 磊 车 丽
陈季香 田一平 鹿学军
孙 敏 李 燕 辛敏裕
张金苏 刘大军 董爱义

清華大學出版社
北 京

内容简介

本书是一本针对高等学校的物理学及其相关专业的专业英语教材。内容主要包括：专业英语的特点、词法、语法、翻译、阅读等专业英语知识的介绍；运动学、质点力学、刚体的定轴转动、振动与波动、电磁学、热学、光学、原子物理、量子力学、相对论等内容的选读，并附有文中出现的物理专业词汇的英汉对照词汇表，便于学生课前有针对性地查阅和掌握；对物理学专业英语中的一些常用表达，在每课中也以专题的形式进行了详尽汇总和举例说明，包括工具、仪器、设备、结构、原料、精度、性能、形状、方位、用途、状态、原因、结论等24个专题。最后在附录部分汇总了物理专业所开设课程名称的英文翻译、教务教学管理词汇、物理学一、二级学科名称英文词汇、数学符号及其英语表达、常用物理基本常数、常用汉英物理学专业词汇表、常用英汉物理学专业词汇表，供读者使用时查阅。

本教材适用于高等学校的物理及其相关专业的本科生、硕士生专业英语教学，也可用于理工科大学物理学的双语教学，也可作为科技工作者和物理学从业人员的参考书和工具书。

图书在版编目（CIP）数据

物理学专业英语简明教程/仲海洋，王轶卓，田莹主编．—北京：清华大学出版社，2022.11
物理类专业系列教材
ISBN 978-7-302-62040-2

Ⅰ．①物…　Ⅱ．①仲…　②王…　③田…　Ⅲ．①物理学－英语－高等学校－教材　Ⅳ．①O4

中国版本图书馆CIP数据核字(2022)第191951号

责任编辑：朱红莲
封面设计：傅瑞学
责任校对：欧　洋
责任印制：刘海龙

出版发行：清华大学出版社
网　　址：http://www.tup.com.cn，http://www.wqbook.com
地　　址：北京清华大学学研大厦A座　　**邮　　编**：100084
社 总 机：010-83470000　　**邮　　购**：010-62786544
投稿与读者服务：010-62776969，c-service@tup.tsinghua.edu.cn
质量反馈：010-62772015，zhiliang@tup.tsinghua.edu.cn
印 装 者：三河市君旺印务有限公司
经　　销：全国新华书店
开　　本：185mm×260mm　　**印　　张**：17.75　　**字　　数**：429千字
版　　次：2022年11月第1版　　**印　　次**：2022年11月第1次印刷
定　　价：58.00元

产品编号：096099-01

前言

在人类追求真理、探索未知世界的过程中，物理学展现了一系列科学的世界观和方法论，深刻影响着人类对物质世界的基本认识、人类的思维方式和社会生活，是人类文明发展的基石，在人才的科学素质培养中具有重要的地位。

随着科学的发展进入21世纪，物理学及其相关领域的国际教育与研究合作交流更加频繁，专业英语早已经是国际上交流的主要语言。国家教育部也提出了高等学校各专业逐步使用英文教材，以培养和增强学生阅读英文专业文献的能力。具备良好的物理学专业英语水平和使用能力是每个物理学及其相关领域的学生和从业人员必备的素质。

本书编者在物理学专业英语教学、理工科大学物理学的双语教学和从事科研和国际交流活动的过程中发现：尽管物理学专业英语在教学和科研工作中经常使用，但是却没有一本合适而全面的参考教材和可供查阅的工具书。很多专业英语课程的学习方式仅仅停留在复印一些英文材料进行阅读、解释和翻译的阶段。学生学习后只是提高了一些专业词汇量和阅读翻译水平，而在使用专业英语交流时，则力不从心。

本书编写的目的就是希望能够通过这本教材的编写，总结一些专业英语和双语教学方面的经验，并将这方面的笔记、资料系统地汇编起来，以各个专题的形式总结出来，供广大物理学及其相关专业的本科生、硕士生、博士生作为教材使用，也适合物理学及其相关领域的科技工作者和从业人员在使用专业英语进行学习和交流时翻阅参考。

作者在教材编写的过程中突出了以下主要特色：①内容选择注重实用性，不仅可以作为教材使用，也可以作为工具书进行查阅参考；② 内容全面精练，例句的选择专业性强，并且恰当贴切；③ 以专题的形式进行介绍或进行汇总，便于独立学习和查询参考；④对于专业英语阅读内容中的专业词汇，在每篇后的词汇表中以中英文对照形式给出；⑤使学习者通过专业英语的学习培养和拓展实际应用英语的能力，进而掌握相应的方法和技巧。

本书共24课，分为6个单元，各单元之间既相互独立也有一定联系。每一课都包括三个部分：第一部分主要介绍有关物理学专业英语方面的知识，通过大量的例句、实例和范文介绍了包括专业英语的特点、词法、语法、句法、翻译、阅读等方面的内容；其中在专业英语的翻译部分，主要介绍了专业词汇、独立主格结构、名词化结构、分隔结构和长句的翻译；第二部分为专业英语的阅读内容，选择了关于运动学、质点力学、刚体的定轴转动、振动与波动、电磁学、热学、光学、原子物理、量子力学、相对论等专业文献的选读，为了读者查阅和节省查字典的时间，在每篇之后按照文中专业词汇出现的顺序编写了英汉对照词汇表。第三部分汇总了物理学专业英语中的一些常用表达，共24个实用的专题内容供参考查阅，包括工具、仪器、设备、结构、原料、精度、性能、形状、方位、用途、状态、原因、结论方面内容；尤其在每个单元的头一课中，汇总的都是在物理实验方面涉及的工具、仪器和设备等实用的内容。附录部分共汇总了7个方面的内容:物理专业所开设课程名称的英文翻译、教务教学管理词汇、物理学一、二级学科名称英文词汇、数学符号及其英语表达、常用物理基本常数、常用汉英物理学专业词汇表、常用英汉物理学专业词汇表。

本书第一、第二单元由仲海洋、李磊、孙敏编写;第三、第四单元由王轶卓、车丽、陈季香编写;第五、第六单元由田莹、付姚、田一平编写。练习和附录部分由鹿学军、李燕、辛敏裕、张金苏、刘大军、董爱义编写。夏文文、杨帆参加了全书的校对工作,深表感谢。

我们还参考了国内外出版物(见本书后面的"参考文献")中的部分观点和内容,在此谨向这些编著者致以真诚的谢意。此外,我们还选用了互联网上的部分资料,由于这些资料涉及面广,选用时作了大量修改,在此不便一一注明,谨向有关人士深表谢意。

由于作者水平有限,错误和不足是难免的,请读者和同行多指教,以利于再版时修正和改进。

仲海洋

2022年9月

目 录

Unit One

LESSON 1

1.1 物理学专业英语的特点

1.1.1 物理学专业英语简介

从体裁上讲，英语主要有文学文体、政论文体、应用文体和科技文体四大类，物理学专业英语属于科技英语文体的一部分。

物理学专业英语把英语和物理学专业知识紧密结合起来，用专业语言来说明客观存在的事物或事实。专业英语的叙述要求客观、真实、明确、简洁，并大量使用科学术语和符号，也经常借用图表和插图来说明专业内容。这就决定了专业英语的文体风格：概念准确、表达正式、陈述客观、逻辑性强、专业性强、严谨周密、行文简练、重点突出、句式严整、少有变化、常用前置性陈述，即在句中将主要信息尽量前置，通过主语传递主要信息。

专业英语的语言特点表现在词汇、句法和修辞三个方面。

1.1.2 物理学专业英语的词汇特点

1. 纯科技词汇

所谓纯科技词汇是指那些只用于某个专业或学科的专门词汇或术语，如 hydroxide(氢氧化物)、diode(二极管)、isotope(同位素)等。随着物理学及其相关领域的发展，新学科、新分支的产生，这样的词汇层出不穷，其词义精确而狭窄，针对性极强。除少数术语是新造的词外，绝大多数术语都是在原有旧词的基础上，靠借用、加前后缀或合成等手段构成，其中尤以合成这种手段最为常见。专业英语中很多新术语都是用这几种手段组成的。阅读专业性强的文献，就要首先了解该领域的专门词汇和术语。

2. 通用科技词汇

通用科技词汇即不同专业都要经常使用的那些词汇，数量较大，如表示时间、空间、方位、方向、尺寸、形状、面积、体积等方面的词汇以及一批要求固定介词的形容词和动词。如 accumulate，accuracy，capital，cell，charge，current，load，intense，motion，operation，potential，pressure，react，reflection，resistance，revolution，tendency 等词汇均属于多门学科使用的半专业词汇。这类词的使用范围比纯科技词汇广、出现频率高，但在不同的专业里有较为稳定的词义。如 power 一词在物理学中的词义为“电”“电力”“动力”“电源”“功率”等，在数学中的词义为“乘方”“功效”“幂”等。

3. 派生词汇

派生词指在已有的词汇上加前、后缀，或以词根生成，或以构词成分形成的新词。这种词汇在物理学专业文献中占有很大的比重。例如，由前缀 hydro-，hyper-，hypo-和 inter-构成的词条在物理学专业英语中就非常多；以表示学科的后缀-logy，-ics 和表示行为、性质、状态等的后缀-tion，-sion，-ance，-ence，-ment 构成的词汇在专业英语文献中俯拾皆是。

前缀的特点：加前缀构成的新词只改变词义，不改变词类。

intergalactic　　银河间的　　inter＋galactic

submicroscopic　　亚微观的　　sub＋microscopic

另外，前缀有固定的意义。如：

multi-表示"多"　　multimedia 多媒体

hyper-表示"超级"　　hyperelastic 超弹性的

后缀的特点：加后缀构成的新词可能改变也可能不改变词义，但一定改变词类。

infinity　无穷远(名词)＝infinite＋y(infinite 是形容词)

quantization　量子化(名词)＝quantize＋ation(quantize 是动词)

有些词加后缀后，语音或拼写可能发生变化。

electrification＝electrify＋cation

accelerator＝accelerate＋or

4. 复合词汇

所谓复合词，就是指两个或两个以上的词组合在一起构建的新词。一般来说，复合词分为复合名词、复合动词、复合形容词和复合副词。有些词已经在词义上发生了变化，从意思上看不出复合的痕迹，但是从词形上仍然能够看出最初复合的痕迹。复合词常以连字符"-"连接两单词构成，或采用短语构成，但有的则去掉连字符形成一个单一的词。

以下是物理学中的一些加连字符复合词的例子：

by-pass 旁路，cathode-ray tube 阴极射线管，change-over switch 换向开关，current-carrying conductor 载流导体，digital-to-analogue conversion 数模转换，fine-adjustment 微调、细调，flat-bottomed flask 平底烧瓶，fly-back 回扫，full-wave rectifier 全波整流器，half-wave rectification 半波整流，infra-red ray 红外线，inverse-square law 平方反比定律，ion-pair 离子偶、离子对，push-button switch 按钮开关

值得一提的是，专业英语中-proof 使用的比较多，它是一个构词成分，意思是"防……的，能抗……的""完全地或成功地抵制……的；不能穿透……的"。这个构词成分能构成许多有用的单词。例如：

a water-proof material　　防水材料

a bullet-proof glass　　防弹玻璃

a sound-proof room　　隔音室

a bomb-proof shelter　　防空洞

a fire-proof material　　防火材料，耐火材料

a fail-proof method　　万无一失的方法

a fool-proof instrument　　操作十分简便的仪器

此外还有 acid-proof 抗酸的、耐酸的，corrosion-proof 抗腐蚀的、防腐蚀的，mar-proof 耐磨损的，ozone-proof 耐臭氧的等。

以下是一些无连字符复合词的例子：

joulemeter 焦耳计，kilowatt 千瓦特、千瓦，loudspeaker 扬声器，output 输出，ratemeter 率计、率表，supersaturation 过饱和

5. 混成词

混成词指由最少两个词语或词语的一部分结合而成的词语，该新词语的意义和读音集组成部分而成。一般而言，混成词是新词，诸如 motel 汽车旅馆（motor＋hotel），smog 烟雾（smoke＋fog），brunch 早午餐（breakfast＋lunch）和 cyborg 生化人（普遍代指 cybernetic organism 自动化生物）等混成词。

以下是一些物理学中混成词的例子：

potentiometer (potential＋meter)	分压器、电位器、电势差计、电位差计
radioisotope (radioactive＋isotope)	放射性同位素
radionuclide (radioactive＋nuclide)	放射性核素
radiotherapy (radioactive＋therapy)	放射疗法
sonometer (sonic＋meter)	弦音计
spectrometer (spectrum＋meter)	光谱仪、分光计
thermocouple (thermal＋couple)	温差电偶、热电偶
telesat(telecommunication＋satellite)	通信卫星

6. 缩略词

缩略词的构成有两种：将英语中较长的单词取其首部或者主干构成与原来单词同义的短单词，称为压缩或省略；将组成词汇的短语的各个单词的首字母拼接为一个字符串，称为缩写。

1) 压缩或省略

一些单词比较长，难于记忆和拼写，通过压缩或省略的方法把它们压缩成一个短小的单词，或者仅仅取其头部，或仅取关键音节。

maths (mathematics)数学，lab (laboratory)实验室，plane (airplane)飞机，ft (foot/feet)英尺，cpd (compound)化合物

2) 缩写

利用词的第一个字母代表一个词构成的缩略词，就叫作首字母缩略词。常见的有以下三类。

① 通常情况下以小写字母出现，并且已经作为常规词汇。

m (metre)公尺，cm (centimeter)厘米，g (gram)克，fm (frequency modulation)调频

scr (silicon-controlled rectifier)可控硅整流器，p. s. i. (pounds per square inch)磅每平方英寸，radar(radio detecting and ranging 无线电探测与定位)雷达，laser(light amplification by stimulated emission of radiation 受激发射光放大器)激光

② 以大写字母出现，有的具有主体发音音节，有的仅为字母缩写。

TV(television)电视，CD(compact disk)激光唱盘，CAD(computer-assisted design)计

算机辅助设计，IT(information technology)信息技术，IDD(international direct dial) 国际直拨电话，PVC(polyvinyl chloride)聚氯乙烯，FRP(fiber glass reinforced plastic)玻璃钢，DNA(deoxyribonucleic acid)脱氧核糖核酸，F(fluorine)氟，U(uranium)铀，CATV(cable television)有线电视，CD-ROM(compact disk-read-only memory)光盘只读存储器(也就是光驱)，GHG(greenhouse gas)温室气体

③ 有的缩略词还可以和其他词连用，如：

E-mail (electric mail)电子邮件，H-bomb (hydrogen bomb) 氢弹，CO_2 (carbon dioxide)二氧化碳

7. 骈词

专业英语中有不少骈词，它们由同义词、近义词构成，两词前后顺序基本固定，中间由 and 连接，在语气、语调及语义方面有特殊的修辞效果。一般有以下三类：

1) 头韵类

如 effective and efficient, integral and indispensable, part and parcel, queries and questions 等。

It is clear that idealized model method provides an *effective and efficient* answer to a number of complex problems. 很明显，理想模型方法是解决许多复杂问题的一种快速高效的办法。

During the last twenty years, holography has become an *integral and indispensable* part of physics. 在最近二十年中，全息技术成为物理学不可或缺的部分。

Mechanics is *part and parcel* of physics. 力学是物理学的组成部分。

If there are *queries and questions* with laser, do not hesitate to contact us. 若对激光器有疑问，请立即和我们联系。

2) 尾韵类

如 first and most (foremost), wear and tear 等。

Calculus is the *first and most* common method of solving problems in physics. 微积分方法是最首要、最常用的解决物理问题的方法。

First and foremost, we should tackle the problem of energy sources. 首先，我们应该解决能源问题。

The *first and most* important step for learning university physics is to have a good command of calculus. 对大学物理学习来说，第一也是最重要的一步是学好微积分。

Regular maintenance of instruments reduces much needless *wear and tear*. 定期维护仪器可以大大减少不必要的磨损。

3) 无韵类

如 each and every, leaps and bounds, pure and simple, trial and error, ways and means 等。

This type of product manufacturing control ensures that consistent rare earth quality is achieved for *each and every* lot delivered to the labs. 这种产品制造控制方法，会保证发往实验室的每批稀土质量是稳定的。

Since the quantum was put forward, it has grown by *leaps and bounds* into a remarkably prosperous branch of physics. 自量子理论提出以来,它迅速发展为一个非常繁盛的物理学分支。

The operating error was due to carelessness *pure and simple*. 操作失误纯粹归因于粗心大意。

In general, the design procedure is not straightforward and will require *trial and error*. 一般来说,设计过程不是一帆风顺的,而是需要反复的试错。

The object of studying thermodynamic processes was to work out *ways and means* of improvements of the efficiency of heat engines. 研究热力学过程是为了找出改进热机效率的各种方法。

8. 专业词汇的隐喻

隐喻是把某种比喻意义从一个事物传给另一事物,即在两个本质不同的事物之间进行的暗含比较,是以相似和联想为基础的。

在物理领域,隐喻已经成为十分重要的形象思维手段和认知工具,词汇中充斥着大量的隐喻。原子结构理论的创立者丹麦物理学家 Henrik David Bohr(玻尔)把肉眼看不见的原子的内部结构想象成一个"太阳系",称为 miniature solar system(微型太阳系),从大的宏观概念直至小的微观概念。与这种视微观世界为宏观世界的放大化隐喻相对应,专业词汇中也存在视宏观世界为微观世界的缩小化隐喻,如天文学家创造了 light-year(光年)这一术语作为计算星球之间距离的单位,用 crab nebula(蟹状云)来指称银河系中的某一强大射电源。当科学家发现了新的事物及其特点、规律时,他们在很多情况下不会任意杜撰或"发明"新的词汇来表示新概念,而是在原有词汇的基础上给新的概念和知识命名。这其中相当一部分是隐喻,即概念与概念之间的类比。

从已知到未知,从具体到抽象,由此及彼,相互类推,就会形成一条"隐喻链"。同一词语应用在不同学科中建构了一系列的科学概念名称,展现出隐喻思维的线索。例如从 wave(波)类推到 sound wave(声波),再到 light wave(光波);到了电子时代,又有了 radio wave(无线电波),electromagnetic wave(电磁波),microwave(微波),ultrasonic wave(超声波)以及 long/medium/short wave(长/中/短波),及其到了量子时代的 matter wave(物质波)等。这些概念的建构一脉相承,都建立在同一喻体之上。

1.2 专业英语阅读

1.2.1 Introduction(引言)

We begin our study of the physical universe by examining objects in motion. The study of motion, whose measurement, more than 400 years ago gave birth to physics, is called kinematics.

Much of our understanding of nature comes from observing the motion of objects. In this chapter we will develop a description for the motion of a single point as it moves through space. Although a point is a geometrical concept quite different from everyday objects such as footballs and automobiles, we shall see that the actual motion of many objects is most easily described as the motion of a single point (the "center of mass"), plus the rotation of the object about that point. Postponing a discussion of rotation, let us begin here with a description of a single point as it moves through space.

1.2.2 Space and Time(时间与空间)

Kinematics is concerned with two basic questions, "Where?" and "When?". Though the questions are simple, the answers are potentially quite complicated if we inquire about phenomena outside our ordinary daily experiences. For example, the physics of very high speeds, or of events involving intergalactic distances or submicroscopic dimensions, is quite different from our common-sense ideas. We will discuss these interesting subjects in later chapters. For the present we shall adopt the space and time of Newton—those concepts we gradually developed as a result of our everyday experiences.

Space is assumed to be continuously uniform and isotropic. These two terms mean that space has no "graininess" and that whatever its properties may be, they are independent of any particular direction or location. In the words of Isaac Newton, "Absolute space, in its own nature, without relation to anything external, remains always similar and unmovable." Every object in the universe exists at a particular location in space, and an object may change its location by moving through space as time goes on. We specify the location of a particular point in space by its relation to a frame of reference.

Time, according to Newton, is also absolute in the sense that it "flows on" at a uniform rate . We cannot speed it up or slow it down in any way, in Newton's words, "Absolute, true, and mathematical time, of itself, and from its own nature, flows equably without relation to anything external, and by another name is called duration." Time is assumed to be continuous and ever advancing, as might be indicated by a clock.

Space and time are wholly independent of each other, though it is recognized that all physical objects must exist simultaneously in both space and time.

Remarkably, many of these traditional ideas turn out to be naive and inconsistent with experimental evidence. The world is just different from the picture we form from our common-sense, intuitive ideas. Space and time, by themselves, are concepts that are difficult (or perhaps impossible) to define in terms of anything simpler. However, we can measure space and time in unambiguous ways. We define certain operations by which we obtain numerical measurements of these quantities using rulers and clocks, based upon standard units of space and time.

For many years, our standard of time was based on astronomical observations of the earth's rotation. Because of the variations in the earth's rotation, in 1967 the 13th General Conference on Weights and Measures, attended by 38 nations, adopted an atomic standard for time.

Similarly, our former standard of length was the distance between two marks on platinum-iridium bar kept at Sevres, France. In 1960, the fundamental length standard was redefined in terms of the wavelength of light emitted during a transition between two atomic energy levels.

The standard units of time and length may be described as follows:

An interval of time. The fundamental unit is the **second** (s), which by international agreement is defined as the duration of 9 192 631 770 periods of radiation corresponding to the transition between the two lowest energy levels in the atomic isotope cesium 133.

An interval of length. The fundamental unit is the **meter** (m), which is defined independently of the time interval. Before 1983, by international agreement the meter was defined as exactly 1 650 763.73 wavelengths of the orange light emitted from the isotope krypton 86. In November 1983, the length standard was defined as the distance that light travels in a vacuum in 1/299 792 458 second.

Certain older units of length are still occasionally used.

$$1 \text{ angstrom}(\text{Å}) = 10^{-10} \text{ m} \quad 1 \text{ micron } (\mu \text{ or } \mu\text{m}) = 10^{-6} \text{ m}$$

Glossary

universe	宇宙	object	物体
measurement	测量	kinematics	运动学
motion of objects	物体的运动	center of mass	质心
space and time	时空	phenomena	现象
intergalactic	银河间的	submicroscopic	亚微观的
dimension	尺度	subject	研究的对象
uniform	均匀的	isotropic	各向同性的
continuously	连续地	graininess	颗粒性
direction	方向	location	位置
specify	规定	frame of reference	参考系
simultaneously	同时地	inconsistent with	与……不一致
define/definition	定义	meridian	子午线
general conference on weights and measures	国际计量大会	atomic standard	原子标准
former standard of length	长度原标准	platinum-iridium	铂铱合金
transition	跃迁	atomic energy level	原子能级
isotope cesium	铯同位素	krypton	氪
vacuum	真空	angstrom	埃

1.3 专业英语常用表达法-1 常用工具

toolbox 工具箱	nail hammer 羊角锤、拔钉锤
handsaw 手锯	sledge hammer 大锤、双手锤
ball-pane hammer 球头锤	pick 镐
ball hammer 圆头锤	double-bladed axe 双刃斧
axe hammer 斧锤	cutting nippers 剪钳、老虎钳

nipper pliers　尖嘴钳,剪丝钳
crowbar　铁锹、撬杠
nail　铁钉
coping saw　弓形锯
chisel　凿子
hand plane　刨子
screw　螺丝钉
screwdriver　螺丝刀
gimlet　手钻
scoop　铲子
triangle　三角板
protractor　量角器
curved ruler　曲尺
adjustable triangle　可调节三角板
T-square　丁字尺
angle square　角尺
drafting machine　平移角尺
dividers　两脚规
tape measure　卷尺
scissors　剪刀
wrench/spanner　扳手
adjustable spanner　活动扳手
double offset ring spanner　梅花扳手
inner hexagon spanner　内六角扳手
pipe spanner　管子扳手
connection cover cutting pliers　剥线钳
wire-cutting pliers　克丝钳
insulated pliers　绝缘钳
metal wire pliers　剪线钳
electric drill　电钻
hollow drill　空心钻
percussion drill　冲击钻
electric (soldering) iron　电烙铁
stopwatch　秒表、跑表、停表

LESSON 2

2.1　物理学专业英语的名词化结构

2.1.1　名词化

物理学专业英语的两个显著特点就是广泛使用名词化结构和大量使用被动语态。名词化是指词性作用的名词性转化,比如起名词作用的非谓语动词和与动词同根或同形的名词,也包括一些形容词来源的名词。这些词可以起到名词的作用,也可以表达谓语动词或形容词所表达的内容,常伴有修饰成分或附加成分,可构成短语。名词化结构指的是大量使用名词和名词词组,即在其他功能和题材的文章里用动词、形容词等词类充当某种语法成分,而在专业英语里往往会转化为由名词充当这种语法成分。

专业英语的“科学性”和“说理性”是名词化大量存在的理据。因为这种结构既可减少句子或分句的出现,又能包容大量的信息,并能反映科学内容的严肃性和客观性。下面就是一个名词化的例子:

(1) We can assume a freely falling body moves in one dimension under constant acceleration if we neglect air resistance.

在专业英语中则通常说成

(2) The motion of a freely falling body can be assumed to be motion in one dimension under constant acceleration by negligence of air resistance.

在这组例子中，名词化发生在两个地方。首先例(1)中的谓语动词转化为(2)中的名词主语，其次，例(1)中的状语从句的谓语动词 neglect 转化为名词结构介词短语来充当状语。经过名词化处理的例(2)由含有两个主谓结构的复合句变成了只含一个主谓结构的简单句，从而使句子的结构更加精练严谨，也由于使用抽象名词替代原来的人称代词做主语而使句子的语体更加正式。

2.1.2 名词化的分类

动词的名词化具有四种形式：

1）动作名词

在四种形式中，动作名词的名词性最强，而动词性最弱。动作名词用于一般性地叙述一个事实或概念，指出动作和技术的特点，而不强调动作本身的进行过程和时间。

Analysis of projectile motion is surprisingly simple if the following three assumptions are made.

2）动词性名词

动词性名词的名词性稍弱一些，而动词性略强。用于把概念和动作过程联系起来的描述，时间性也不强。

The *analyzing* of projectile motion is surprisingly simple if the following three assumptions are made.

3）动名词

动名词的动词性更强一些。在强调动作过程，概括地叙述一般行为，而不是特定行为，并且时间性也不强时，就可以使用动名词。

Analyzing projectile motion is surprisingly simple if the following three assumptions are made.

4）动词不定式

动词不定式的名词性最弱，动词性最强。通常用动词不定式表达某一次有时间性的特定动作或过程。

Projectile motion of the ball needs *to be analyzed* now if the following three assumptions are made.

2.1.3 名词化结构的构成

名词化结构的应用使得整个句子的结构便于写作修辞，也使得词句负载信息的容量增加。经常使用的名词化结构有：

1）名词(行为名词)＋介词＋名词

在此结构中，若“介词＋名词”构成的介词短语在逻辑上是行为名词的动作对象或动作的发出者，行为名词的含义在深层中转换或变异，使原来的名词变为动词，构成了动宾或主谓的关系。

The acceleration of the car is due to the force applied on it. 车的加速是由于外力的作用。

此句的 The acceleration of the car＝The car accelerates.

In the case of all freely falling bodies, gravity is essential in *the change of the velocity*. 所有作自由落体运动的物体，速度的改变都离不开重力。

其中 the change of the velocity＝the velocity changes.

2）介词＋名词（行为名词）

在此结构中，行为名词的动作意义相对完整，与句中的其他部分之间存在着一定的逻辑关系，能起到时间状语、原因状语、条件状语和让步状语等作用。

A rigid body can change its position *by translation or rotation*. 刚体位置的改变可以通过平移或转动实现。

3）谓语动词＋行为名词（＋介词短语）

此结构可以将宾语（介词宾语）转换成谓语。

Kepler's laws have *found application for* the exploration of the planets. 开普勒定律已经被用来探索行星。

此句可改变为 People have applied Kepler's laws to explore the planets.

此结构中谓语动词 find 含义空泛，只起语法作用，翻译时可以不译。类似的动词有：do, keep, have, make, take, pay, show, perform, offer 等。又如：

Friction *offers resistance to* the movement of the block. 摩擦力阻碍木块的运动。句中 offer 几乎不表示什么意义，只起连接作用。此句可以改为：Friction resists the movement of the block.

4）与动词构成固定搭配

名词化结构与动词构成固定搭配的常用形式为：动词＋动词名词化结构＋介词名词化结构。这种搭配大量地以一个动词短语的形式出现，约定俗成。例如：

make use of 利用，do research for 研究，lay emphasis on 强调，pay attention to 注意，等等。

5）行为名词＋短语/从句

在此结构中行为名词可以译成动词，与后面的成分一起构成汉语的动宾结构。

I have a *doubt* whether the instrument works well. 我怀疑这仪器是否运行良好。

此句可以改换成 I doubt whether the instrument works well or not.

6）名词＋名词（行为名词）

在此结构中，名词在表层结构上是前置定语，但在翻译过程中，其深层结构的内在含义可以译成动宾词组，行为名词转换成谓语。如：

heat conduction 导热，rust prevention 防锈，performance examination 性能检验

2.1.4 名词化的功能

名词化具有可以使语篇简洁、客观、量化、正式、严密等多种功能。

1）简洁功能

简洁明了是物理学专业英语的重要特点之一，要求以精练的语言传达大量的信息，而名词化的使用可以使句子省去一些不必要的词语，使得结构紧凑，具有较强的可读性。例如：

Kepler's laws had been developed. For this reason, Newton could discover his laws of motion.

以上两句可用名词化结构分别表达为 the development of Kepler's laws 和 for Newton to discover his laws of motion。

若再把这两个名词化结构按句法要求组织起来,则是:

The development of Kepler's laws makes it possible for Newton to discover his laws of motion.

所以,利用名词化结构,不仅可把两个句子合二为一,使语言更简洁、更精练,而且可把更多的信息结构融合成逻辑关系明确的整体。

2)客观功能

英语中用动词体现过程,动词须有参与者,有时还须有补语。而名词化则使本来由动词体现的过程转而由名词体现,这个过程变成另一过程的参与者,与旧过程相关的参与者就可以省略掉,从而使表达显得客观、真实,避免主观因素和主观色彩。

If we substitute some rolling friction for sliding friction, we can considerably reduce the friction.

改用名词化结构后可以去掉句子的主语 we,提高了客观的程度。

The substitution of some rolling friction for sliding friction results in a very considerable reduction in friction.

3)量化功能

专业英语有时需要使用表示数量的词来表示某人做了多少事情或某物占有多少分量等,但是动词和形容词是不能被量化的,这时名词化就成了可采取的有效手段。例如:

Last year 17 major changes and improvements were made toward making the lab even more perfect.

这个句子使用了数字加动词 change 和 improve 的名词化形式,使表达准确、严谨。如果采用 Last year we changed 17 major things and improved the lab even more perfect. 则显得不够严谨,数量的概念不是很强。

4)正式功能

专业英语要求使用较为正式的语言,适当使用一些名词化结构,可以增加语篇的正式程度。

The engineers are confident about the motion of the "Shenzhou" spaceship from the very beginning.

这个句子中,使用动词 move 的名词化形式 motion,使整个句子显得很正式。如果改成 The engineers are confident about how the "Shenzhou" spaceship moves...则正式程度大为降低。

5)严密功能

由于名词化结构中大量使用抽象名词,因此可以借助于抽象思维的逻辑性和概念化使科技文章的表达更确切,更严密。此外,在使用上,句子信息结构最复杂、最重要的部分往往是名词化结构。

We begin our *study of* the physical universe by examining objects *in motion*. The *study of* motion, whose measurement, more than 400 years ago *gave birth to* physics, is called kinematics. Much of our *understanding of* nature comes from observing *the motion of* objects.

2.2 专业英语阅读

2.2.1 Vectors(矢量)

Many quantities in physics have magnitude and direction. **Vectors** are quantities with magnitude and direction. Examples include velocity, acceleration, momentum, and force. Quantities with magnitude but no associated direction—for example, distance and speed—are called **scalars.**

A vector is represented graphically by an arrow drawn in the same direction as that of the vector, and with a length that is proportional to the magnitude of the vector. When the magnitude of a vector is given, its unit must also be given.

Two vectors are defined to be equal if they have the same magnitude and the same direction. Graphically, this means that they have the same length and are parallel to each other. A consequence of the definition is that moving a vector so that it remains parallel to itself does not change it. Vectors do not depend on the coordinate system used to represent them (except for position vectors, which are introduced later).

2.2.2 Properties of Vectors(矢量的性质)

In comparing vectors and performing other mathematical operations such as addition and subtraction, we may translate vectors anywhere in the coordinate space for convenience. We must be careful, however, to preserve their magnitudes and directions with respect to the axes.

Vector Addition(矢量加法)

Two vectors are added graphically by placing the tail of one, $\boldsymbol{B}$, at the head of the other, $\boldsymbol{A}$ (Fig. 2-1). The resultant (or net) vector, $\boldsymbol{C}=\boldsymbol{A}+\boldsymbol{B}$, extends from the tail of $\boldsymbol{A}$ to the head of $\boldsymbol{B}$. This is the so-called **head-to-tail method.**

An equivalent way of adding vectors, called the **parallelogram method**, is to move $\boldsymbol{B}$ so that it is tail-to-tail with $\boldsymbol{A}$. The diagonal of the parallelogram formed by $\boldsymbol{A}$ and $\boldsymbol{B}$ then equals the resultant vector $\boldsymbol{C}$, as shown in Fig. 2-2.

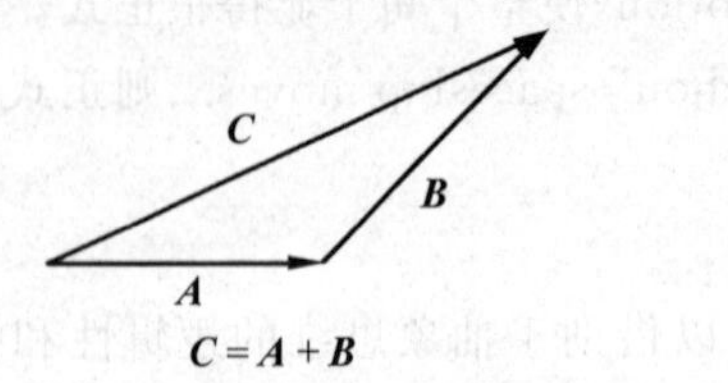

Fig. 2-1 Head-to-tail method of vector addition

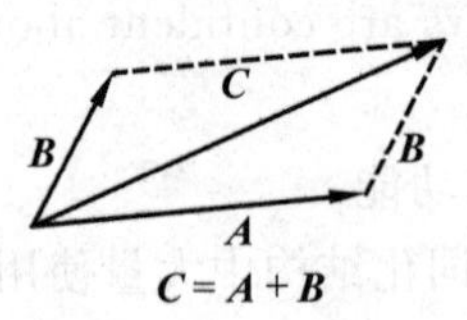

Fig. 2-2 Parallelogram method of vector addition

The vectors have the mathematical property of "obeying the **commutative law** in addition".

$$\boldsymbol{A}+\boldsymbol{B}=\boldsymbol{B}+\boldsymbol{A} \tag{2-1}$$

Vector Subtraction（矢量减法）

We subtract vector $\boldsymbol{B}$ from vector $\boldsymbol{A}$ by adding $-\boldsymbol{B}$ to $\boldsymbol{A}$. The result is shown in Fig. 2-3. Note that vector addition or subtraction can be done only when vectors are in the same unit.

Scalar Product (Dot Product)（标积/点积）

The scalar product of any two vectors is defined as a scalar quantity equal to the product of the magnitudes of the two vectors $\boldsymbol{A}$ and $\boldsymbol{B}$ and the cosine of the angle ϕ that is included between the directions of $\boldsymbol{A}$ and $\boldsymbol{B}$.

That is, the scalar product (or dot product) of $\boldsymbol{A}$ and $\boldsymbol{B}$ is defined by the relation

$$\boldsymbol{A} \cdot \boldsymbol{B} = AB\cos\phi \tag{2-2}$$

where ϕ is the angle between $\boldsymbol{A}$ and $\boldsymbol{B}$ as in Fig. 2-4. A is the magnitude of $\boldsymbol{A}$, and B is the magnitude of $\boldsymbol{B}$, Note that $\boldsymbol{A}$ and $\boldsymbol{B}$ need not have the same unit.

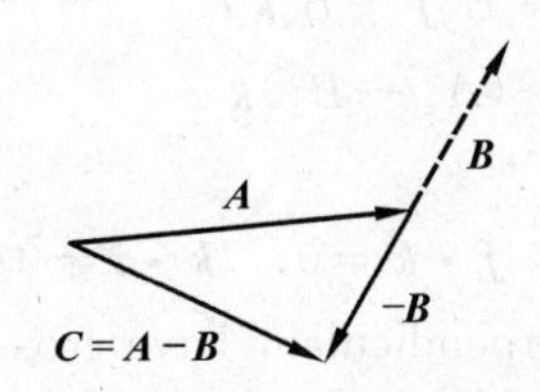

Fig. 2-3 Vector subtraction

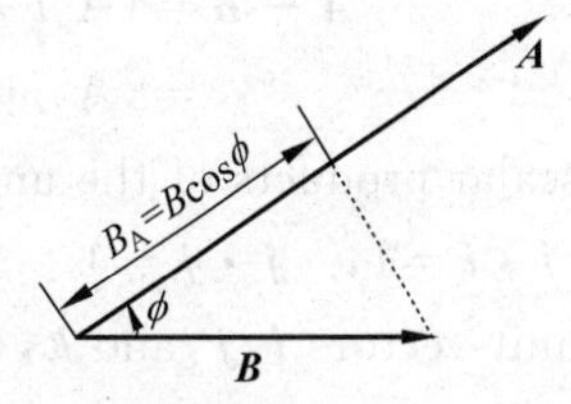

Fig. 2-4

Cross Product (Vector Product)（叉积/矢积）

The cross product of two vectors $\boldsymbol{A}$ and $\boldsymbol{B}$ is defined to be a vector $\boldsymbol{C}=\boldsymbol{A}\times\boldsymbol{B}$ whose magnitude equals the area of the parallelogram formed by the two vectors, as in Fig. 2-5(a). The vector $\boldsymbol{C}$ is perpendicular to the plane containing $\boldsymbol{A}$ and $\boldsymbol{B}$ in the direction given by the right-hand rule, that is, as your right-hand fingers curl from the direction of $\boldsymbol{A}$ toward the direction of $\boldsymbol{B}$, the direction of $\boldsymbol{A}\times\boldsymbol{B}$ is given by your thumb (Fig. 2-5(b)). If ϕ is the angle between the two vectors and $\boldsymbol{n}$ is the unit vector that is perpendicular to each in the direction of $\boldsymbol{C}$, the cross product of $\boldsymbol{A}$ and $\boldsymbol{B}$ is

$$\boldsymbol{C} = \boldsymbol{A} \times \boldsymbol{B} = (AB\sin\phi)\boldsymbol{n} \tag{2-3}$$

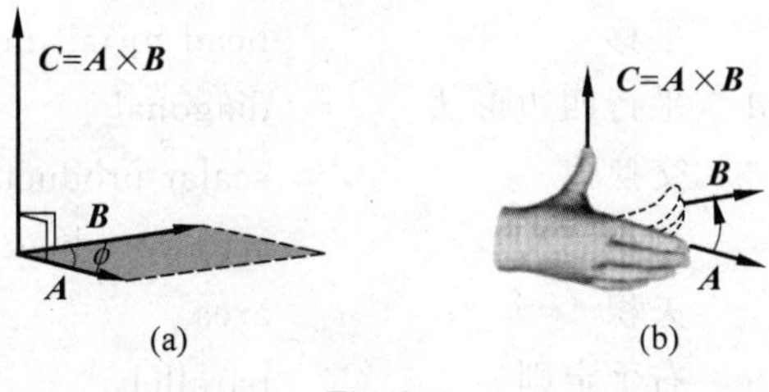

(a) (b)

Fig. 2-5

If $\boldsymbol{A}$ and $\boldsymbol{B}$ are parallel, $\boldsymbol{A}\times\boldsymbol{B}$ is a zero vector.

2.2.3 Unit Vectors(单位矢量)

A **unit vector** is a dimensionless vector with unit magnitude. Unit vectors that point in the positive x, y, and z directions are convenient for expressing vectors in terms of their rectangular components. They are usually written as $\boldsymbol{i}$, $\boldsymbol{j}$ and $\boldsymbol{k}$, respectively (Fig. 2-6). For example, the vector $A_x\boldsymbol{i}$ has a magnitude $|A_x|$ and points in the positive x direction if A_x is positive (or the negative x direction if A_x is negative). A general vector $\boldsymbol{A}$ can be written as the sum of three vectors, each of which is parallel to a coordinate axis

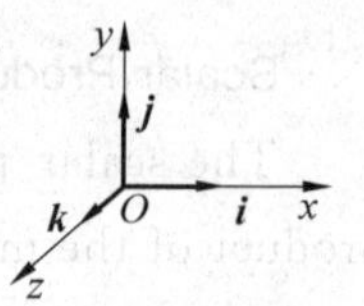

Fig. 2-6

$$\boldsymbol{A} = A_x\boldsymbol{i} + A_y\boldsymbol{j} + A_z\boldsymbol{k} \tag{2-4}$$

The addition of two vectors $\boldsymbol{A}$ and $\boldsymbol{B}$ can be written in terms of unit vectors as

$$\begin{aligned}\boldsymbol{A} + \boldsymbol{B} &= (A_x\boldsymbol{i} + A_y\boldsymbol{j} + A_z\boldsymbol{k}) + (B_x\boldsymbol{i} + B_y\boldsymbol{j} + B_z\boldsymbol{k}) \\ &= (A_x + B_x)\boldsymbol{i} + (A_y + B_y)\boldsymbol{j} + (A_z + B_z)\boldsymbol{k}\end{aligned} \tag{2-5}$$

The scalar products of the unit vectors are

$$\boldsymbol{i}\cdot\boldsymbol{i} = 1,\quad \boldsymbol{j}\cdot\boldsymbol{j} = 1,\quad \boldsymbol{k}\cdot\boldsymbol{k} = 1,\quad \boldsymbol{i}\cdot\boldsymbol{j} = 0,\quad \boldsymbol{j}\cdot\boldsymbol{k} = 0,\quad \boldsymbol{k}\cdot\boldsymbol{i} = 0 \tag{2-6}$$

The unit vectors $\boldsymbol{i}$, $\boldsymbol{j}$, and $\boldsymbol{k}$, which are mutually perpendicular, have cross products given by

$$\boldsymbol{i}\times\boldsymbol{j} = \boldsymbol{k},\quad \boldsymbol{j}\times\boldsymbol{k} = \boldsymbol{i},\quad \text{and}\quad \boldsymbol{k}\times\boldsymbol{i} = \boldsymbol{j} \tag{2-7}$$

$$\boldsymbol{i}\times\boldsymbol{i} = \boldsymbol{j}\times\boldsymbol{j} = \boldsymbol{k}\times\boldsymbol{k} = \boldsymbol{0} \tag{2-8}$$

Glossary

vector	矢量	magnitude	大小
velocity	速度	acceleration	加速度
momentum	动量	scalar	标量
proportional to	正比于	parallel	平行
position vector	位置矢量	coordinate system	坐标系
resultant/net vector	合矢量	addition	加法
subtraction	减法	equivalent	等价的
translate	平移	head-to-tail method	三角形法
parallelogram method	平行四边形法	diagonal	对角线
commutative law	交换律	scalar product	标积
dot product	点积	cross product	叉积
vector product	矢积	area	面积
right-hand rule	右手定则	parallel	平行
unit vector	单位矢量	unit magnitude	单位大小
dimensionless	无量纲的	respectively	分别地

2.3 专业英语常用表达法-2 仪器设备

apparatus 装置、设备、器械
appliance 器具、电器
device 装置
equipment 设备
facilities 设备、用具
gear 用具、装置
instrument 仪器、仪表
machine 机器、机械
machinery 机器设备、机械
mechanism 机械装置
plant 成套机械、装置
realia 教具、教学用品
set (成套)设备、仪器
tool 工具
unit 装置、组件、元件
a battery supply set 电池供电设备
a device for regulating temperature 控制温度的装置
a hand-operated tool 手动工具
a remote-control gear 遥控装置
acoustic analytical instrument 声分析仪
acoustical instrument 声学仪器
adiabatic apparatus 绝热装置
adjusting instrument 调节仪器、调节装置
all-purpose instrument 多用工具、万能仪表
altitude instrument 高度仪
an air-conditioning equipment for 供……用的空调设备
an instrument for measuring the spectra 测量光谱的仪器
arc-suppressing apparatus 灭弧装置
beat measuring apparatus 拍频测试仪
bolometer 辐射热测定器
bolometric instrument 辐射热量计
calorimeter 量热计
cathode ray apparatus 阴极射线仪器
chromatographic instrument/chromatograph 色谱仪
current-measuring instrument 测流仪器
depth-measuring instrument 测深仪
detecting instrument 检测仪器、探测仪器
dial instrument 指针式仪表、有刻度的仪表
digital measuring instrument 数字式测量仪
displaying instrument 指示仪器
double-scale instrument 双标度仪表
double-range instrument 双量程仪表
dynamometer 功率计、电力测功仪、测力计、动力计
echo-sounding instrument 回声探测仪

eddy current instrument　涡流仪器
educational instrument　教学仪器
electric instrument　电工测量仪表、电表
electrical appliance　电器用具
electroacoustical instrument　电声(测试)仪器
electromagnetic acoustical instrument　电磁声学仪器
electronic measuring instrument　电子测量仪器
electronic test instrument　电子试验(测试)仪器
electrostatic acoustical instrument　静电声学仪器
electrostatic instrument　静电式仪表
electrostatic measuring instrument　静电式电表、静电式测量仪
electrothermic instrument　热电式仪表
fine measuring instrument　精密测量仪器、精密量具
first-order instrument　一阶仪器
flow instrument　流量计
humidity-measuring instrument　湿度测量仪
instructional instruments　教学仪器
insulation test instrument　绝缘(电阻)测试仪器
laboratory apparatus　实验仪器(装置)
laboratory instrument　实验室仪器
laser distance-measuring instrument/laser range finder　激光测距仪
level instrument　位面计、水平仪
levelling instrument　水准器、水平尺、测平仪
measuring instrument　测量仪表、测量仪器
metrologic instrument　计量仪器
needle instrument　指针型仪器
optical instrument　光学仪器
photomicrographic apparatus　显微照相装置
portable instrument　便携式仪器
power plant/unit　动力装置/机组
precise instrument　精密仪器
radio instrument　无线电仪器
research instrument　试验设备、研究设备
resistance instrument　电阻式仪表
scientific apparatus/instrument　科学仪器/仪表
scientific experiment package　科学实验装置
sensing instrument　灵敏仪表、灵敏元件
spraying apparatus　喷雾器
supersonic thickness meter/gauge　超声测厚仪

surveying instrument 测量仪器、测绘仪器

test instrument 试验工具、测试设备

testing instrument 试验仪器

the latest research equipment 最新研究设备

visual instrument 目视仪器

X-ray diffraction instrument X射线衍射仪

LESSON 3

3.1 物理学专业英语的语法特点

3.1.1 常用的动词时态

专业英语在时态运用上有限，尽管英语的动词有 16 种时态，但在专业英语中常见的只有四种：一般现在时、一般过去时、一般将来时和现在完成时。

1）一般现在时

在物理学英语文献资料中用得最多的时态是一般现在时，用以表述无时间性的科学定义、定理、公式、现象、过程等。究其原因可能是科学家和物理工作者都想表明他们所说的、所写的都是真理性的，都不受时间的限制。即使是叙述一个已完成的实验，或者是叙述一个将要做的实验，也大都使用一般现在时，意在表明其他人在任何时候都可以重复这样的实验而得到同样的结果。主要有以下三种用法。

（1）表示一般叙述过程。如：

As the electrons move, the surface charge density increases until the magnitude of the internal field equals that of the external field, giving a net field of zero inside the conductor.

If we now imagine the surface to shrink to zero like a collapsing balloon, until it essentially encloses a point, the charge at the point must be zero.

（2）叙述客观事实或科学定理。如：

A good electrical conductor contains charges (electrons) that are not bound to any atom and are free to move about within the material.

The electric field is zero everywhere inside the conductor.

Work done on a particle equals the change in its kinetic energy.

（3）表达通常或习惯发生的行为。如：

Much of our understanding of nature comes from observing the motion of objects.

The solutions of kinematic equations are usually obtained quite easily in a direct fashion using integral calculus.

2）一般过去时

在提到以前叙述过的事情、叙述物理学发展史时，经常使用过去时态。如：

The result is the same as Eq. (7-1), which was calculated directly from Coulomb's law.

The first capacitor was the Leyden jar, a glass container lined inside and out with gold foil. It was invented at the University of Leyden in the Netherlands by eighteenth-century experimenters who, while studying the effects of electric charges on people and animals, got the idea of trying to store a large amount of charge in a bottle of water. An experimenter held up a jar of water in one hand while charge was conducted to the water by a chain from a static electric generator. When he reached over to lift the chain out of the water with his other hand, he was knocked unconscious. Benjamin Franklin realized that the device for storing charge did not have to be jar-shaped and used foil-covered window glass, called Franklin panes.

3）一般将来时

在表示将来发生的行为或情况，或表达假设条件时，常用到将来时态。如：

In this chapter we shall use the energy concept in our study of electricity.

As we shall see in subsequent chapters, the concept of electric potential is of great practical value.

If the uncharged conductor is grounded, what will be the charge distribution and electric field near the slabs?

If a conductor is placed in an external field, the charges in the conductor will redistribute themselves, in the meantime, the external field will also be altered by the charge redistribution of the conductor.

4）现在完成时

表示已取得的成果或完成的工作。如：

We have stated that φ and $\boldsymbol{E}$ are equivalent descriptions of electric fields, and have determined how to calculate φ from $\boldsymbol{E}$.

This problem has already been solved graphically.

So far our use of Gauss's law has been confined to the situation in which no dielectric was present.

3.1.2 使用介词的侧重点

尽管介词为英语各种文体所共用，但侧重点有所不同。如在文学英语中 with 多表示"产生某种生理或心理状态的原因(pale with anger)"，而在科技英语中更为常见的是用以表示"工具、手段"。如：

The capacitor is sprayed with aluminum to prevent rust.

Nowadays labs are built with large cranes and modern equipment.

3.1.3 并列成分的使用

在专业英语中，并列成分得到了大量的使用，下面的例句中都使用了并列的成分：

The story of the quantum is the one of a confused and groping search for knowledge *conducted by* scientists of many lands on a front wider than the world of physics had ever seen before, *illumined by* flashes of insight, *aided by* accidents and guesses, and *enlivened by* coincidences such as one would expect to find only in fiction.

Since the beginning of the present century, an increasing degree of uncertainty about the wave theory of light has been developing *due to* the discovery of the photoelectric effect, the quantum theory, and a number of allied phenomena, and *due also to a* detailed investigation into the nature and physical properties of the ether. Also, during this period a great deal has been learned about the structures of the atom which is not altogether limited to a wave picture of light. Hence the photon, or a light corpuscle, has been postulated.

In announcing that "every particle of matter in the universe attracts every other particle with a force inversely proportional to the square of the distance between the two particles", *in showing that* the one universal and comparatively simple law governs not only the motion of the planets round the sun and of the satellites round their planets but, probably, also the relative motions of all the heavenly bodies, Newton gave to the world a truth the importance of which in all branches of human thought can hardly be overestimated. Of value to science, of course, from microphysics to macrophysics. But consider the effect on man's concept of nature and of his relations thereto of realizing, indeed, of having proven to him, for the first time that the physical universe is governed by law, not by caprice; and if the physical universe, why not the biological universe, even the moral universe.

3.1.4 经常使用的复合修饰语

复合修饰语的使用是为了使行文简洁,它可以由名词、形容词、副词、介词短语、从句等充当。在专业英语中,经常使用复合修饰语,通常包括以下三种情况:

1) 一个以上的修饰语同时修饰同一个中心词。如:

sound and light waves 声波和光波,*particle and wave* duality 波粒二象性

"*Absolute*, *true*, *and mathematical* time, of itself, and from its own nature, flows equably without relation to anything external." 绝对的、真实的、数学的时间,就其本身而言,就其本性而言,均匀流逝,与外界任何事情都无关。

2) 一个修饰语同时修饰一个以上的中心词。如:

analyze the *efficiency and performance* of machinery 分析机械的运行情况及其效率

3) 一个以上的修饰词同时修饰一个以上的中心词。如:

Scientists always check statements and make experiments *carefully and objectively* to verify them. 科学家总是谨慎而客观地核对各种说法,并以同样的态度通过实验来加以验证。

3.1.5 广泛使用被动语态

专业英语中的谓语约三分之一是被动语态。这是因为物理文献侧重叙事推理,强调客观准确。第一、二人称使用过多时,会造成主观臆断的印象,因此尽量使用第三人称叙述,采用被动语态,例如:

Attention must be paid to the working temperature of the machine. 应当注意机器的工作温度。

而很少说：You must pay attention to the working temperature of the machine. 你们必须注意机器的工作温度。

此外，如前所述，专业文章将主要信息前置，放在主语部分。这也是广泛使用被动态的主要原因。观察并比较下列两段短文的主语：

We can store electrical energy in two metal plates separated by an insulating medium. We call such a device a capacitor, or a condenser, and its ability to store electrical energy capacitance. It is measured in farads. 电能可储存在由一绝缘介质隔开的两块金属极板内。这样的装置称为电容器，其储存电能的能力称为电容。电容的测量单位是法(拉)。

变为如下被动语态句：

Electrical energy can be stored in two metal plates separated by an insulating medium. Such a device is called a capacitor, or a condenser, and its ability to store electrical energy capacitance. It is measured in farads.

这一段短文中各句的主语分别为：electrical energy, such a device, its ability to store electrical energy 和 it(capacitance)。它们都包含了较多的信息，并且处于句首的位置，非常醒目。四个主语完全不同，避免了单调重复，前后连贯，自然流畅，充分表明被动结构可取得简洁客观的效果。

3.1.6 大量使用从句

专业英语对客观事物的描述要求准确、完整，因而各种从句用得较多，但层次清楚。以下是一些从句的例子：

1) 状语从句

The body possesses a definite store of potential energy *while it is in the elevated position*.

2) 定语从句

Switching circuits are the ones *that perform logical functions*.

3) That 引导的从句

在句子中可以起主语、宾语、补语等作用。如：

For example, in electric circuits it is often assumed *that the relations between voltages and currents are linear*.

Stated in words, Ohm's law says *that the steady current through any portion of an electric circuit equals the potential difference across that portion of the circuit divided by the resistance of that portion of the circuit*.

4) 疑问词从句

由疑问词 what, where, how, which, who, whose, whether 等引导的从句，在句子中可以起主语、宾语、补语、同位语等作用。如：

In order to determine *whether a system is performing properly and ultimately* to control the system performance, the engineer must know *what the system is doing* at any instant of time.

The question *whether the atom can or cannot be split up* is known to have interested scientists from ancient times.

3.1.7 大量使用后置定语

大量使用后置定语也是科技文章的特点之一。常见的结构有以下五种：

1）介词短语

The forces *due to friction* are called frictional forces. 由于摩擦而产生的力称为摩擦力。

2）形容词及形容词短语

In radiation, thermal energy is transformed into radiant energy, *similar in nature to light*. 热能在辐射时，转换成性质与光相似的辐射能。

3）副词

The force *upward* equals the force *downward* so that the balloon stays at the level. 向上的力与向下的力相等，所以气球就保持在这一高度。

4）单个分词，但仍保持较强的动词意义

The heat *produced* is equal to the electrical energy *wasted*. 产生的热量等于浪费了的电能。

5）定语从句

The molecules exert forces upon each other, *which depend upon the distance between them*. 分子相互间都存在着力的作用，这些力的大小取决于它们之间的距离。

关于定语从句，会在以后的章节中进一步讨论，在此不再详述。

3.1.8 大量使用非限定动词结构

专业英语文章要求行文简练，结构紧凑。为此，往往使用分词短语代替定语从句或状语从句；使用分词独立结构代替状语从句或并列分句；使用不定式短语代替各种从句；使用介词+动名词短语代替定语从句或状语从句。这样既可将句子缩短，又比较醒目。试比较下列各组句子。

A direct current is a current which flows always in the same direction.

A direct current is a current flowing always in the same direction.

直流电是一种总是沿同一方向流动的电流。

When heat radiates from the earth, it causes air currents to rise.

Radiating from the earth, heat causes air currents to rise.

热量由地球辐射出来时，使得气流上升。

A body can move uniformly and in a straight line if there is no cause to change that motion.

A body can move uniformly and in a straight line, there being no cause to change that motion.

如果没有改变物体运动的原因，那么物体将作匀速直线运动。

Vibrating objects produce sound waves, and each vibration produces one sound wave.

Vibrating objects produce sound waves, each vibration producing one sound wave.

振动的物体产生声波,每一次振动产生一个声波。

There are different ways which change energy from one form into another.

There are different ways of changing energy from one form into another.

将能量从一种形式转变成另一种形式有各种不同的方法。

When the radio waves are made to correspond to each sound in turn, messages are carried from a broadcasting station to a receiving set.

By making the radio waves correspond to each sound in turn, messages are carried from a broadcasting station to a receiving set.

使无线电波依次对每一个声音作出相应变化时,消息就由广播电台传递到接收机。

3.2 专业英语阅读

3.2.1 Linear Motion with Constant Acceleration(匀加速直线运动)

A very common and simple type of one-dimensional motion occurs when the acceleration is constant, or uniform. Consequently, the velocity increases or decreases at the same rate throughout the motion.

For motion along a straight line, we choose a rectangular coordinate system that is oriented so that one of the axes (for example, the x-axis) is along the line. Then, components of position vector, velocity, and acceleration lie along this direction, and the y and z components are zero. The vector equations then become scalar equations. If the velocity is v_0 at time $t=0$, and v at some later time t, the corresponding acceleration is

$$a=\frac{\Delta v}{\Delta t}=\frac{v-v_0}{t-0}=\frac{v-v_0}{t} \tag{3-1}$$

A Case in Point—Freely Falling Bodies

It is well-known that all objects, when dropped, will fall toward the earth with nearly constant acceleration. In the idealized case, where air resistance is neglected, such motion is referred to as **free fall**.

We shall denote the acceleration due to gravity by the symbol $\boldsymbol{g}$. The magnitude of $\boldsymbol{g}$ decreases with increasing altitude. Furthermore, there are slight variations in g with altitude. The vector $\boldsymbol{g}$ is directed downward toward the center of the earth. At the earth's surface, the magnitude of $\boldsymbol{g}$ is approximately 9.80 m/s^2.

If we neglect air resistance and assume that the gravitational acceleration does not vary with altitude, then the motion of a freely falling body is equivalent to motion in one dimension under constant acceleration. Therefore our kinematic equations for constant acceleration can be applied. We shall take the vertical direction to be the y axis and call y positive upward. With this choice of coordinates, we have $a=-g$. The negative sign simply indicates that the acceleration is downward. We get the following expressions:

$$v = v_0 - gt \quad (3\text{-}2)$$

$$y - y_0 = v_0 t - \frac{1}{2}gt^2 \quad (3\text{-}3)$$

$$v^2 - v_0^2 = -2g(y - y_0) \quad (3\text{-}4)$$

You should note that the negative sign for the acceleration is already included in these expressions. Therefore, when using these equations in any free-fall problem, you should simply substitute $g = 9.80\ \text{m/s}^2$.

3.2.2 Projectile Motion(抛体运动)

Anyone who has observed a football in motion (or, for that matter, any object thrown in the air) has observed projectile motion. This very common form of motion is surprisingly simple to analyze if the following three assumptions are made: (1) the acceleration due to gravity, $\boldsymbol{g}$, is constant over the range of motion and is directed downward, (2) the effect of air resistance is negligible, and (3) the rotation of the earth does not affect the motion. With these assumptions, we shall find that the path of a projectile, which we call it trajectory, is always a parabola.

If we choose our reference frame such that the y direction is vertical and positive upward, then $a_y = -g$ (as in one-dimensional free fall) and $a_x = 0$ (since air friction is neglected). Furthermore, let us assume that at $t = 0$, the projectile leaves the origin $(x_0 = y_0 = 0)$ with a velocity v_0, as in Fig. 3-1.

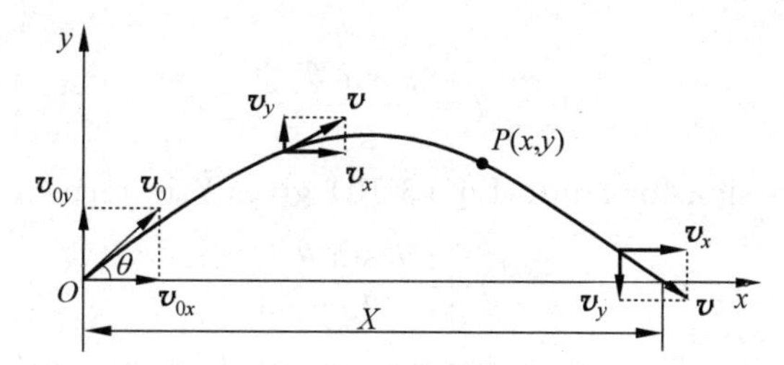

Fig. 3-1 The parabolic trajectory of a projectile that leaves the origin with a velocity v_0

If the vector makes an angle with the horizontal, then the initial velocity and acceleration have components

$$v_{0x} = v_0 \cos\theta \quad (3\text{-}5a)$$

$$v_{0y} = v_0 \sin\theta \quad (3\text{-}5b)$$

$$a_x = 0, \quad a_y = -g \quad (3\text{-}6)$$

Notice that v_x does not depend on v_y and vice versa: The horizontal and vertical components of projectile motion are independent. The velocity components and coordinates for the projectile at any time t are

$$v_x = v_{x0} = v_0 \cos\theta = \text{constant} \quad (3\text{-}7)$$

$$v_y = v_{y0} - gt = v_0 \sin\theta - gt \quad (3\text{-}8)$$

$$x = v_{x0} t = (v_0 \cos\theta) t \tag{3-9}$$

$$y = v_{y0} t - \frac{1}{2} g t^2 = (v_0 \sin\theta) t - \frac{1}{2} g t^2 \tag{3-10}$$

From Eqs. (3-9) and (3-10), we conclude that projectile motion is the superposition of two motions: (1) the motion of a freely falling body in the vertical direction with constant acceleration and (2) uniform motion in the horizontal direction with constant velocity.

The Total Flight Time(总飞行时间)

The total flight time T is obtained by setting $y=0$ in Eq. (3-10)

$$y = (v_0 \sin\theta) T - \frac{1}{2} g T^2 = 0, \quad T > 0$$

The flight time of the projectile is thus

$$T = \frac{2 v_0 \sin\theta}{g} \tag{3-11}$$

Horizontal Range and Maximum Height(射程与最大高度)

There are two special points that are interesting to analyze: the peak with Cartesian coordinates labeled $(X/2, Y)$ and the point with coordinates $(X, 0)$. The distance X is called the **horizontal range** of the projectile, and Y is its **maximum height.**

We can determine the maximum height, Y, reached by the projectile by noting that at the peak, $v_y = 0$. Therefore, Eq. (3-8) can be used to determine time t it takes to reach the peak

$$t = \frac{v_0 \sin\theta}{g}$$

Substituting this expression for t into Eq. (3-10) gives Y in terms of v_0 and θ

$$Y = \frac{v_0^2 \sin^2\theta}{2g} \tag{3-12}$$

The range, X, is the horizontal distance traveled in the total flight time T. Using Eq. (3-9) and noting that $x = X$ at $t = T$, we find that

$$X = v_0 \cos\theta \cdot \frac{2 v_0 \sin\theta}{g} = \frac{2 v_0^2 \sin\theta \cos\theta}{g}$$

Since $\sin 2\theta = 2\sin\theta\cos\theta$, X can be written in the form

$$X = \frac{v_0^2 \sin 2\theta}{g} \tag{3-13}$$

The maximum value of $\sin 2\theta$ is unity, which occurs when $2\theta = 90°$. Therefore, we see that X is maximum when $\theta = 45°$, as you would expect if air friction is neglected.

Glossary

one-dimensional	一维的	corresponding	对应的
freely falling body	自由落体	air resistance	空气阻力

acceleration due to gravity	重力加速度	altitude	高度
vertical direction	竖直方向	negative sign	负号
projectile	抛体	trajectory	轨迹
assumption	假设	negligible	可忽略的
rotation	转动	air friction	空气摩擦
parabola	抛物线	parabolic trajectory	抛物线轨迹
initial	初始的	horizontal	水平的
superposition	叠加	flight time	飞行时间
horizontal range	射程	maximum height	最大高度

3.3 专业英语常用表达法-3 状态 状况 技术设备的更新

1. 表示"状态、状况"等的词汇

at a complete standstill 处于完全停顿状态

at high temperature 在高温下

at/in/on/under... 处于……状态

at low concentration 在低密度(浓度)下

at rest 处于静止状态

at room temperature 在室温下

at the same electric potential 在相同电势下

at work 正在工作(运行)

bring/put...into/to... 使处于……状态

have a high density 具有高密度

in a good state of repair 维修良好、状态良好

in a stable state 处于稳定状态

in a state of disorder 处于杂乱(无序)状态

in a state of unstable equilibrium 处于非稳定平衡状态

in a suspended state 处于悬挂/暂停状态

in danger 处于危险状态、受到危险

in equilibrium 处于平衡状态

in motion 处于运动状态

in operation 正在运行、在运行(工作)中

in progress 正在进行、在进行中

in suspension 吊着、悬浮着、悬着

in vacuum 在真空状态下

on fire 着火、在燃烧

on test 正在经受试验/测试

on the decrease 正在减少

on the increase 正在增长

under construction 正在建造中

under control　处于受控状态
under development　正在研制之中
under discussion　正在讨论中
under heat treatment　在热处理中
under investigation　在调查中
under study　在研究中
under way　正在进行

2. 表示“过时、陈旧、淘汰”等的词汇

obsolete　荒废的、陈旧的
old　陈旧的、古老的
old-fashioned　老式的、过时的
out-of-date　老式的、过时的
timeworn　陈旧的、老化的
behind the times　过时、落伍
fall into disuse　淘汰
out of repair　失修
in disrepair　年久失修
in poor condition　状况不佳

3. 表示“更新、换代”等的词汇

remove　更换、升级
renew (renewal)　更新
renovate (renovation)　革新、改造
replace　取代、替换
introduce (introduction)　引进、纳入
install (installation)　安装、安置
update　使现代化、更新
upgrade　使升级、改良
in one's place　代替
in place of　代替
take the place of　代替

以下是一些具体的例子：

An object *at rest* will remain *at rest* and an object *in motion* will continue *in motion* with a constant velocity unless it experiences a net external force. 任何物体都保持静止的或沿一条直线作匀速运动的状态，除非作用在它上面的力迫使它改变这种状态。

Both Fermi-Dirac and Bose-Einstein become Maxwell-Boltzmann statistics *at high temperature* or *at low concentration*. 费米-狄拉克统计和玻色-爱因斯坦统计在高温或者低密度的情况下都会变为麦克斯韦-玻尔兹曼统计。

When no motion of charge occurs within the conductor, the conductor is *in electrostatic equilibrium*. 当导体内的电荷没有运动时，导体就处于静电平衡状态。

If an object is instantaneously *at rest*, is the force on it necessarily zero? 如果一个物体在某一瞬间是静止的，那么它所受到的外力是否必然为零呢？

Because the two spheres are connected by a conducting wire, the entire system is a single conductor and all points must be *at the same electric potential*. 由于这两导体球被导线连接，所以整个系统构成一个导体，因而所有各点处于电势相同的状态。

If a body is acted on by a number of forces and still remains stationary, the body is said to be *in equilibrium*. 如果一个物体受到数个力的作用而仍保持静止状态，我们说该物体处于平衡状态。

If the electric field were not zero, free charges in the conductor would accelerate *under the action of the electric force*. 如果电场不为零,导体内的自由电荷在电场力的作用下会加速运动。

The pig-iron comes out of the blast-furnace *in a molten state*. 生铁呈熔化状态流出高炉。

Metal which is hardened by cold-working may *be brought back to its original state/condition* by annealing. 通过冷加工而变硬的金属可通过退火恢复到它原来的状态。

The engine should be *under normal working conditions* during the test. 在试验时发动机应处于正常运行状态。

The surface of the liquid is *under a tension*. 液体的表面处在张力作用下。

Thus *renewal* of equipment began. 就这样设备开始更新了。

These were *obsolete* and had to be replaced with digital technology. 这些(设备)已经陈旧不堪,将被淘汰而改用数字技术(设备)。

Many old facilities have been *updated* since 2005. 自 2005 年以来许多陈旧的设施已被更新换代。

When we arrived here we found a very congested *old* system which needed total *renovation*. 当我们到这里时,我们发现系统陈旧又拥挤不堪,需要进行彻底改造。

They had to *remove* many lines and exchanges which were over 20 years old and completely *out-of-date*. 他们必须把已经用了 20 多年的、完全过时的线路和电话交换台予以更换。

This system will most probably be *upgraded* in the long run to a completely fiber-optic cable route. 这套系统最终将会全部升级,改用光纤线路。

LESSON 4

4.1 物理学专业英语的修辞特点

物理学专业英语的主要作用是提供事实、数据、假设、理论、实验等信息,运用概念、判断、推理,系统论述自然科学。其内容往往是一种客观的叙述,用于描述事实、记录实验,阐明规律或探讨理论,比如经常对环境、仪器设备进行描写,对实验过程和工艺流程进行描写,对工作原理、实验结果、性能、用途、条件等进行说明。不论是科学文章、科普文章、原版专业书籍,还是技术文本,都需要把科学道理说清楚。专业英语具有平铺直叙、简洁、确切的特点,注重的是科学事实和逻辑概念等,因而其文体十分严谨,修辞比较单调。为了具有科学性、准确性、客观性和抽象性等风格特点,专业英语经常运用如下几种修辞方式。

4.1.1 多使用逻辑语法词表示各种关系

表示原因的逻辑语法词如 because, due to, owing to, as a result of, for...

表示语气转折 but, however, nevertheless, yet, otherwise...

表示限制 only, if only, except, besides, unless...

表示逻辑顺序 so, thus, therefore, furthermore...

表示假设 suppose, assuming, provided, providing...

4.1.2 使用一些词语表示过程或顺序

表示先后顺序、承上启下的词语,常用的有:first, first of all, second, secondly, third, thirdly, next, then, before, prior to, after, after that, last, lastly, finally 等,还可用 for one thing, for another, also, in addition, on top of that, more than that, furthermore 等。此外,还有如 when, where, which, and, by the time 等起连接作用的词语也经常被采用。

One must judge a model by how well it meets two standards. *First*, is the model based upon the best experimental evidence that we have? *Second*, is it good enough to allow us to predict what would happen in a new situation that has not been investigated before? 评判一个模型必须根据它在多大程度上符合下面两个标准:①该模型是不是以我们所获得的最佳实验证据为其依据?②它是不是足以使我们能预测在至今尚未研究过的新条件下将会发生什么情况?

We have decided not to import the complete plant. *For one thing*, it is expensive; *for another* its reliability is open to question. 我们已决定不进口这套成套设备了。首先,它价格昂贵;其次,它的可靠性也有问题。

Following are a set of procedures in making a scientific discovery. A problem must *first of all* be recognized. *Secondly*, existing knowledge and theories about the problem must be studied. On this basis, *the third step* is to collect factual data relevant to the problem. *Next* comes the most important step of all: the formation of a hypothesis. *After* developing a hypothesis, *the next step* is testing it. *The last step* is reporting the discovery. 科学发现的步骤是这样的:首先是确认课题,接着是研究与该课题有关的、现有的学说。在此基础上,第三步就是收集与该课题相关的资料。第四步最重要,就是形成假设。紧接着就是对该假设加以验证。最后就是把这一科学发现公之于众。

4.1.3 下定义

下定义是为了对某一科学概念或事物的本质和特征有清晰的介绍而采用的一种描述方法。一般说来定义语言都遵循一定的格式,即:被定义的事物=被定义事物的种属、类别+被定义事物的特殊属性(如果被定义的事物是可数名词,要用单数形式,并冠以不定冠词 a 或 an)。常用 be, be called, be known as, be defined as, be referred to as...等结构。如:

<u>A vernier caliper</u> is <u>a measuring instrument</u> <u>consisting of an L-shaped frame with...</u>

被定义事物　　被定义事物的种属　　被定义事物的特殊属性

定义通常采用关系从句(限制性或非限制性关系从句,多使用关系代词 who, which, whom, that, whose 等)或非限定动词结构,也可以采用括号或者破折号对所定义的对象进行解释,如:

The particles making up atoms are often referred to as elementary particles or fundamental particles. 构成原子的粒子通常称为基本粒子。

A force is an effect of attraction or repulsion between two bodies or particles. 力是物体或粒子之间的吸引或排斥效应。

另外,我们还会见到一句话定义后,又补充有例证、对比、说明等较长的定义形式。如:

Latent heat is the heat needed to change the state of matter of a material. While the latent heat is given to the material, its temperature remains constant. 潜热是改变物质的物态所需的热量。当物质吸收潜热时，其温度保持不变(恒定)。

4.1.4 分类

物理学中的事物错综复杂、多种多样，为研究、分析、讨论的方便，往往要对事物进行分类，找出它们之间的共同点，说明它们之间的区别。

对事物进行分类，需要三个方面的信息：①被分类的一组事物；②被分成的不同类型；③分类的标准或依据。

专业英语文本中常见的分类叙述有以下几种。

1) 完整的分类叙述：有关文字叙述直接提供上述三方面的信息。如：

Waves may be divided into two types based on the direction of the particles' oscillations relative to the wave velocity: transverse waves and longitudinal waves.

从本句可以看到：①被分类的事物为 waves；②被分成的类型为两种，即 transverse waves and longitudinal waves；③分类依据为 the direction of the particles' oscillations。

2) 不完整的分类说明：略去上述三条信息中的有关分类的标准或依据。如：

Quantities in physics can be classified as vectors or scalars.

3) 完全分类：根据某一标准把分类事物的全部归到一定数目的类型之中，加以分析讨论。如：

All the different forces observed in nature can be classified as members of one of four fundamental forces. These four fundamental forces are the gravitational force, the electromagnetic force, the strong nuclear force and the weak nuclear force. For example, the weight of the body is the gravitational force.

4) 不完全分类：在分类时，只提及有讨论价值的几种类型，而略去其余无关紧要的，或没有代表性的事物。如：

According to their shapes, capacitors can be classified into several categories: parallel-plate capacitors(平行板电容器), cylindrical capacitors(圆柱形电容器), spherical capacitors (球形电容器), etc.

在此例句中，没有说明电容器到底可以分成几种类型，只举了三种常见类型，所以我们称为"不完全分类"。

5) 多级分类：根据某一分类原则把一大类事物分成若干"子类"，然后再根据另一个分类原则把某一"子类"再分成若干更小的类别，更小的类别还可以继续分类，以此类推。我们把这种逐层分类法称为多级分类。如：

Synthesis techniques for nanomaterials, in general, can be divided into two broad categories as chemical methods and physical methods. Under chemical methods different routes, viz., colloidal, capping, cluster formation, sol-gel, electrochemical, etc., are being followed. Physical methods mostly used are molecular beam epitaxy, ionised cluster beam, liquid metal ion source, consolidation, sputtering and gas aggregation of monomers.

6）隐含分类：在上述五种分类中，我们可以直接使用分类词，如 divide，classify 等。但有时文章中不出现这种直接提供分类信息的词语，这种情况我们称为隐含分类。如：

In physics sometimes we are concerned with the properties of substances, which may be either physical properties or chemical properties. Physical properties of a substance are those which can be observed or measured without changing it into another substance; for example, its color, form, hardness and its melting and boiling points. Chemical properties of a substance are those which describe its behavior when it changes into other substances, e.g. when it burns or is destroyed by acids.

虽然在此作者没有使用分类词语，但同样为我们提供了有关分类的三方面的信息：①被分类的事物，properties of substance；②被分成的类型，physical properties 及 chemical properties；③分类依据，whether the properties involve any change from one substance to another substance，or more substances.

4.2 专业英语阅读

4.2.1 Circular Motion（圆周运动）

Motion along a circular path, or a segment of a circular path, is called **circular motion.**

If a particle moves along a circular arc, the direction from the particle toward the center of the circle is called the centripetal direction.

Angular Velocity and Angular Acceleration（角速度与角加速度）

Fig. 4-1 illustrates a particle undergoes circular motion about O. As the particle moves along the circle from point A to the point B in a time interval Δt, it moves through an arc length Δs and the radius vector sweeps out an angle $\Delta\theta$, which equals the angular displacement. We define the average angular velocity as the ratio of this angular displacement to the time interval Δt:

$$\omega = \frac{\Delta\theta}{\Delta t} \tag{4-1}$$

The **instantaneous angular velocity**, ω, is defined as the limit of the ratio in Eq. (4-1) as Δt approaches zero:

$$\omega = \lim_{\Delta t \to 0} \frac{\Delta\theta}{\Delta t} = \frac{d\theta}{dt} \tag{4-2}$$

Angular velocity has units of rad/s, or s^{-1}, since radians are not dimensional. We shall take ω to be positive when θ is increasing (counterclockwise motion) and negative when θ is decreasing (clockwise motion).

Note that angular velocity is a vector. Its direction follows the **right-hand rule**, perpendicular to the plane the particle moves in. That is, the four fingers of the right hand are wrapped in the direction of the moving particle, and the extended right thumb points in the direction of ω (Fig. 4-2). Angular velocities, when they add vectorially, follow parallelogram rule for vectors.

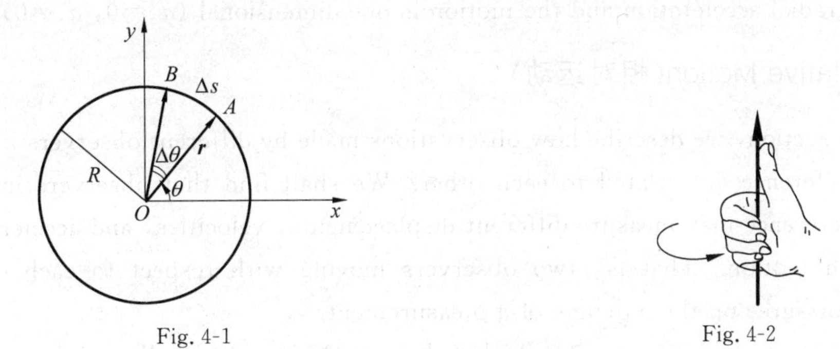

Fig. 4-1 Fig. 4-2

If the instantaneous angular velocity changes, the particle has an angular acceleration. In analogy to linear acceleration, the **angular acceleration $\boldsymbol{\alpha}$** is defined as

$$\boldsymbol{\alpha}=\lim_{\Delta t\to 0}\frac{\boldsymbol{\omega}_{t+\Delta t}-\boldsymbol{\omega}_t}{\Delta t}=\lim_{\Delta t\to 0}\frac{\Delta\boldsymbol{\omega}}{\Delta t}=\frac{\mathrm{d}\boldsymbol{\omega}}{\mathrm{d}t} \tag{4-3}$$

Angular acceleration has units of rad/s^2, or s^{-2}. Note that α is positive when ω is increasing in time and negative when ω is decreasing in time.

Circular Motion with Varying Speed（变速圆周运动）

Let us consider the motion of a particle along a circular path where the velocity changes both in direction and in magnitude. In this situation, the velocity of the particle is always tangent to the path; however, the acceleration vector $\boldsymbol{a}$ is now at some angle to the path. As the particle moves along the curved path, the direction and magnitude of the acceleration vector, $\boldsymbol{a}$, may change from point to point. This vector can be resolved into two component vectors: a radial component vector, $\boldsymbol{a}_\mathrm{n}$, and a tangential component vector, $\boldsymbol{a}_\mathrm{t}$. That is, the total acceleration vector, $\boldsymbol{a}$, can be written as the vector sum of these component vectors:

$$\boldsymbol{a}=\boldsymbol{a}_\mathrm{n}+\boldsymbol{a}_\mathrm{t} \tag{4-4}$$

The **tangential acceleration** arises from the change in the speed of the particle, and its magnitude is given by

$$a_\mathrm{t}=\frac{\mathrm{d}v}{\mathrm{d}t} \tag{4-5}$$

The **radial or normal acceleration** is due to the time rate of change in direction of the velocity vector and has a magnitude given by

$$a_\mathrm{n}=\frac{v^2}{R} \tag{4-6}$$

Since $\boldsymbol{a}_\mathrm{n}$ and $\boldsymbol{a}_\mathrm{t}$ are perpendicular component vectors of $\boldsymbol{a}$, it follows that $a=\sqrt{a_\mathrm{n}^2+a_\mathrm{t}^2}$. As in the case of uniform circular motion, $\boldsymbol{a}_\mathrm{n}$ always points toward the center of the circle. The direction of $\boldsymbol{a}_\mathrm{t}$ is either in the same direction as $\boldsymbol{v}$ (if the speed v is increasing) or opposite $\boldsymbol{v}$ (if the speed v is decreasing).

Note that in the case of uniform circular motion, where v is constant, $a_\mathrm{t}=0$ and the acceleration is always radial. Furthermore, if the direction of $\boldsymbol{a}$ does not change, then there is no radial acceleration and the motion is one-dimensional ($a_n=0$, $a_t\neq 0$).

4.2.2 Relative Motion(相对运动)

In this section, we describe how observations made by different observers in different frames of reference are related to each other. We shall find that observers in different frames of reference may measure different displacements, velocities, and accelerations for a particle in motion. That is, two observers moving with respect to each other will generally not agree on the outcome of a measurement.

Suppose a person on a moving vehicle (observer A) throws a ball straight up in the air according to his frame of reference, as in Fig. 4-3(a). According to observer A, the ball will move in a vertical path. On the other hand, a stationary observer B, standing on the ground, will see the path of the ball as a parabola, as illustrated in Fig. 4-3(b).

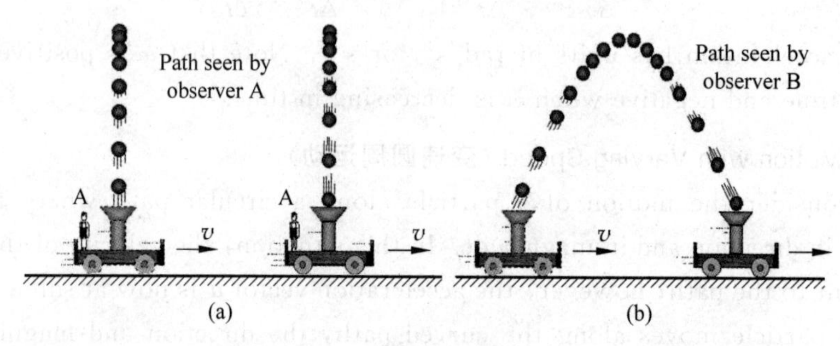

Fig. 4-3

(a) Observer A in a moving vehicle throws a ball upward and sees a straight-line path for the ball;
(b) A stationary observer B sees a parabolic path for the same ball

In a more general situation, consider a particle located at the point P in Fig. 4-4. Imagine that the motion of the particle is being described by two observers, one in reference frame S, fixed with respect to the earth, and the other in reference frame S', moving to the right relative to S with a constant velocity $\boldsymbol{u}$. (Relative to an observer in S', S moves to the left with a velocity $-\boldsymbol{u}$.)

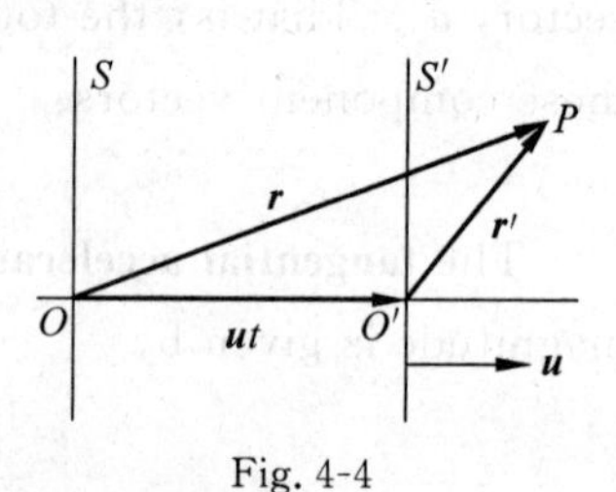

Fig. 4-4

We label the position of the particle with respect to the S frame with the position vector $\boldsymbol{r}$ and label its position relative to the frame S' with the vector $\boldsymbol{r}'$, at some time t. If the origins of the two reference frames coincide at $t=0$, then the vectors $\boldsymbol{r}$ and $\boldsymbol{r}'$ are related to each other through the expression

$$\boldsymbol{r}=\boldsymbol{r}'+\boldsymbol{u}t \tag{4-7}$$

If we differentiate Eq. (4-7) with respect to time and note that $\boldsymbol{u}$ is constant, we get

$$\frac{\mathrm{d}\boldsymbol{r}}{\mathrm{d}t}=\boldsymbol{u}+\frac{\mathrm{d}\boldsymbol{r}'}{\mathrm{d}t} \tag{4-8}$$

$$v = u + v' \tag{4-9}$$

where v' is the velocity of the particle observed in the frame S' and v is the velocity observer in the S frame. Eqs. (4-7) and (4-9) are known as **Galilean transformation equations.** They relate the coordinates and velocity of a particle in the earth's reference frame to those measured in a frame of reference in uniform motion with respect to the earth.

Glossary

circular motion	圆周运动	centripetal	向心的
arc length	弧长	angular displacement	角位移
instantaneous angular velocity	(瞬时)角速度	radian(s)	弧度
dimensional	有量纲的	counterclockwise	逆时针
clockwise	顺时针	circle	圆
center of a circle	圆心	vectorially	矢量地
angular acceleration	角加速度	tangential acceleration	切向加速度
resolve	(矢量)分解	uniform circular motion	匀速圆周运动
radial	径向的	radial/normal acceleration	法向加速度
perpendicular	垂直的	observer	观察者
outcome	结果	measurement	测量
stationary	静止的	differentiate	微分
Galilean transformation	伽利略变换		

4.3 专业英语常用表达法-4 组成 构成 成分 质量 标准

1. 表达"组成、构成、成分"的常用名词

component 成分、组分、零件

composition 合成、组成、成分

constitution 构造、组成

constituent 成分、要素

content 内容、含量

formation 形成、构成

ingredient 成分、配料

make-up 组成、构造

2. 表达"组成、构成、成分"的常用词组

component assembly 零件装配

shock wave formation 激波系

component part 成分、组分、零件

microscopic constitution 显微组织

component wire (电缆)芯线

structural constitution 结构成分

bath composition 电解液成分

component (of) velocity 分速度

composition of forces 力的合成

component sine waves 正弦波分量

alloying ingredient 合金的组分

pressure component 分压力

crystal formation 结晶、晶体生成

composition metal 合金

formation of n-p-n junction N-P-N 结结构

composition of radiance 辐射谱

moisture content 湿度、含水量

formation of image　成像
chemical constitution　化学成分(结构)
isotopic constitution　同位素成分
pulse formation　脉冲的形成
molecular constitution　分子结构(构成)

以下是一些具体的例子：

The factory produces *components* for aircraft. 这个工厂生产飞机构件(部件)。

The *composition* of cast-iron is different for different purposes. 铸铁的成分随用途的不同而有所不同。

The carbon *content* of wrought-iron is very low. 熟铁的碳含量很低。

Ferrite and carbon are the *constituents* of mild steel. 铁氧体和碳是软钢的成分。

The moisture *content* of the cylinder increased. 汽缸的水分含量增加了。

The chemical *composition* of water remains constant whether it is in solid, liquid or gaseous state. 无论处于固态、液态或气态，水的化学成分都保持不变。

3. 表达"组成、构成、成分"的常用动词和短语

be combined into...　组成、构成
compose　组成、构成
constitute　组成、构成
form　组成、构成
make up　组成、构成
be composed of...　由……组成
be made up of...　由……组成
be sprayed with...　用……喷镀
consist of...　由……组成
comprise　由……组成
contain　包含、包括

以下是一些具体的例子：

be sprayed with a metal coating　喷镀一层金属

be sprayed with a coating of paint　喷镀一层涂料

A hydrogen atom *consists of* a single electron moving round a single proton. 一个氢原子是由一个电子绕着一个质子旋转构成的。

All substances on the earth, whether gaseous, liquid or solid, *are made up of* atoms. 地球上所有的物质无论它们是气态的、液态的或是固态的，都是由原子组成的。

This physical instrument *is composed of* several different parts. 这台物理仪器由几个不同的部件组装而成。

Steel *is composed of* iron and a number of other elements. 钢是由铁和若干种其他元素构成的。

Fermi-Dirac statistic is a part of the science of physics that describes the energies of single particles in a system *comprised of* many particles that obey the Pauli Exclusion Principle. 费米-狄拉克统计是物理学的一部分，它用来描述由遵守泡利不相容原理的多粒子所构成系统的单个粒子的能量。

The technician *made up* a bottle of dilute sulfuric acid. 技术人员配置了一瓶稀硫酸。

Brass *consists of* copper and zinc. 黄铜由铜和锌构成。

The atmosphere *comprises* a number of gases. 大气由许多种气体构成。

The electro-magnet *comprises* a soft iron core shaped like a horseshoe. 电磁铁包括一块马蹄形的软铁芯。

Cast-iron *is made up of* about six different substances. 铸铁大约由六种不同的物质构成。

Sometimes two different types of machines *are combined into* a complex one. 有时两台不同的机器组成一台复杂的机器。

Ferrite and carbon *make up* mild steel. 铁氧体和碳构成软钢。

The alloy *contains* 10% nickel and 8% iron. 这合金含有10%的镍和8%的铁。

The rare earths *comprise* the series of elements in the sixth row of the periodic table stretching from lanthanum to ytterbium. 稀土(元素)包括元素周期表第六行中的从镧到镱的一系列元素。

4. “质量高,优质”的常用表达

常用 fine, high, good, better, best, top, first-class, first-rate 等词,也可说

standard quality　符合标准的质量

choice quality　精选的质量

a fine quality of...　优质的……

a good quality stop watch　一块高质量的停表

products of quality　优质产品

quality concrete　高级/优质混凝土

be superior in quality　质量好

have quality　质量好

of good quality　上等的、优质的

5. “符合质量要求”的常用表达

reach the (official) standard　达到(官方)标准

maintain the quality　保持合格质量

meet the (minimum) standard　达到(最低)标准

be up to the standard (grade, category)　达标

come up to the standard　达标

make the grade　达到(理想)标准、合乎(质量)要求

6. “质量一般”的常用表达

fair quality　质量还行

acceptable quality　可接受的质量

above the average quality　高于平均质量水平

7. “基本够格,稍有问题”的常用表达

be borderline　基本够格

have a borderline result　结果勉强够格

8. “质量次、差”的常用表达

除 bad, poor, inferior 等词外,还可说

off quality　品质差

below standard　低于标准

not up to standard　未达标

below the average quality　低于平均质量水平

out of grade　未达标

of poor quality　劣等的、劣质的

Unit Two

LESSON 5

5.1 物理学专业英语中的比较

在物理学中，经常需要对所研究的事物或采用的方法进行比较说明。比较是人们认识客观事实的一个重要途径，它利用事物之间的相似或不同之处，通过类比使人们理解或接受那些不易理解或接受的东西。英语中的比较结构内容丰富，形式多样，并有其独特的表现形式，所以正确理解和表达比较结构及其相应翻译在专业英语学习中很有必要。通过比较可以发现被描述物之间的相同点和不同点，能对抽象的概念，如事物的性质、特点、功能、形状、尺寸等，有较为具体的了解。下面列举几种常见的比较和对比。

5.1.1 同等比较

表示两对象在性质、特征等方面相似或相同。其结构形式有：

1) as...as （和……一样）

The lightning rod is as good as any other. 这避雷针与其他任何一种避雷针一样好。

2) be identical with, be the same as, be equal to （和……相同）

The TV set uses the same capacitor as that one does. 这台电视机用的电容器和那台一样。

The rating of this engine is equal to that of Toyota motor. 这台发动机的额定值与那台丰田发动机的额定值等同。

3) no more(less)...than... （和……一样不……，不亚于……）

Rubber is no more conductor than glass. 橡胶和玻璃一样都不是导体。

类似的表达结构还有：

As (Just as)..., so... （和……一样，……也……）

The same thing is true of... （同样……也……）

Like (Similarly, Likewise, In the same way)... (同样……)

A and B have/share several things in common. A 和 B 有几个共同点。

A and B are alike/identical/similar/the same (in design/in every respect/in most respects). A 和 B(在设计上/在每一方面/在大多数方面)相像/一致/类似/一样。

A like/as B has/have/is/are C. A 和 B 一样，都是 C。

A bears some close resemblance(s) to B. A 和 B 有一些相似之处。

There is little resemblance between A and B. A 和 B 没有相似之处。

A do/does/be…, so do/does/be B. （A……,B 也……）

A do/does/be not…, nor do/does/be B. （A……, B 也不……）

（As ＋从句/短语）＋主句＋as 从句/短语 （和……一样……）

这些句型还可以与一些副词连用来表示程度,例如 exactly, just, precisely, more or less, almost, approximately, about, practically 等。

5.1.2 差等比较

差等比较包括两方面:"优等比较"和"劣等比较"。前者表示"A 优于 B",后者表示"A 劣于 B"。其结构形式有:

1）…than… （比更……一些）

The carbon content of cast steel is greater than that of mild steel. 铸铁的碳含量比软钢高。

2）…not as/so …as （和……不一样,不如……）

Water doesn't have a boiling point as low as alcohol. 水的沸点不如酒精低。

3）be superior to /be inferior to （优于……/劣于……）

The power of the new engine is superior to that of the old one. 这台新发动机的功率优于旧的。

5.1.3 有额比较

表示具体比较结果,即"额"是多少。其结构形式有:

1）额＋比较级＋than （比……多,是……）

Sound travels nearly three times faster in copper than in lead. 声音在铜中的传播速度几乎是铅中的 3 倍。

2）额＋as＋原级＋as, 额＋the size/weight/price/number/age/that… （是多少……）

Jupiter is 1500 times as large as/the size of the Earth. 木星是地球的 1500 倍大。

The effort is two thirds as great as the resistance. 作用力是阻力的 2/3。

3）表示减少或增加意义的动词(decrease, increase, rise, exceed, grow, raise, expand, go up 等)＋额（增加/减少了……）

The production of various picture tubes has increased three times as against 2003. 各种显像管的产量比 2003 年增加了 3 倍。

The new invention makes it possible that small amounts of electricity can increase hydrogen production from waste water 4 times. 这项新发明使得低耗电废水提取氢的产量提高 4 倍成为可能。

The technicians tried hard to make the balloon expand its size 6 times. 技师们尽最大努力使得气球的体积膨胀了 6 倍。

4）表示倍数意义的动词(double, treble, quadruple) ＋宾语/表语 （增加了……倍）

Production quadruples this year. 今年产量增加了 3 倍。

5.1.4 递进比较

这种比较表示一种不断递进的语义关系。其结构形式有：

1) the＋比较级＋句子/句子＋比较级，the＋比较级＋句子 （越……，越……）

The thicker the wire is，the more freely it can carry current. 导线越粗，导电就越容易。

A subject weighs less，the farther it gets from the surface of the earth. 物体离地球表面越远重量越轻。

2) 比较级＋and＋比较级 （越来越……）

More and more new knowledge is discovered in physics. 越来越多的物理学新知识被发现。

More and more scientists today work for government sponsored bodies. 今天越来越多的科学家为政府支持的机构工作。

3) never...enough，never too...，can't...over/too （越……越……，怎么也不为过）

You can never be careful enough when operating this instrument. 操作这台仪器越小心越好。

The significance of the invention can't be overestimated. 这个发明的意义不管怎样估计都不会过高。

5.1.5 终极比较

表示某物在特定范围内显得最突出，或某一动作在一定情况下达到最高程度。其结构主要用形容词和副词的最高级。

The kilometer is the largest metric unit of measurement. 千米是最大的米制衡量单位。

注意：终级比较和差等比较也可表示完全相同的意思，只是说话者看问题的角度不同而已。这时候的差等比较结构中往往带有 other 或 else。如：

Strontium titanate is far better as a dielectric than any other ones in the group. 作为电介质，钛酸锶比其他电介质强得多。

5.1.6 选择性比较

1) more...than，less...than，better...than，not so much...as （与其……不如……，不是……倒是……，不但没有……反而……）

His explanation about the Hall effect was more confusing than clarifying. 他对于霍尔效应的解释不但没使人明白，反而使人更糊涂了。

This material is not so much a conductor as a semiconductor. 与其说这种材料是导体不如说是半导体。

2) prefer...to...，prefer to...，rather than... （喜欢……不喜欢……）

Physicists usually prefer idealized models to real objects when dealing with problems. 在处理问题时，物理学者通常喜欢采用理想模型取代实际物体。

He prefers to carry out experiments rather than theoretical induction. 他喜欢动手做实

验，不喜欢理论推理。

5.1.7 表示 A≠B

即 A 与 B 不相同，有差异，其常用句式为：

A is not as/so＋形容词 as B.

A is (totally/completely/entirely/quite) different from B (in every way/in every respect/in its shape/that 从句).

A (totally/completely/entirely/quite) differs from B (in every way/in every respect/in its shape/that 从句).

A can be distinguished from B (by its shape).

A unlike/as distinct from/as against B has/have C/ is/are C.

例句：

Heat waves are different/differ from light waves only in their wavelengths. 热波与光波的差异只是波长不同。

Iron, unlike glass, is a good conductor of electric current. 铁不同于玻璃，它是电流的良导体。

5.2 专业英语阅读

5.2.1 Newton's Laws of Motion(牛顿运动定律)

The purpose of classical mechanics is to provide a connection between the acceleration of a body and the forces acting on it. Keep in mind that classical mechanics deals with objects that are large compared with that dimensions of atoms ($\approx 10^{-10}$ m) and move at speeds that are much less than the speed of light (3×10^{8} m/s).

In this chapter, we describe Newton's three laws of motion and begin using them to solve problems involving objects in motion and at rest.

5.2.2 Newton's First Law(牛顿第一定律)

A modern wording of **Newton's first law of motion** is:

An object at rest will remain at rest and an object in motion will continue in motion with a constant velocity (that is, constant speed in a straight line) unless it experiences a net external force (or resultant force).

In simpler terms, we can say that when the resultant force on a body is zero, its acceleration is zero. From the first law, we conclude that an isolated body (a body that does not interact with its environment) is either at rest or moving with constant velocity.

Inertial Frames(惯性系)

Newton's first law is sometimes called **law of inertia**, and it applies to objects in an inertial frame of reference.

An **inertial frame of reference** is one in which an object, subject to no force, moves with constant velocity. That is, a reference frame in which Newton's first law is valid is called an **inertial frame.**

In effect, Newton's first law defines an inertial frame of reference. A reference frame that moves with constant velocity relative to the distant stars is the best approximation of an inertial frame. The earth is not an inertial frame because of its orbital motion about the sun and rotational motion about its own axis. However, in most situations we shall assume that the earth is an inertial frame.

Thus, if an object is in uniform motion ($\boldsymbol{v}=$ constant), an observer in one inertial frame (say, one at rest with respect to the object) will claim that the acceleration and the resultant force on the object are zero. An observer in any other inertial frame will also find that $\boldsymbol{a}=0$ and $\boldsymbol{F}=0$ for the object. According to the first law, a body at rest and one moving with constant velocity are equivalent. Unless stated otherwise, we shall unusually write the laws of motion with respect to an observer "at rest" in an inertial frame.

5.2.3 Newton's Second Law(牛顿第二定律)

Newton's second law answers the question of what happens to an object that has a nonzero resultant force acting on it.

Newton's second law of motion is:

The time rate of change of momentum of an object is equal to the resultant external force acting on the object.

$$\sum \boldsymbol{F}=\frac{\mathrm{d}\boldsymbol{p}}{\mathrm{d}t}=\frac{\mathrm{d}(m\boldsymbol{v})}{\mathrm{d}t} \tag{5-1}$$

where $\boldsymbol{p}$ is the momentum of the object as the product of the mass, m, and the velocity, $\boldsymbol{v}$

$$\boldsymbol{p}=m\boldsymbol{v} \tag{5-2}$$

Eq. (5-1) is the most general form of Newton's second law, which is valid in any inertial frame of reference.

If m is treated as a constant, then Eq. (5-1) can be expressed

$$\sum \boldsymbol{F}=\frac{\mathrm{d}\boldsymbol{p}}{\mathrm{d}t}=\frac{\mathrm{d}(m\boldsymbol{v})}{\mathrm{d}t}=m\frac{\mathrm{d}\boldsymbol{v}}{\mathrm{d}t}=m\boldsymbol{a} \tag{5-3}$$

which can be stated as the acceleration of an object is directly proportional to the resultant force acting on it and inversely proportional to its mass.

5.2.4 Newton's Third Law(牛顿第三定律)

Newton's third law states that *if two bodies interact, the force exerted on body* 1 *by body* 2 *is equal to and opposite the force exerted on body* 2 *by body* 1. That is,

$$\boldsymbol{F}_{12}=-\boldsymbol{F}_{21}$$

This law is equivalent to stating that forces always occur in pairs, or that a single isolated force cannot exist. The force that body 1 exerts on body 2 is sometimes called the **action force**, while the force of body 2 on body 1 is called the **reaction force.** Either force can be labeled the action or reaction force. The action force is equal in magnitude to the reaction force and opposite in direction. In all cases the action and reaction forces act on different objects.

5.2.5 The Four Fundamental Forces(四种基本力)

All the different forces observed in nature can be explained in terms of four basic interactions that occur between elementary particles.

The Gravitational Force(引力)—the force of mutual attraction between objects

Newton's law of universal gravitation states that *every particle in the universe attracts every other particle with a force that is directly proportional to the product of the masses of the particles and inversely proportional to the square of the distance between them.*

$$F = \frac{Gm_1m_2}{r^2} \tag{5-4}$$

where $G = 6.67 \times 10^{-11}\ \text{N} \cdot \text{m}^2/\text{kg}^2$ is the **universal gravitational constant.**

The Electromagnetic Force(电磁力)—the force between electric charges

This class of forces includes electric and magnetic forces. The electric force binds atoms and molecules in compounds to form ordinary matter. It is much stronger than the gravitational force. Coulomb's law states that the magnitude of the electrostatic force between two charged particles is

$$F_e = \frac{kq_1q_2}{r^2}$$

where $k = 9 \times 10^9\ \text{N} \cdot \text{m}^2/\text{C}^2$ is the **Coulomb constant.**

The Strong Nuclear Force(强核力)—the force between subatomic particles

This force holds the nucleus of an atom together. All nuclei except those of hydrogen contain electrically neutral neutrons and positively charged protons. The charged protons repel each other. It is the strong nuclear force that counteracts the repulsive electrical interactions. Note that the nuclear force is extremely short range and its strength decreases very rapidly outside the nucleus.

The Weak Nuclear Force(弱力)—the force between subatomic particles during certain radioactive decay processes

The weak nuclear force is a **short-range force** that is responsible for a common form of radioactivity called beta decay, in which a neutron in a radioactive nucleus is transformed into a proton while ejecting an electron and an essentially massless particle. The weak force play no direct role in the behavior of ordinary matter, but it is very important in interactions among fundamental particles.

Glossary

at rest	静止	approximation	近似
net external force/resultant force	合外力	inertial frame of reference	惯性参考系
inertia	惯性	act on＝exert	(力)作用于
unless stated otherwise	除非另有说明	nonzero	非零的
rate of change	变化率	mass	质量
directly proportional to	正比于	momentum	动量
inversely proportional to	反比于	interact	相互作用
action force	作用力	opposite	相反、相对
reaction force	反作用力	isolated	孤立的
gravitational force	引力	universal gravitational constant	万有引力常数
electromagnetic force	电磁力	electric charge	电荷
atom	原子	molecule	分子
compound	混合物、化合物	bind	约束
Coulomb's law	库仑定律	electrostatic force	静电力
charged particle	带电粒子	strong nuclear force	强力
subatomic	亚原子的	hydrogen	氢
nucleus (*pl*. nuclei or nucleuses)	原子核	neutron	中子
proton	质子	repel	排斥
repulsive	排斥的	counteract	抵抗
strength	强度	weak nuclear force	弱力
short-range force	短程力	radioactivity	放射性
radioactive decay	放射性衰变	beta decay	β衰变
electron	电子	massless	无质量的
fundamental particles	基本粒子		

5.3 专业英语常用表达法-5 力学常用实验仪器

pendulum wire 摆线	stopwatch 秒表
surface tension tester 表面张力测定仪	torsion pendulum 扭摆
dynamic meter 测力计	viscometer 黏度计
ballistic pendulum 冲击摆	gas density balance 气体密度秤
simple pendulum 单摆	balance/scale 天平
poise/weight 砝码	standard line meter 线纹米尺
(running)time meter/timer 计时器	pressure vacuum meter 压力真空表
Jolly balance 焦利秤	vernier caliper 游标卡尺
micrometer 螺旋测微仪、千分尺	pendulum weight/bob 摆锤

standard meter 标准米尺
tweezers 镊子
gear 齿轮
torsion balance 扭力天平、扭秤
ticker-tape timer 纸带打点计时器
air track 气垫导轨
spring balance 弹簧秤
barometer 气压计
analytical balance 分析天平
balance reading glass 天平读镜
psychrometer 干湿计
pressure meter/gauge 压力计
measuring cup/graduate 量杯
Young's modulus tester 杨氏模量测定仪
densimeter 密度计、比重计
balance spring/hair spring 游丝
centrifugal pendulum/centrifugal governor/centrifugal speed regulator 离心调速器
graduate, graduated/measuring volumetric cylinder 量筒
physical/compound pendulum 复摆、物理摆
target collision experiment facility 碰撞打靶实验仪
moment of inertia detector 转动惯量测定仪
new type of simple pendulum experiment facility 新型单摆实验仪
coefficient of liquid surface tension detector 液体表面张力系数测定仪
shear modulus and moment of inertia experiment facility 切变模量与转动惯量实验仪
general experimental sound velocity measurement instrument 声速测量综合实验仪
viscosity of liquid drop-ball tester 落球法液体黏滞系数测定仪
simple harmonic motion and spring stiffness coefficient experiment facility 简谐振动与弹簧劲度系数实验仪

LESSON 6

6.1 物理学专业英语中的举例和列举

6.1.1 举例

举例是专业英语中常见的表达方式。举例是作者为了更形象、更生动地揭示所表达的内容，帮助读者理解而采取的一种方法，起到进一步解释说明的作用。举例可以使抽象的内容更加具体、形象、生动，便于理解、记忆和运用。

在专业英语中，举例一般用于作者下完定义或写明某一观点之后。有时作者使用浅显的信息进一步解释抽象的概念、观点，使文章通俗易懂，表达清晰。

举例时常用到如下句型：

For example... 例如……

For example 既可以放在句首，也可以放在句中。放在句首时，要大写，后面有逗号；放在句中时，前后都有逗号，形似插入语；也可以放在破折号后面，因为破折号本身表示解释说明的意思。For example 没有连接词的作用，不能连接两个分句。

e.g. ... 例如……

e. g. 是拉丁语 exempli gratia 的缩写，意为 for example，但不如 for example 正式，使用时一般放在括号内。

take... as an example　以……为例

take... for example　以……为例

For instance...　例如……

与 for example 用法一样，但 for instance 比 for example 举例更具体些。

common/typical examples are...　常见的/典型的例子是……

an even more dramatic example is...　一个更生动的例子是……

examples of which are...　属于此种类型的例子有……

use an example to show...　用一个例子表明……

an example of...　一个……的例子

furnishes a good example of...　提供了一个……的好例子

like...　像……

afford an illustration of...　提供了一个……的例子

a case in point　恰当的例子

as an example　例如、举例来说

as an illustration　作为实例

as demonstrated by...　例如

by way of example　例如、举例来说

illustrated with...　用……举例说明

including　包含、包括

involving　包括

like　类似

range over　涉及、包括

say　比如说

something like...　有点像……

specifically　具体地

such　诸如此类的

such as　例如、诸如

take the case of...　以……为例

to illustrate by an example 举例说明

以下是一些具体的例子：

Waves transport energy and momentum through space without transporting matter. As a water wave moves across a pond, *for example*, the molecules of water oscillate up and down, but do not cross the pond with the wave. 波动只传播能量和动量而不传播媒质。例如，当水波穿过水池时，水分子上下振动，但是不随波穿过水池。

Take the spring system *as an example*. A block attached to a spring on a frictionless track moves in simple harmonic motion. 以弹簧系统为例，一个与弹簧相连、放在光滑轨道上的木块的运动就是简谐运动。

Mechanical waves are waves that disturb and propagate through a medium; the ripple in the water due to the pebble and a sound wave, for which air is the medium, are *examples of* mechanical waves. 机械波是扰动在媒质中传播形成的波，石头落入水中形成的波纹、以空气作媒质的声波都是机械波的例子。

To illustrate this, we *take* temperature *for example*. 为了阐明此理，我们以温度为例说明。

Another method that can be used to obtain the potential due to a finite continuous charge distribution is to start with the definition of the potential difference given by Eq. (8-7). If $\boldsymbol{E}$ is known or can be obtained easily (*e.g.* from Gauss's law), then the line integral of $\boldsymbol{E} \cdot \mathrm{d}\boldsymbol{r}$ can be evaluated. 还有另外一种方法可以用来求有限连续分布电荷产生的电势，那就是利用方程(8-7)定义的电势差。如果电场 $\boldsymbol{E}$ 已知或者可以容易求得(例如利用高斯定律求得)，那么 $\boldsymbol{E} \cdot \mathrm{d}\boldsymbol{r}$ 的线积分就可以计算出来了。

When, *for instance*, we hit a nail with a hammer, we expect the nail to move. 比如，当我们用锤子敲打钉子时，我们可预知钉子要移动。

The mutual inductance depends on the geometrical arrangement of the two circuits. *For instance*, if the circuits are far apart, the flux through circuit 2 due to the current I_1 will be small and the mutual inductance will be small. 互感与两个回路的几何形状位置有关。例如，如果两个回路相距较远，电流 I_1 在回路 2 产生的通量就会小，从而互感就会小。

Electromagnetic waves are a special class of waves that do not require a medium in order to propagate; it is the varying electromagnetic field that propagates through space, *examples of which are* light waves and radio waves. 电磁波是一种特殊种类的波，它的传播不需要媒质；它是变化的电磁场在空间的传播，例如光波和无线电波。

An even more dramatic example of longitudinal waves is sound waves in air. 空气中的声波是纵波的一个更生动的例子。

Many machines vibrate because they have rotating parts that are not in perfect balance (observe a washing machine in the spin cycle *for an example*). 许多机器都会振动，那是因为它们的转动部分未在精确的平衡位置，作为一个例子，你可以观察一下内部旋转时的洗衣机。

A common example of the conversion of work into heat is movement with friction. 功转化为热的一个常见的例子就是摩擦运动。

Some common examples of adiabatic processes include the expansion of hot gases in an internal combustion engine, the liquefaction of gases in a cooling system, and the compression stroke in a diesel engine. 绝热过程的一些常见的例子包括：内燃机内热气的膨胀、冷却系统内的气体液化、还有柴油机内的压缩冲程。

We'll *use an example to* show that if the heat-engine statement is false, then the refrigerator statement is false. 我们可以用一个实例来说明，如果关于热机的说法是错误的，那么关于冰箱的说法也是错误的。

Another example of such a forbidden process would be if all of the air in your room spontaneously moved to one half of the room, leaving a vacuum in the other half. 另外一

个不可能发生的过程的例子是：你房间内的空气自动聚集到房间的一半里边，而另一半成了真空状态。

The result shows that the capacitance is proportional to the length of the cylinders and also depends on the radii of the two cylindrical conductors. *As an example*, a coaxial cable consists of two concentric cylindrical conductors of radii a and b separated by an insulator. 结果表明，电容正比于两个圆柱形导体的半径和长度。半径分别为a和b的两个同轴导体，中间用绝缘体隔开形成的同轴电缆就可以作为一个例子。

A nonconducting material, *such as* air, glass, paper, or wood, is called a dielectric. 不导电的物质，比如空气、玻璃、纸、木头，叫作电介质。

A water wheel, turned by water flowing over a weir, *furnishes a good example of* potential energy being turned into useful mechanical energy. 由流水过坝转动的水轮提供一个说明位能正在转变成有用的机械能的例子。

A city electric-power plant *affords an illustration of* complicated transformations of energy. 城市发电厂为能量转换复杂关系的说明提供了例子。

6.1.2 列举

物理学专业英语中经常会用到多项内容的列举，例如同一对象呈现出的多种特性、变化，支持同一结论的各种依据、原因、理论、方法等。各项内容之间的关系也不尽相同，可以是并列、递进或对比等。这些内容之间的关系可以通过使用过渡词语来更为准确、清楚地表达，同时增加了内容的条理性，不会使某些方面被忽略。这些过渡词语让我们把两个观点更好地衔接在一起，使行文或口语表达紧凑、流畅、条理清晰。

以下就是一些专业英语中常见的表示列举的过渡词语。

(1) 表示并列、顺序与附加关系的

additionally 另外、加之
afterward 后来、以后
again 再次
alike 相似地、同样地
all the same 同样
also 也、而且、此外
analogous to 类似地
and 而且、亦、另外
another 另外的、另一
as well as 也、以及
aside from(this) 除……以外
at first 首先、最初
at the same time 同时
besides (that) 此外
by the same token 同样地
concurrently 同时
conjointly 相连地
coupled with 连同
equally important 同样重要
first of all 首先
following this 之后
for a start 首先
for one thing 首先、一则
in addition (to) 除此之外
in like manner 同样地
in similar fashion 相似地
in the first place 起初、首先
in the same way 同样地
later on 后来
likewise 同样地
meanwhile 同时
next 其次

on top of that 最后 so forth 等等

once again 再次 so on 等等

once more 再一次 subsequently 随后、后来

or 或者 then 然后

over again 再一次 to begin with 首先

plus 并且 together with 连同

similarly 类似地 what is more 而且

simultaneously 同时

Basically… Similarly… As well…

Both… and…

Either… or…

First… Second… Third… Next…

For one thing…, for another…

In the first place… Also… Lastly…

In the first place… Just in the same way… Finally…

Neither… nor…

Not only… but also…

To be sure… Additionally… Lastly…

例如：

Whether a physical model is good or not should be judged by how well it meets three standards. *First*, is the model similar with the original object in some specific aspects? *Second*, is it rational to neglect the difference between the model and the original object? *Finally*, is the model good enough to allow us to predict what would happen in a new situation that has not been investigated before? 一个物理模型是否完善应该通过其是否满足三个标准来判定。首先，这个模型是否与原物在某些具体方面相似？其次，对原物与模型间的差异进行忽略是否合理可行？最后，这个模型是否完善到可以使我们在尚未研究过的新环境下对未来将要发生的情况进行预测？

(2) 表示递进关系的

even more 更 more important 更重要的是

further 此外、进一步 moreover 此外、加之

furthermore 而且、此外 still 更

Clearly… Then…

Generally… Furthermore… Finally…

In the first place… In the second place… Lastly…

In the first place… Pursuing this further… Finally…

On the one hand…, on the other hand…

(3) 表示转折、让步与对照关系的

alternatively 作为选择、可选择的 as opposed to 与……相对比

although 尽管、虽然 but 但是

by comparison 比较起来
by way of contrast 两相对比
compared to 与……相比
contrarily 反之、相反地
conversely 相反地
despite 不管、尽管
even so 虽然如此
even though 即使
however 然而
in contrast 与此相反
in spite of 尽管
in turn 反之、反过来
instead 代替、也可以
Not... but... 不是……,而是……
Not only... but also... 不但……,而且……
Sometimes... But not always... 有时……,但不总是……
nevertheless 仍然、不过
nonetheless 尽管如此(仍然)
notwithstanding 虽然、尽管
on the contrary 正相反
otherwise 否则、要不然
rather than 不是……而是……
rather 恰恰相反
still 仍然
whereas 然而、反之、却
yet 然而

6.2 专业英语阅读

6.2.1 Work and Energy(功和能)

Work and **energy** are important concepts in physics as well as in our everyday lives. In physics, a force does work if its point of application moves through a distance and there is a component of the force in the direction of the velocity of the force's point of application. For a constant force in one dimension, the work done equals the force component in the direction of the displacement times the displacement.

Energy is closely associated with work. When work is done by one system on another, energy is transferred between the two systems. For example, when you do work pushing a swing, chemical energy of your body is transferred to the swing and appears as kinetic energy of motion or as gravitational potential energy of the earth-swing system.

6.2.2 Work(功)

In dot-product notation, the element work $\mathrm{d}A$ done by a constant force $\boldsymbol{F}$ on a particle over a displacement $\mathrm{d}\boldsymbol{r}$ is

$$\mathrm{d}A = \boldsymbol{F} \cdot \mathrm{d}\boldsymbol{r} \tag{6-1}$$

The work done on the particle as it moves from point 1 to point 2 is

$$A = \int_1^2 \boldsymbol{F} \cdot \mathrm{d}\boldsymbol{r} \tag{6-2}$$

Work is a scalar quantity that is positive if $\mathrm{d}\boldsymbol{r}$ and $\boldsymbol{F}$ have the same signs and negative if they have opposite signs. The SI unit of work and energy is the **joule** (J), which equals the product of a newton and a meter:

$$1\ \text{J} = 1\ \text{N} \cdot \text{m}$$

A convenient unit of work and energy in atomic and nuclear physics is the electron volt (eV):

$$1\ \text{eV} = 1.6 \times 10^{-19}\ \text{J}$$

Commonly used multiples are keV (10^3 eV) and MeV (10^6 eV). The work required to remove an electron from an atom is of the order of a few eV, whereas the work needed to remove a proton or a neutron from an atomic nucleus is of the order of several MeV.

Work Done by a Variable Force（变力的功）

In Fig. 6-1 we plot a constant force F_x as a function of position x. The work done on a particle whose displacement is Δx is represented by the area under the force versus-position curve, indicated by the shading in Fig. 6-1.

Many forces vary with position. For example, a stretched spring exerts a force proportional to the distance it is stretched. And the gravitational force the earth exerts on a spaceship varies inversely with the square of the distance between the two bodies. We can approximate a variable force by a series of constant forces (Fig. 6-2). The work done by a variable force is then

$$A = \lim_{\Delta x_i \to 0} \sum_i F_x \Delta x = \text{area under the } F_x\text{-versus-}x \text{ curve} \tag{6-3}$$

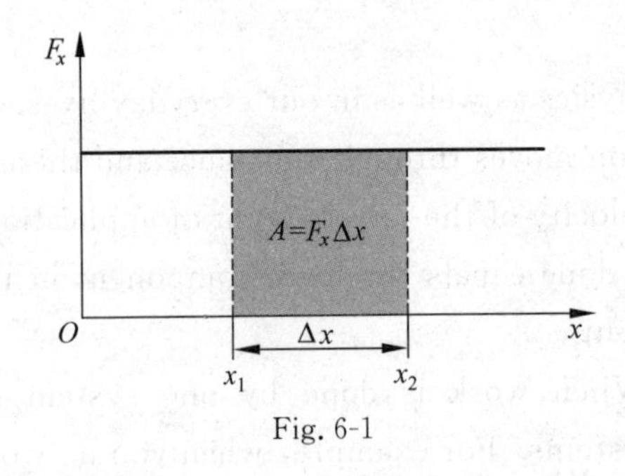

Fig. 6-1

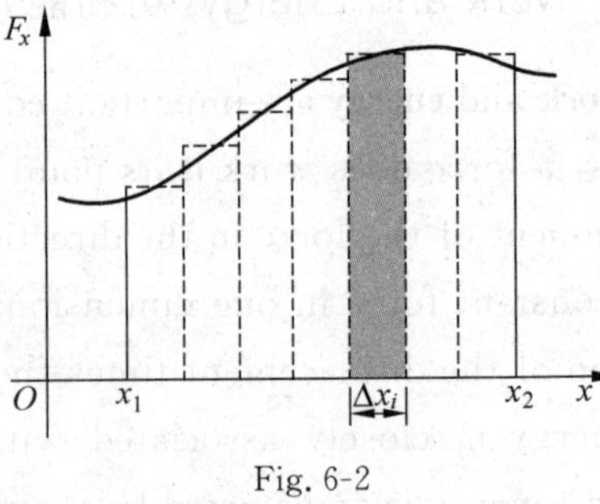

Fig. 6-2

This limit is the integral of F_x over x. So the work done by a variable force F_x acting on a particle as it moves from x_1 to x_2 is

$$A = \int_{x_1}^{x_2} F_x \mathrm{d}x = \text{area under the } F_x\text{-versus-}x \text{ curve} \tag{6-4}$$

For each displacement interval Δx_i, the force is essentially constant. Therefore the work done equals the area of the rectangle of height F_{xi} and width Δx_i. As was shown earlier, this work equals the change in kinetic energy for this displacement interval (if the force is the net force). The total work done is the sum of the areas over all displacement intervals. It follows that the total work equals the change in kinetic energy for the entire displacement. Thus, $A_{\text{total}} = \Delta E_{\text{k}}$ holds for variable forces as well as for constant forces.

6.2.3 Conservative Force and Potential Energy（保守力和势能）

The total work done on a particle equals the change in its kinetic energy. But we are often interested in the work done on a system consisting of two or more particles. Often, the work done by external forces on a system does not increase the total kinetic energy of the system, but instead is stored as **potential energy**—energy associated with the configuration of the system.

Potential Energy(势能)

Consider lifting a barbell of mass m to a height h. The work done by the gravitational force on the barbell starts at rest and ends at rest. Because the kinetic energy of the barbell does not change, we know the total work done on the barbell is zero. That means the work done by the force of your hands on the barbell is $+mgh$. Now consider the barbell and the earth to be a system of two particles (You are not the part of this system). The external forces on the earth-barbell system are the gravitational attraction you exert on the earth, the force your feet exert on the earth, and the force mg exerted by your hands on the barbell (Fig. 6-3). The barbell moves, but the motion of the earth is negligible, so only the force exerted on the barbell by your hands does work on the system. The total work done on the earth-barbell system by all forces external to the system is mgh. This work is stored as potential energy, which is energy associated with the position of the barbell relative to the earth. That is energy associated with the configuration of the earth-barbell system. This kind of energy is called **gravitational potential energy.**

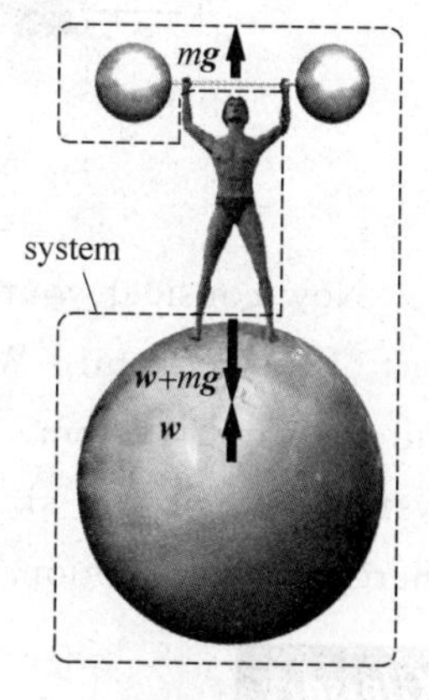

Fig. 6-3

Another system that stores energy associated with its configuration is a spring. If you stretch or compress a spring, energy associated with the length of the spring is stored as potential energy. Consider the spring shown in Fig. 6-4 as the system. You compress the spring, pushing it with equal and opposite forces $\boldsymbol{F}_1$ and $\boldsymbol{F}_2$. These forces sum to zero, so the net force on the spring remains zero. Thus, there is no change in the kinetic energy of the spring. The work you do on this system is stored not as kinetic energy, but as **elastic potential energy.** The configuration of this system has been changed, as evidenced by the change in the length of the spring. The total work done on the spring is positive because both $\boldsymbol{F}_1$ and $\boldsymbol{F}_2$ do positive work.

Conservative Force (保守力)

When you ride ski lift to the top of a hill of height h, the work done by gravity on you is $-mgh$ and the work done by the lift on you is $+mgh$ independent of the shape of the hill. The total work done by gravity on you during the round trip up and down the hill is zero, independent of the path you take. The force of gravity, exerted by the earth on you is a **conservative force.**

A force is conservative if the total work it does on a particle is zero when the particle moves around any closed path, returning to its initial position. From Fig. 6-5 we see that this definition implies that: The work done by a conservation force on a particle is independent of the path taken as the particle moves from one point to another.

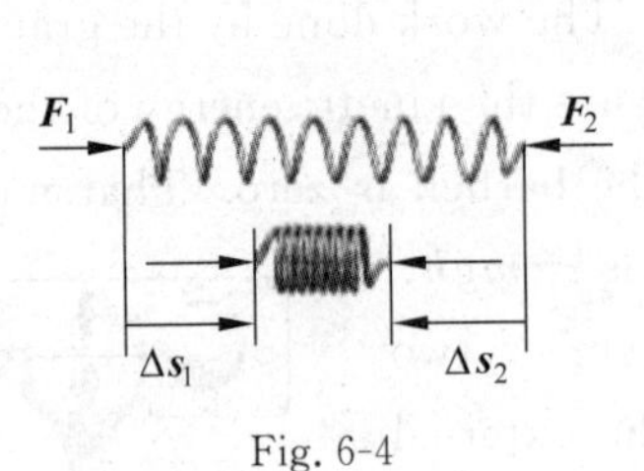

Fig. 6-4

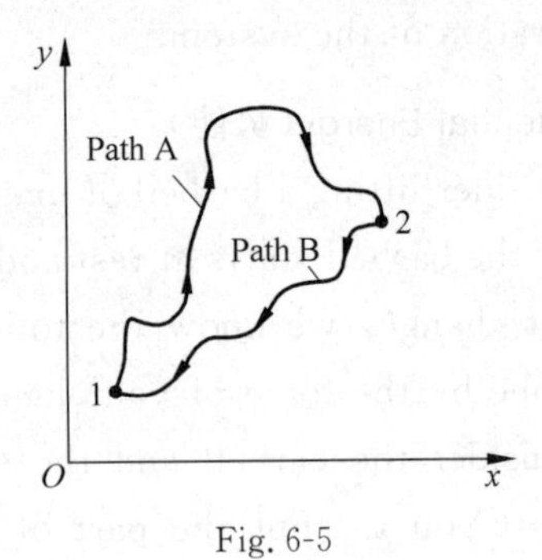

Fig. 6-5

Now consider yourself and the earth to be a two-particle system. (The ski lift is not part of this system.) When a ski lift raises you to the top of the hill, it does work *mgh* on the you-earth system. This work is stored as the gravitational potential energy of the system. When you ski down the hill, this potential energy is converted to the kinetic energy of your motion.

Glossary

work	功	point of application	作用点
joule	焦耳	notation	符号
eV	电子伏特	conservative force	保守力
potential energy	势能	positional function	位置函数
gravitational force	重力	elastic potential energy	弹性势能
independent of	与……无关		

6.3 专业英语常用表达法-6 原料 用途 功能 性能和特点

1. 表示原料

除用 have,(be) made of,with 外,主要用“名词(或形容词) +名词”这种结构:

a brick structure 砖结构	cut-off glass screen 玻璃隔热屏障
a concrete core 混凝土芯	a steel mast 钢柱
cast glass 压铸玻璃	a stone wall 石墙
rough-cast glass 毛玻璃	a wooden frame 木框
a glass screen 玻璃屏	tools made of iron 铁制工具

以下是一些具体例子:

The new greenhouse *has a mild steel structure*, with glass beneath it. 这座新暖房(用的)是低碳钢结构,结构下面是玻璃。

A curved screen *is made of* steel and glass. 弧形隔热屏障是用钢和玻璃制的。

Up in the roof here, we have a 6mm thick *cast glass*. 在这里屋顶上用的是 6 mm 厚的压铸玻璃。

Nearly 4500 square meters of space, under one low, *glass roof*. 近 4500 m^2 的面积的都是在一个低矮的玻璃屋顶之下。

Locomotives *are built of* steel, and airplanes *of* aluminum. 火车头由钢制成,而飞机由铝制成。

We want *such materials as can bear high temperature and pressure*. 我们需要能耐高温高压的材料。

2. 表示用途、功能

(1) 用 for 或动词不定式

for reference　供参考

a manual *for* technician　技工手册

solve the equation *for* x　解方程求 x

use the symbol Pb *for* lead　用符号 Pb 表示铅

What is the tool *for*? 这工具是做什么用的?

This is the very instrument *for* the experiment. 这是最适合做这个实验的仪器。

The institute is an international centre *for* physics research. 该机构是物理学研究的国际中心。

The key lab is primarily a research establishment but also it is there *to welcome the public*. 这个重点实验室主要是一个科研机构,但同时也面向公众。

(2) 用 the function (或 purpose) is+动词不定式

The rough-cast glass has several functions, and probably the main one *is to hide the internal structure*. 毛玻璃有好几个作用,主要的可能是为了把内部结构遮盖起来。

At the end we have a clear glass, and *its function* there *is* to give you a view out of the lab. 在末端用的是透明玻璃,它的作用是使你看得见实验室外边的情景。

(3) 用下列词语

be designed+动词不定式(或 for 短语)

help+动词不定式

have the function of

serve the purpose of

例如:

The building *is designed primarily for* experiment facilities rather than people. 这座建筑是供实验仪器而不是供人们使用的。

The cast glass *helps to express* the internal configuration of the lab. 压铸玻璃有助于显露出实验室的内部结构。

It also *serves the purpose of* giving access over the roof for maintenance and for cleaning the glass. 它还起到这样的作用,使(工人)能走上屋顶,进行维修保养、清洗玻璃。

3. 表示性能和特点

介绍仪器、设备等的性能特点是专业英语的重要功能之一,这在说明书、操作指南、产品鉴定会、学术会议等场合是常用的。其中有一些属通用词汇，对各行各业都适用，如 feature，characteristic，performance 等名词，major，main，leading，striking，unique 等形容词，以及 feature，characterize 等动词。作为口头表达，常用 The idea is…。例如：

The instrument *has several features of*… 该仪器有如下特点……

This system *features* simplicity of operation as its major design objectives. 操作简便是在设计本系统时考虑到的主要特点之一。

Operational flexibility and durable service life *feature* the equipment. 操作灵便，经久耐用是本设备的两大特色。

Micrometer *are characterized by* extreme accuracy，high reliability as well as simplicity and ease of control. 螺旋测微仪不仅操作简易，而且精密度高，性能可靠。

The rare earths *are characterized by* a partially filled $4f$ shell that is shielded from external fields by $5s^2$ and $5p^6$ electrons. 稀土（元素）的特征为：由于被 $5s^2$ 和 $5p^6$ 壳层的电子屏蔽了外场，$4f$ 壳层仅被部分填充。

Every *character* is of practical value and is based on long experience in building this type of lidar. （本产品的）每一个特点都有其实用价值，这是（本公司）在制造这类激光雷达中长期经验积累的结果。

The idea of the device is to allow the operator to perform handling and loading with the greatest convenience. （设计）这个装置的想法是使操作者能十分方便地进行装卸。

The whole idea is that ever greater attention must be paid to a better working environment. 总的想法是，必须更加重视创制一个更为良好的工作环境。

LESSON 7

7.1 物理学专业英语中的结果和结论

7.1.1 结果

在专业英语中，表示过程或计算的结果时，时态可采用一般过去时，语态可采用主动语态或被动语态。但是，被动语态的使用频率高于主动语态的使用频率。如果采用主动语态进行表达，那么，可以采用这一句型：主语＋宾语或由 that 引导的宾语从句。如：

Our final equation revealed an important general property of simple harmonic motion：the total mechanical energy in simple harmonic motion is proportional to the square of the amplitude. 最后的方程揭示了简谐运动的一个重要特点：简谐运动的总机械能与振幅的平方成正比。

展示研究结果，常用词汇有：show，result，present 等。

常见的表达方式有：…was(were)…；We found…；There was…等。

常见的用 that 从句表示结果的有：

The results indicated that … 结果表明……

The results showed that … 结果显示……

The results demonstrated that ... 结果证明了……

The results revealed that ... 结果揭示了……

常见的由 It 引导的 that 从句有：

It was shown that ... 结果显示……

It can be seen that ... 可以看出……

It was found that ... 结果发现……

It was discovered that ... 发现……

It was concluded that ... 结论为……

It has been demonstrated that ... 结果已经证明……

It was clarified that ... 结果明确为……

It was revealed that ... 说明/阐明……(结果)，揭示了……

It is considered that ... 考虑到……

It was confirmed that ... 可以确信……

It is suggested that ... 提出(……结果)

It was supposed that ... 推测(……结果)，假定(……结果)

It has become apparent that ... 已经明显表明……

也可以用 to be 后面的表语成分表示测定、计算等的结果。例如

When the measured value of ε_0 and defined value of μ_0 are put into Eq. (15-1), the speed of electromagnetic waves is found *to be* about 3×10^8 m/s, the same as the measured speed of light. 当把 ε_0 的测定值和 μ_0 的定义值代入方程(15-1)时，得到的电磁波的速度结果约为 3×10^8 m/s，这与测量所得到的光速相同。

For paramagnetic and diamagnetic materials, the magnetization is found *to be* proportional to the applied magnetic field that produces the alignment of the magnetic dipoles in the material. 对于顺磁质和抗磁质，磁化的程度正比于使磁介质产生磁偶极取向排列的外加磁场的强度。

The theoretical equation was proved *to be* correct by the experimental results. 实验的结果证实了理论方程的正确。

7.1.2 结论

结论句一般置于正文的最后，通常用于总结主要的研究成果、提出的独到见解或建议、表明作者的观点和看法、陈述主要结论等。专业文章的结论语句简洁、突出重点。专业英文通常用完整的句子来表达结论，动词时态用一般现在时或现在完成时。在写作时通常直接写结论，也可用一些句型引出结论。介绍结论常用的词汇有：summary, sum up, introduce, conclude, as a result, in short, in a word, in brief, therefore 等。常用的英文表达方式有：

...is probably... ……或许是……

...is... ……是……

Our conclusion is that... 我们的结论是……

This study shows that... 研究表明……

This study suggests that… 研究提出……

This study confirms that… 研究确信……

These observations support… 这些观察支持(……结论)

The findings indicate that… 结果表明……

The results indicate/show that… 结果表明……

This paper concludes that… 这篇文章的结论……

This study/investigation/research leads the author(s) to conclusion that… 通过本研究作者得到的结论为……

The research enables us to conclude that… 通过研究我们得到的结论为……

It is concluded that… 结论为……

Thus, it can be concluded that… 所以,可以得出结论……

Draw the following conclusion… 得到如下结论……

The results agree with … 这些结果与……一致

The experiment shows that… 实验表明……

It is suggested that… 可以推想(……结论),可以料想(……结论)

These results will/can be significant for… 这些结果对……来说会/可能有意义

In conclusion, … 最后/总之,……

In general, … 总体来说,……

To conclude, … 总之,……

In brief, … 简而言之,……

In a word, … 总之,……

In short, … 简而言之,……

In summary, … 概括来说,……

By and large, … 大体上/基本上……

All in all, … 总而言之,……

Generally speaking, … 总体来说,……

On the whole, … 大体上……基本上……

For the above-mentioned reasons, … 基于上述原因,……

As we have seen, … 正如我们所知,……

As already mentioned, … 正如已提及的一样,……

To sum up, … 总之/总而言之,……

To summarize, … 概括而言,……

We can easily come to the conclusion that… 我们很容易就可以得到这样的结论,……

What we can be sure of is that… 可以肯定的是……

Now it is believed that… 由此我们相信……

It is almost certainly true to say that… 几乎可以准确地说……

Clearly, … 显然……

It may be possible to (do something)… 可能(发生……情况)

Therefore, we can find that... 因此,我们发现……

It is quite clear that... 很明显……,很清楚……

It is well-known that... 众所周知……

Now it comes to conclusion that... 可以得出结论……

There is no doubt that... 无疑……

以上句型中,措辞不同,表明对结论的肯定程度不同。以下为具体例子:

We can *draw the following general conclusion*: For any system the most probable macroscopic state is the one with the greatest number of corresponding microscopic states, which is also the macroscopic state with greatest disorder and the greatest entropy. 我们可以得到如下一般结论:任何系统的最概然宏观态就是具有最大数目对应微观态的状态,也就是具有最大无序和最大熵的宏观态。

To illustrate this further, *we can make the conclusion of* boundary conditions for Half-wave Loss... 为了进一步说明,我们可以把半波损失的边界条件的结论归纳为……

Comparing Eq. (20-23) and Eq. (20-24), we can *conclude* that when simple harmonic waves propagate, for each element of the medium, kinetic energy equals potential energy at any time. 把方程(20-23)和方程(20-24)相比较,我们得到的结论为:当简谐波传播时,媒质中的每一质元的动能和势能在任何时刻都相等。

Clearly, this inelastic collision is an irreversible process. 显然,这种非弹性碰撞是不可逆过程。

Now I would like *to sum up* what I said in a few sentences. 现在我想用几个句子来总结一下我讲过的内容。

The long and short of it is that the experimental results coincide with the theoretical speculations to a certain extent. 总之,实验结果与理论推测在一定程度上相吻合。

Thus the conclusion we can draw is that the kind of materials is not suitable to be used under high temperature. 因此我们可以得出结论,这种材料不适合用在高温环境中。

In general, all the characteristics of the ferromagnetic material come from the special internal structure of it. 总之,铁磁材料的特性都源自其内部的特殊结构。

7.2 专业英语阅读

7.2.1 Rigid Body(刚体)

In order to study some problems easily, we used to imagine some bodies as **rigid bodies**, which are idealized models of particles system. If the body is nondeformable or the separations between all pairs of particles in the body remain constant, we may look it as a rigid body, which consists of a large number of particles, each with its own velocity and acceleration.

7.2.2 Law of Rotation of a Rigid Body about a Fixed Axis(刚体定轴转动定律)

To set a top spinning, you twist it. In Fig. 7-1, a disk is set spinning by the forces $\boldsymbol{F}_1$ and $\boldsymbol{F}_2$ exerted at the edges of the disk in the tangential direction. The directions of these forces are important. If the same forces are applied in the radial direction (Fig. 7-2), the disk will not start to spin.

Fig. 7-1　　Fig. 7-2

Fig. 7-3 shows a particle of mass m attached to one end of a massless rigid rod of length r. There is an axis perpendicular to the rod and passing through its other end, and the rod is free to rotate about this axis. Consequently, the particle is constrained to move in a circle of radius r. A single force $\boldsymbol{F}$ is applied to the particle as shown. Applying Newton's second law to the particle and taking components in the tangential direction gives $F_t = ma_t$. We wish to obtain an equation involving angular quantities. Substituting $r\alpha$ for a_t and multiplying both sides by r gives

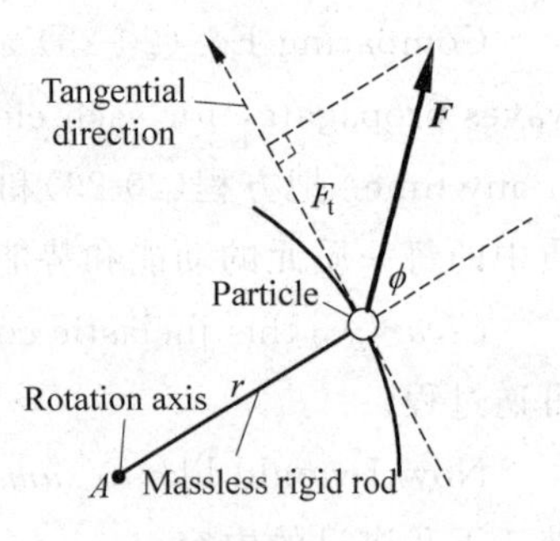

Fig. 7-3

$$rF_t = mr^2\alpha \tag{7-1}$$

The product rF_t is the **torque** M associated with the force. That is,

$$M = F_t r \tag{7-2}$$

Substituting into Eq. (7-1) gives

$$M = mr^2\alpha \tag{7-3}$$

A rigid object that rotates about a fixed axis is just a collection of individual particles, each of which is constrained to move in a circular path with the same angular velocity ω and acceleration α.

$$M_{i\text{-net}} = m_i r_i^2 \alpha$$

Applying a rigid object that rotates about a fixed axis is just a collection of individual particles, each of which is constrained to move in a circular path with the same angular velocity ω and acceleration α. Applying Eq. (7-1) to the ith of these particles gives

$$M_{i\text{-net}} = m_i r_i^2 \alpha$$

where $M_{i\text{-net}}$ is the torque due to the net force on the i th particle. Summing both sides over all particles gives

$$\sum M_{i\text{-net}} = \sum m_i r_i^2 \alpha = \left(\sum m_i r_i^2\right)\alpha = J\alpha \tag{7-4}$$

The sum in the term on the right is the object's **moment of inertia** J for the axis of rotation.

$$J = \sum_i m_i r_i^2 \tag{7-5}$$

In the past we saw that the net force acting on a system of particles is equal to the net force acting on the system because the internal forces (those exerted by the particles within the system on one another) cancel in pairs. The treatment of internal torques exerted by the particles within a system on one another leads to a similar result, that is, the net torque acting on a system equals the net external torque acting on the system. We can thus write Eq. (7-4) as

$$M_{\text{net,ext}} = \sum M_{\text{ext}} = J\alpha \tag{7-6}$$

So *torque acting on a rigid body is proportional to its angular acceleration, and the proportionality constant is the moment of inertia*. This is **law of rotation of a rigid body about a fixed axis**, which is analog of Newton's second law for linear motion, $\sum \boldsymbol{F} = m\boldsymbol{a}$.

7.2.3 Torque and Angular Momentum(力矩和角动量)

Torque is expressed mathematically as the cross product (or vector product) of $\boldsymbol{r}$ and $\boldsymbol{F}$

$$\boldsymbol{M} = \boldsymbol{r} \times \boldsymbol{F} \tag{7-7}$$

Fig. 7-4 shows a particle of mass m moving with a velocity $\boldsymbol{v}$ at a position $\boldsymbol{r}$ relative to the origin O. The linear momentum of the particle is $\boldsymbol{p} = m\boldsymbol{v}$. The **angular momentum** $\boldsymbol{L}$ of the particle relative to the origin O is defined to be the cross product of $\boldsymbol{r}$ and $\boldsymbol{p}$

$$\boldsymbol{L} = \boldsymbol{r} \times \boldsymbol{p} \tag{7-8}$$

If $\boldsymbol{r}$ and $\boldsymbol{p}$ are in the xy plane, as in Fig. 7-4, $\boldsymbol{L}$ is parallel with the z axis and is given by $\boldsymbol{L} = \boldsymbol{r} \times \boldsymbol{p} = mvr\sin\phi\boldsymbol{k}$. Like torque, angular momentum is defined relative to a point in space.

Fig. 7-5 shows a particle of mass m attached to a circular disk of negligible mass in the xy plane with its center at the origin. The disk is spinning about its axis with angular speed ω. The speed v of the particle and its angular speed are related by $v = r\omega$. The angular momentum of the particle relative to the center of the disk is

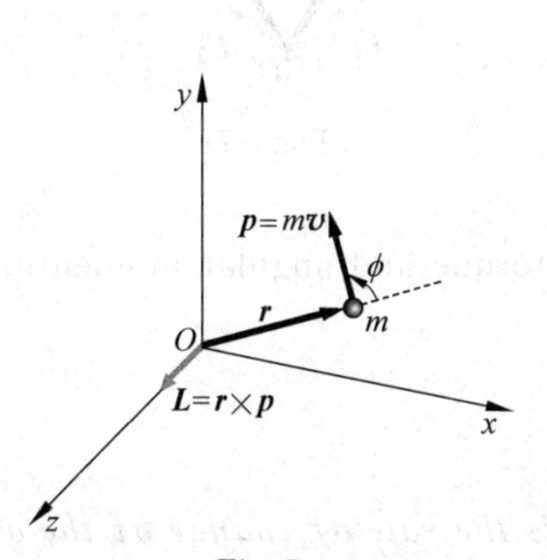

Fig. 7-4

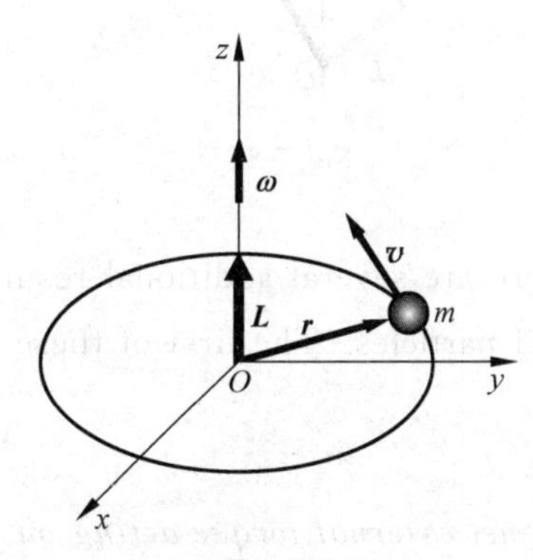

Fig. 7-5

$$\boldsymbol{L} = \boldsymbol{r} \times \boldsymbol{p} = \boldsymbol{r} \times m\boldsymbol{v} = rmv\sin 90^\circ \boldsymbol{k}$$
$$= rmv\boldsymbol{k} = mr^2\omega\boldsymbol{k} = mr^2\boldsymbol{\omega}$$

The angular momentum vector is in the same direction as the angular velocity vector. Since mr^2 is the moment of inertia for a single particle about the z axis, we have

$$\boldsymbol{L} = mr^2\boldsymbol{\omega} = J_z\boldsymbol{\omega}$$

This result does not hold for the angular momentum about a general point on the z axis. Fig. 7-6 shows the angular momentum vector $\boldsymbol{L}$ for the same particle attached to the same disk but with $\boldsymbol{L}'$ computed about a point on z axis that is not at the center of the circle. In this case, the angular momentum is not parallel to the angular velocity vector $\boldsymbol{\omega}$, which is parallel with z axis.

In Fig. 7-7, we attach a second particle of equal mass to the spinning disk. The angular momentum vector $\boldsymbol{L}_1'$ and $\boldsymbol{L}_2'$ are shown relative to the same point O'. The total angular momentum $\boldsymbol{L}'_1 + \boldsymbol{L}'_2$ of the two-particle system is again parallel to the angular velocity $\boldsymbol{\omega}$. In this case, the axis of rotation, the z axis, passes through the center of mass of the two-particle system, and the mass distribution is symmetric about this axis. Such an axis is called a **symmetry axis.** For any system of particles that rotates about a symmetry axis, the total angular momentum (which is the sum of the angular momenta of the individual particles) is parallel to the angular velocity and is given by

$$\boldsymbol{L} = J\boldsymbol{\omega} \tag{7-9}$$

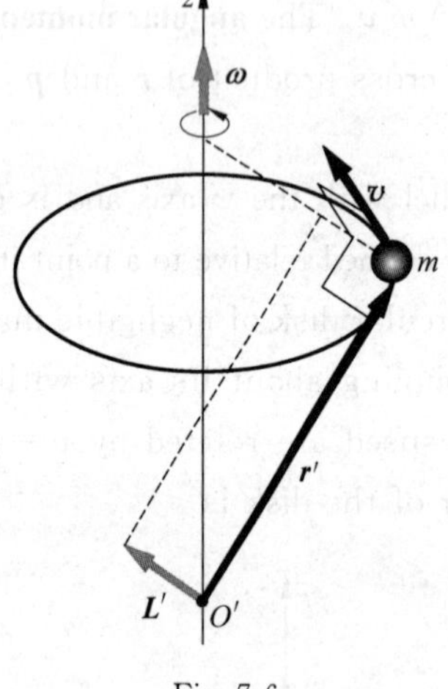

Fig. 7-6

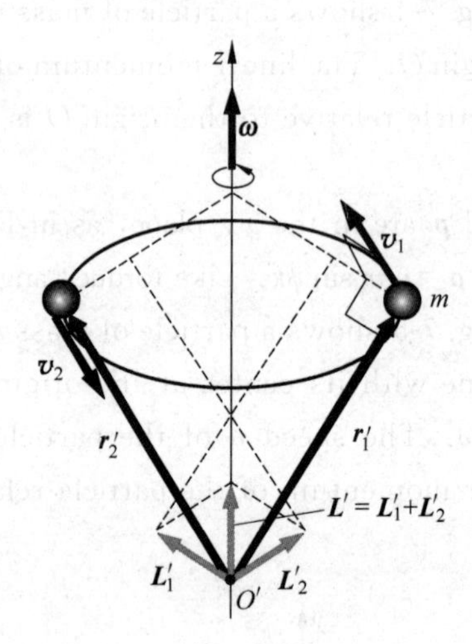

Fig. 7-7

There are several additional results concerning torque and angular momentum for a system of particles. The first of these is

$$\boldsymbol{M}_{\text{net,ext}} = \frac{\mathrm{d}\boldsymbol{L}_{\text{sys}}}{\mathrm{d}t} \tag{7-10}$$

The net external torque acting on a system equals the rate of change of the angular momentum of the system. This is called **theorem of angular momentum.**

Glossary

torque	力矩	axis of rotation	旋转轴
moment of inertia	转动惯量	proportionality constant	比例常数
analog	类似、相似	is proportional to	与……成正比
mass distribution	质量分布	symmetry axis	对称轴
angular momentum	角动量	theorem of angular momentum	角动量定理

7.3 专业英语常用表达法-7 形状 面积 体积 容量

1. 基本形状

triangle 三角形	right triangle 直角三角形
isosceles triangle 等腰三角形	equilateral triangle 等边三角形
obtuse angle 钝角	acute angle 锐角
right angle 直角	square 正方形
rectangle 矩形	parallelogram 平行四边形
trapezoid 梯形	rhombus 菱形
polygon 多边形	quadrilateral 四边形
pentagon 五边形	hexagon 六边形
octagon 八边形	circle 圆形
semicircle 半圆	ellipse 椭圆
cylinder 圆柱	cone 圆锥
cube 正方体/立方形	cuboid 长方体
sector 扇形	cone 锥形
curve 曲线	cylinder 圆柱(筒)形
sphere 球形	hemisphere 半球形
ellipsoid 椭面、椭圆形	oblong 长方形
pyramid 棱锥形	

2. 描述形状常用形容词

circular 圆形的、环状的	conical 锥形的
(criss-)cross 十字形的	cubical 立方形的
curved 曲线的、弯曲的	cylindrical 圆柱形的
depressed 凹陷的	ellipsoidal 椭圆的
flat (扁)平的	half-round 半圆的
half-moon-shaped 半月形的	heart-shaped 心形的
hemispherical 半球形的	hollow 空心的
hyperbolic 双曲线的	I-shaped 工字形
oblong 长方形的	pointed/sharp 尖的

pyramidal 棱锥形的、金字塔形的
round/rounded 圆的
spherical 球形的
square 正方形的
straight 直(线)的
triangular 三角形的
rectangular 矩形的
semicircular 半圆形的
spindle-shaped 纺锤形的
solid 实心的
star-shaped 星形的
U-shaped U形的

3. 表示形状常用词组

a curved surface 曲面
a pointed roof 尖形屋顶
a round box 圆盒
triangular compasses 三角规
T-socket T形套管
I-steel 工字钢
herring bone pattern 人字形
Y-connection Y形连接
U-pipe U形管
A-frame A形架
O-ring O形环
T-track 锤形径迹
V-belt 三角皮带
clamp 弓形夹具
Y-pipe 叉形管
a flat cable 扁平电缆
rectangular timber 方木
a spherical mirror 球面镜
T-plate T字板
I-bar 工字铁
cross-wire 十字线
herring bone gear 人字齿轮
X-tube X形管
Z-beam Z字梁
C-washer C形垫圈
S-wrench S形扳手
U-bolt 马蹄螺栓
set square 三角板
U-steel 槽钢
X-type 交叉形

4. 表示形状常用词语结构

be＋adj.(＋in shape/form) (形状)是……的,是……形的
be shaped like... 形状像……
have the shape of... 具有……形状
(be) in the shape of... 呈……的形状,以……的形式
take the shape of... 呈……的形状
以下是一些具体例子:
This plate is *curved*/*flat*. 这块板是弯的/平的。
This rod is *pointed*/*rounded* at one end. 这根棒一头是尖/圆的。
This line is *straight*/*curved*. 这条线是直的/弯的。
What *shape* is the top of the rocket? 火箭顶部是什么形状?
The lab building is built *in the shape of* a letter "L"。这座实验室造成字母L形。
The top of the planetarium *is like a snail in shape*. 天文馆的顶部形状像一只蜗牛。
The coil *is rectangular in shape*. 这线圈是矩形的。
The flask *resembles a pear in shape*. 这烧瓶的外形像只梨。
The curves *are hyperbolic in form*. 这些曲线是双曲线。

A wheel *has the same shape as* the letter "O"。车轮的形状与字母O相同。

The curve of normal distribution *is shaped like* a bell. 正态分布曲线的形状像一口钟。

A safety-pin *is so shaped that* it cannot easily prick you. 安全别针做成不易伤人的形状。

The simplest wave form *is a sinusoidal* curve. 最简单的波形是正弦曲线。

In weightlessness, a drop of liquid *takes a spherical shape/takes the shape of a sphere*. 在失重状态,一滴液体呈球形。

5. 面积的表示法

(1) 通常用 by 表示乘、除及度量关系

The room is 6 meters *by* four. 这个房间长 6 m,宽 4 m。

Multiply the base *by* the height and you have the area of a rectangle. 底乘以高,就可得出长方形的面积。

This is a room 20 *by* 30 feet. = This is a room 20 feet *by* 30. = This is a room 20×30 ft. 这是一个 20 英尺宽 30 英尺长的房间。

(2) 其他表示方法

The room has an area of ten square meters. = The area of the room is ten square meters. =The room is ten square meters in area. =The room covers an area of ten square meters. 这个房间的面积为 10 m^2。

Tianjin is about 11 760.26 square kilometers in the area. 天津面积约为11 760.26 km^2。

China's land area is 9 600 000 sq km. 我国国土面积是 960 万 km^2。

6. 体积、容量的表示法

(1) 常用"基数词+长度名词+ by +基数词+宽度名词+ by +基数词+高度名词"表示立方体尺寸,若长、宽、高的单位一致,则通常省略后两个单位名词。如:

The box is eight inches *by* six *by* five. 这个盒子长、宽、高分别为 8 英寸、6 英寸、5 英寸。

The box measures three feet *by* four *by* two. 这箱子的尺寸是(3×4×2)英尺。

(2) 用 capacity, volume 以及 cube(立方), cubed(自乘二次的), to the third power (……的三次方), liter, cubic meter, cumec, cubic centimeter 等。其结构有:The volume/capacity is...; has a capacity of...; be with a capacity of...等。例如:

The pretreatment works will *have a capacity of* 45 *cumecs*, or 45 cubic meters per second. 这个预处理厂将会有 45 m^3/s 的处理能力。

The *capacity* of the treatment works *is approximately* 45 cumecs. 这个处理厂的处理能力约为 45 m^3/s。

Before the waste joins the sea, it will pass through a large new pretreatment works *with a capacity of* 45 cumecs, or 45 cubic meters per second. 污水在流入海之前,要先通过一个容量为 45 m^3/s 新建的大型预处理厂。

The formula for *the volume of* a cube with edge l is $V=l^3$. (V equals l cubed, or V equals l to the third power) 求边长为 l 的立方体的体积，其公式为 $V=l^3$.

The *volume* measuring accuracy of this system is ±0.25%. 系统容积的测量精度为±0.25%。

LESSON 8

8.1 物理学专业英语在口语体上的特点

专业英语口语体兼有专业英语及一般英语口语体的某些特点。

1) 使用大量的术语以及书面词语

例如常见的用语有 be concerned with, be subjected to, be exposed to, the order of 等。

It is well known that, different from physics, chemistry *is concerned with* the composition, properties and structure of substances and *with* the changes they undergo. 众所周知，与物理学不同，化学是研究物质的成分、(化学)性质及其结构的，还研究这些物质(在化学反应中)的变化。

The sample temperature was controlled by a thermocouple with a precision on *the order of* 0.1℃. 样品温度用热电偶控制，其精度在 0.1℃左右。

2) 重在客观的叙述，较少使用人称代词

为了使内容的叙述更加客观，句子中通常不出现人称代词，如果使用也常用 one 或 we。例如：

There is a steady loss of energy in the cycle. 在循环中能量持续损失。

In the real case *there is* a need for energy compensation. 在实际情况中需要补充能量。

As *one* shakes the end of a stretched rope, the disturbance transfers energy from the hand to the rope and a wave pulse travels down the rope. 摇动绳子的一端，扰动就会把能量从手传递给绳子，一个波脉冲就会在绳子上传播。

If *we* look into the experiment we can see the secret of what is happening. 如果我们审视一下实验，就会明白事情发生的原因。

3) 用名词短语替代从句

口语使用名词短语的目的在于可以使整个句子的结构简化，易于理解。例如：

They enjoy *the newness of the building*. 相当于 They like the building *because it is new*.

These experiments demonstrated *the truth of the theory*. 相当于 These experiments demonstrated *that the theory is true*.

4) 用“主语＋谓语动词的被动态＋动词不定式”替代名词从句。

例如：

An atom *is found to consist of* electrons and some other very small particles. (相当于 *It is found that* an atom consists of electrons...). (据认为)原子是由电子及其他微粒子所

组成。

这就是说,“主语＋谓语动词的被动态＋动词不定式”这种结构由于能使语言简练,在专业口语体中是常见的,能用于这种结构的谓语动词有 know,think,believe,say 等。

Alloy steel *is said to be* a carefully made steel.(相当于 *It is said that* alloy steel *is* a carefully made steel.)合金钢可以说是一种精细冶炼的钢。

5) 对量度表达没有书面语那么确切并具体

专业英语口语在表达量度时通常不像书面语那么具体,有时只是笼统地说“高、矮、大、小、长、短”等。

The tidal range is *of the order of* 10 *meters*. 浪高在 10 m 的数量级。(书面体)

The tide is *as high as* 10 *meters*/10 *meters high*. (口语体)

Some waves have been *of heights of up to* 7 *or* 9 *meters*. 有的浪高达到 7 m 或 9 m。(书面体)

Some waves have been 7 *or* 9 *meters high*. (口语体)

6) 经常使用主动语态

在专业英语书面体中常用被动语态,但在口语体中,会经常用 we 作主语的主动语态。

If the end of a stretched rope *is given* a quick shake, the disturbance transfers energy from the hand to the rope。(书面体)

If *we* give the end of a stretched rope a quick shake, the disturbance transfers energy from the hand to the rope.(口语体)

7) 其他特点

一般地说,专业英语口语体还有不简练、冗余度大、用词不推敲、语法不规范、结构和层次不严密,边说边自我纠正等特点。例如:

The fact that the lab *is designing* primarily for experiment facilities rather than people, *that* makes it particularly special. 实验室的设计主要是供实验设备而不是供人使用的,这就使得它有其独特之处。

这句话是口语体,在语法和用词上不很严格。如果严格规范要求,is designing 要改为 is designed;第二个 that 要去掉。

Internally the structure is very much played down and the various instruments are beginning to dominate and it's very important, I feel, right from the outset, that the instruments should be allowed to dominate internally, especially as the spaces are relatively small and the roof is relatively low. 在实验室内,钢架结构已不占主要地位,突出的是各种仪器。我认为从一开始就要重视使各种仪器在室内居主导地位,这尤其是因为要考虑到室内的空间相对来说是少了些,屋顶也较低了些。

这是一个包含有 6 个子句的长句。因为属于口语体,有几层意思是反复说的,显得不够简练。其中,I feel 是一个表示态度的插入句;right 用以强调 from the outset,意思是:从一开始就……;it is important that... 是一句型,表示愿望、建议等,that 从句中的谓语用 should ＋*v*.(这是口语体),如果是书面语体,可用动词原形,即 It is important that the instruments be allowed...。

Any glassware which has a... a delicate shape, an interesting shape, is seen very clearly against a cast glass background, it's seen in silhouette.

在口语体中,因为是即席讲话,往往不可能做到字斟句酌,于是当讲话人发现用字不当时,可随时加以修正。如这里先用了 has a... a delicate shape(有一个细小的、优美的形状),说后感到 delicate 并不达意,于是又加上了 an interesting shape,这个短语可以说是对前一个短语的修正。

8.2 专业英语阅读

8.2.1 Oscillation(振动)

Oscillation occurs when a system is disturbed from a position of stable equilibrium. There are many familiar examples: boats bob up and down, clock pendulums swing back and forth, and the strings and reeds of musical instruments vibrate. Others, less familiar examples are the oscillations of air molecules in a sound wave and the oscillations of electric currents in radios and television sets.

In this section, we deal mostly with simple harmonic motion, the most basic type of oscillatory motion. Applying the kinetics and dynamics of simple harmonic motion provides the analysis of the oscillatory motion of a variety of interesting systems. In some situations dissipative forces dampen the oscillatory motion, but in other situations driving forces sustain the motion by compensating for the damping.

8.2.2 Description of Simple Harmonic Motion(简谐运动的描述)

A common, very important and very basic kind of oscillatory motion is **simple harmonic motion(SHM)** such as the motion of an object attached to a spring (Fig. 8-1).

The **kinetic equation** for a simple harmonic motion is

$$x(t)=A\cos(\omega t+\varphi) \tag{8-1}$$

where A, ω, and φ are constants. It is a mathematical representation of the position of the particle as a function of the time. For the spring system, x is the displacement of the object from its equilibrium position. In general, x can be any oscillatory quantities.

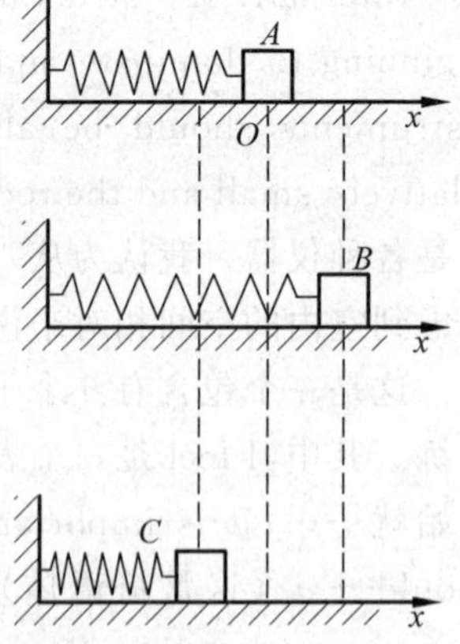

Fig. 8-1 A spring system

There are three property quantities for a simple harmonic motion:

The maximum displacement x_{max} from equilibrium position is called the **amplitude** A. The value of the amplitude depends on initial position and the energy of the system. The constant ω is called the **angular frequency**. It has units of radians per second and dimensions of inverse time, the same as angular speed, which is also designated by ω. *The time it takes for a displaced object to execute a complete cycle of oscillatory motion*—from one extreme to the other extreme and back—is called the **period** T. The unit of the period is second. The reciprocal of the period is the **frequency** ν, which is *the number of cycle per second*

$$\nu = \frac{1}{T} \tag{8-2}$$

The unit of the frequency is the cycle per second (cy/s), which is called a hertz (Hz).

The period T is the shortest time satisfying the relation

$$x(t) = x(t + T)$$

for all t. Substituting into this relation using Eq. (8-1) gives

$$\begin{aligned} A\cos(\omega t + \varphi) &= A\cos[\omega(t + T) + \varphi] \\ &= A\cos(\omega t + \varphi + \omega T) \end{aligned}$$

The cosine (and sine) function repeats in value when the angle increases by 2π, so

$$\omega T = 2\pi \quad \left(\text{or } \omega = \frac{2\pi}{T}\right) \tag{8-3}$$

The frequency is the reciprocal of the period

$$\nu = \frac{1}{T} = \frac{\omega}{2\pi} \tag{8-4}$$

The argument of the cosine function in Eq. (8-1), $\omega t + \varphi$, is called the **phase** of the motion, and the constant φ is called the **phase constant**, which is the phase at $t = 0$. If we have just one oscillating systems, we can always choose $t = 0$ at which $\varphi = 0$. If we have two systems oscillating with the same amplitude and frequency but different phase, we can choose $\varphi = 0$ for one of them. The equations for the two systems are then

$$x_1 = A\cos(\omega t)$$

and

$$x_2 = A\cos(\omega t + \varphi)$$

We can see by inspection that each time t increases by T, the phase increases by 2π and one cycle of the motion is completed.

8.2.3 Damped Oscillations(阻尼振动)

Left to itself, a spring or a pendulum eventually stops oscillating because the mechanical energy is dissipated by frictional forces. Such motion is said to be **damped.** If the damping is large enough, as, for example, a pendulum submerged in molasses, the oscillator fails to complete even one cycle of oscillation. Instead it just moves toward the equilibrium position with a speed that approaches zero as the object approaches the equilibrium position. This type of motion is referred to as **overdamped.** If the damping is small enough that the system oscillates with an amplitude that decreases slowly with time—like a child on a playground swing when Mum stops providing a push each cycle—the motion is said to be **underdamped.** Motion with the minimum damping for nonoscillatory motion is said to be **critically damped.**

8.2.4 Driven Oscillations and Resonance(受迫振动和共振)

To keep a damped system going, mechanical energy must be put into the system. When this is done, the oscillator is said to be **driven** or **forced**. When you keep a swing going by "pumping", that is, by moving your body and legs, you are driving an oscillator. If you put mechanical energy into the system faster than it is dissipated, the mechanical energy increases with time, and the amplitude increases. If you put mechanical energy in at the same rate it is being dissipated, the amplitude remains constant over time. The motion of the oscillation is then said to be in **steady state.**

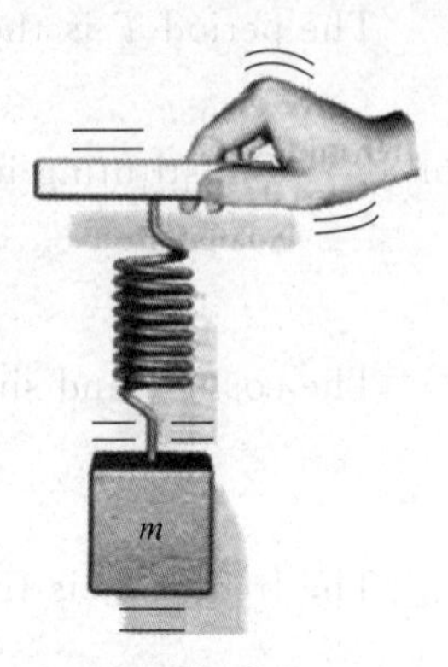

Fig. 8-2

Fig. 8-2 shows a system consisting of an object on a spring that is being driven by moving the point of support up and down with simple harmonic motion of frequency ω. At first the motion is complicated, but eventually steady state motion is reached in which the system oscillates with the same frequency as that of the driver and with a constant amplitude and, therefore, at constant energy. In the steady state, the energy put into the system per cycle by the driving force equals the energy dissipated per cycle due to the damping.

The amplitude, and therefore the energy, of a system in the steady state depends not only on the amplitude of the driving force, but also on its frequency. The **natural frequency** of a oscillator, ω_0, is its frequency when no driving or damping force are present (In the case of a spring, for example, $\omega_0 = \sqrt{k/m}$). If the driving frequency is approximately equal to the natural frequency of the system, the system will oscillate with a relatively large amplitude. For example, if the support in Fig. 8-2 oscillates at a frequency close to the natural frequency of the mass-spring system, the mass will oscillate with a much greater amplitude than it would if the support oscillates at higher or lower frequencies. This phenomenon is called **resonance.** When the driving frequency equals the natural frequency of the oscillator, the energy per cycle transferred to the oscillator is maximum. The natural frequency of the system is thus called the **resonance frequency.**

There are many familiar examples of resonance. When you sit on a swing, you learn intuitively to pump with the same frequency as the natural frequency of the swing. Many machines vibrate because they have rotating parts that are not in perfect balance (observe a washing machine in the spin cycle for an example). If such a machine is attached to a structure that can vibrate, the structure becomes a driven oscillatory system that is set in motion by the machine. Engineers pay great attention to balancing the rotary parts of such machines, damping their vibrations, and isolating them from building supports.

Glossary

oscillation/vibration	振动	oscillatory motion	振动
simple harmonic motion (SHM)	简谐运动	dissipative force	耗散力
		spring	弹簧
damping	阻尼	angular frequency	角频率、圆频率
equilibrium position	平衡位置	period	周期
radian	弧度	reciprocal	倒数
hertz(Hz)	赫兹	argument	辐角
amplitude	振幅	phase angle/(phase constant)	初相、初位相
phase	相、位相、周相		
average value	平均值	dissipate	消耗
damped oscillations	阻尼振动	submerge	浸入
molasses	糖蜜	oscillator	振子
overdamping	过阻尼	critically damped	临界阻尼的
overdamped	过阻尼的	driven oscillation	受迫振动
steady state	稳态	underdamped	欠阻尼的
resonance	共振	critical damped	临界阻尼
natural frequency	固有频率	resonance frequency	共振频率
nonoscillatory	不摆动的、不振动的		

8.3 专业英语常用表达法-8 一致 符合

1. 常用动词

accord	符合、一致	agree	符合、一致
coincide	一致、相符	answer	符合
fit	符合	meet	符合、适合

2. 常用名词

accord	一致、符合	agreement	符合、一致
coincidence	符合、一致	consistence/consistency	一致(性)
correspondence	符合、一致	conformity	符合、一致
congruence/congruency	一致、符合		

3. 常用形容词

consistent	一致的、符合的	accordant	一致的
congruous	一致的、全等的	conformable	一致的、符合的
corresponding	一致的、符合的		

4. 常用词组

answer to 适应、符合

be in agreement with...　与……相一致

be in accord with...　与……相一致

be in conformity with/to...　与……相一致

be/as consistent with...　与……相一致(协调)、符合……

bring...into accord with...　使……与……一致(相调和)

bring...into correspondence with...　使……与……一致起来

conform to...　(使)一致、(使)符合

conform M to N　使 M 与 N 一致

conform with　与……一致、与……符合

correspondence between M and N　M 与 N 相符

correspond to/with...　符合于……

fit in with　符合、适应

fit...into...　使……符合……

fit...like a glove　完全符合……

tally with...　与……相符合、与……一致

例如:

answer a description　与描述相符

observations that *fit* the theory nicely　与理论完全相符的观测

consistence of composition　成分的一致性

consistency check　一致性检验

curve *fit*　曲线拟合

meet a criterion　符合标准

meet the specification　符合规格/规范

The lab *answers to* his description. 这实验室与他的描述相符。

The experimental results *agree/accord with* the theoretical calculations. 实验结果与理论计算相符合。

The actual experimental figures are *in agreement with/in accord with/in conformity with/consistent with* the estimated figures. 实验中的数字与估计数字相符合。

A contact lens *conforms to/is consistent with* the curvature of the eyeball. 接触透镜与眼球的曲率相符。

The theory does not *correspond to* the experimental facts. 这一理论与事实情况不符。

Our results *coincide with* those obtained by other professors. 我们的结果与其他一些教授所取得的结果相一致。

It is shown that the QCT(quasi-classical trajectory) calculated results *agree well with/show a good agreement with* the experimental data for the reaction. 结果表明,对于这种反应,利用准经典轨线法计算的结果与实验获得的数据吻合得很好。

Unit Three

LESSON 9

9.1 物理学专业英语中的长句分析(一)

总的来说,专业英语文献中的句子长度呈现两种趋向:长难句多,短的简单句也多。一个句子长达一百来个单词,构成一个完整段落的现象屡见不鲜。

物理学专业英语用于表达科学理论、原理、规律、概述,以及各事物之间错综复杂的关系,而复杂的科学思维是无法使用简单句来表达的。为了表示严谨、精确的含义,语法结构复杂的长句有较多的应用,而这种严谨周密、层次分明、重点突出的语言手段也就成了专业英语文体又一重要特征。长句由基本句型扩展而成,其方式有增加修饰成分如定语、状语的,有用各种短语如介词、分词、动名词或不定式短语充当句子成分的,也可能是通过关联词将两个或两个以上的句子组合成复合句。从句子结构来看,英语中句子可分简单句,并列复合句和主从复合句。

英语利用形态变化、词序和虚词三大语法手段可构成包孕许多修饰成分或从句的长句,句中各部分顺序灵活多样。通常英语句中的表态部分(如判断和结论)在先,而叙事部分(如事实和描写)在后,汉语则正好相反;英语句中先短后长,"头轻脚重",而汉语也正好相反;英语借助形态变化和连接手段而将句中成分灵活排列,汉语则常按时间和逻辑顺序由先到后、由因到果、由假设到推论、由事实到结论这样排列。

在专业英语的学习中,长句的分析是一个非常重要的基础,无论是专业文献阅读、翻译还是严谨精确地写作,都要从长句分析开始。长句分析既重要也有些难度。然而,无论多长的句子、多么复杂的结构,它们都是由一些基本的成分组成的。只要弄清英语原文的句法结构,找出整个句子的中心内容,理解各层意思,然后分析各层意思之间的逻辑关系,再按汉语的特点和方式表达就可以理解或译出原文了。

9.1.1 长句结构的分析

抓住主干,添枝加叶。所谓抓住主干,就是在理解长句时首先要找到主语、谓语这两个主要成分。添枝加叶,就是在主语、谓语这两个主要成分的基础上,逐个加上各种修饰语,包括定语、状语、补语、非谓语动词结构、各种从句等。比如:

In the engine, the heat required to change the water into steam is produced by the combination of the oxygen gas in the air with the fuel. 在蒸汽机中,把水变成蒸汽所需要的热是由空气中的氧气与燃料结合而产生的。

该句主语 heat 容易确定，它后面紧跟一个动词过去分词 required，这个 required 很容易被误解为谓语，其实真正的谓语是 is produced。

Laser, its creation being thought to be one of today's wonders, is nothing more than a light①*that differs from ordinary lights only in* ②*that it is many times more powerful* and so be applied in fields ③ *that no ordinary light or other substance has ever been able to get in*.

此例是带有两个定语从句和一个宾语从句的主从复合句。主句为"Laser...is nothing more than a light."。定语从句①修饰主句中的名词 light，宾语从句②为介词 in 的宾语，定语从句③修饰名词 fields。另外，主句中又有一个独立主格结构"its creation... today's wonder"作状语。

The property the air has of taking up a great amount of water when heated and giving it out when cooled, is the cause of our clouds and rain. 空气所具有的这种热时吸收、冷时放出大量水汽的性能，是云和雨生成的原因。

该句主语 property 后面省略了一个 that 或 which，用来引导定语从句 the air has，修饰主语的介词短语 of taking... and giving...中又套两个用 when 引导的状语从句，显得句子结构复杂。借助 is 前面这个逗号，使我们找到了这个谓语联系动词。

在进行长句分析时，首先要理解透原文，分析句子结构，弄清语法关系。而对一个句子做结构分析，通常可采取下列步骤：通读全句，确定句子的种类——简单句或复合句，然后对句子的各种成分进行层次分析。

9.1.2 简单句的分析

对于简单句，进行句子分析时，要先找出句子的主要成分——主语、谓语，并进一步判明句中其余部分的语法成分——宾语、定语、状语、补足语、同位语、插入语等。一个简单句中只有一个主语和谓语，主谓结构是句子的骨干（主干）；其他的句子成分，如定语、状语、补语等，则是生长在这个骨干上的次要（枝叶）成分。因此，在分析简单句时，应首先找出主语和谓语，再确定其他成分跟主语和谓语之间的关系以及各次要成分之间的关系。在把原文的意思彻底弄清楚之后，再来考虑如何用中文进行表达。例如：

In order to understand the operation of a transistor, it is first necessary to be familiar with the mechanism of charge transfer across a junction formed between a piece of n-type semi-conductor and a piece of p-type semi-conductor.

分析：在这个句子中。It 是句子的形式主语，真正的主语是 to be familiar with the mechanism; is necessary 是谓语。of charge transfer across a junction 是个介词短语，作 mechanism 的定语；formed 是过去分词，作 junction 的定语；between a piece of n-type semi-conductor 是介词短语，作 formed 的状语。In order to...transistor 是个目的状语，修饰全句。这个句子有三层意思：①为了理解晶体管的工作原理；②首先必须熟悉电荷越过结的机理；③结是在一块 N 型半导体材料和一块 P 型半导体材料之间形成的。

9.1.3 复合句的分析

所谓复合句就是由两个或两个以上完整的子句组合而成的句子。一般来说，英语中一个句子只能有一个谓语，如果出现两个谓语，那么其中一个谓语只能是以从句、并列句或非谓语动词的形式出现。

把两个子句结合在一个句子中，英语需要加上连接词或加上标点符号。而由连接词and，or，but...或标点符号结合的子句，在全句内处于并列的地位，这样的复合句称为并列复合句。这种复合句不过是简单地将两句写成一句而已，所以分析起来并不困难。

另一类复合句中的子句地位不平等，一个子句是主句，其他是从句，这样的复合句称为主从复合句。主从复合句是由"主句＋从句"构成，它是专业英语学习中比较复杂的句子结构。所谓从句是指从属于主句的句子，由从属连词连接。从句的种类有很多，但根据其性质和作用可以分为：名词性从句、形容词性从句（即定语从句）、副词性从句（即状语从句）三大类。

在进行句子分析时，如遇到的长句为复合句，按照我们刚才介绍的方法"抓住主干，添枝加叶"，应先找出主句，然后确定各个从句的性质及其作用，判明各从句之间的关系——并列关系或从属关系，并对各成分进行层次分析。

对于结构复杂，层次纷纭的主从复合句要抓住引导各种从句的连词，弄清各种从句的性质，这样就能掌握全局结构，分清主次。

在复合句中，包含从句较多的长句又可以分为以下几种：

1）以状语从句或含有从句的状语短语开头的长主从复合句

由于状语可以位于句首，因而很多长句是以状语从句或含有从句的状语短语开头的。在从句末尾往往有一逗号，逗号之后就是主句。这样的长句，在结构上与汉语的句子类似。例如：

①Although there exists much experimental knowledge in regard to the behaviour of bodies ②which are not in the conditions ③to which the mathematical theory is applicable, yet it appears ④that the appropriate extensions of the theory ⑤which would be needed in order to incorporate such knowledge within it cannot be made ⑥ until much fuller experimental knowledge has been obtained.

这是一个以 although 引导的让步状语从句开头的复合句。它的从句多达六个：①although 引导的修饰主句谓语的让步状语从句；②which 引导的修饰 bodies 的定语从句；③to which 引导的修饰 conditions 的定语从句；④that 引导的主语从句；⑤which 引导的修饰 extensions 的定语从句；⑥until 引导的修饰 cannot be made 的时间状语从句。

2）以主句开头的长主从复合句

主句位于句首，后面可以接以引导词引导的各种从句，如主语从句、定语从句、状语从句等。例如：

Viscosity is a general term for all those properties of matter ①by virtue of which the resistance ②which a body offers to any change, depends upon the rate ③at which the change is effected.

这是一个以主句开头的复合句。其中包含三个从句：①以 by virtue of which 引导的定语从句；②以 which 引导的修饰 resistance 的定语从句；③以 at which 引导的修饰 rate 的定语从句。

3）并列成分多或长的长复合句

正确判断并列成分是能否对长句进行正确分析的重要因素。这些并列的成分，既可以是并列的分句,也可以是并列的主语、谓语、宾语、定语和状语等。而且,并列成分一般多由 and 等连词连接。在大多数情况下,根据该连词后面那个词或短语的性质到前面去找与之性质相同的词或短语,即可对并列成分做出正确判断。例如:

Such activities as investigating the strength and uses of materials, extending the findings of pure mathematics to improve the sampling procedures used in agriculture or social sciences, and developing the potentialities of atomic energy, are all examples of the work of the applied scientist.

该句中 and 后面是动名词短语 developing the potentialities of atomic energy,and 前面也有两个动名词短语 investigating... 和 extending...,可以肯定这三个动名词短语并列,用作 as 引导的定语从句的主语。

4) 从句中含有从句的长复合句

在有的长句中,某个从句本身有可能也是含有一个或一个以上从句的复合句;甚至这种从句本身又含有从句。遇到这种比较复杂的现象,必须要对每个从句的性质做出正确的分析,找出与它发生关系的词。例如:

A further inference was drawn by Pascal, who reasoned that if this "sea of air" existed, its pressure at the bottom (i. e. sea-level) would be greater than its pressure further up, and that therefore the height of mercury column would decrease in proportion to the height above the sea-level.

此句中,who 引导的定语从句为一复合句,含有一个以 that 引导的宾语从句。而这个 that 引导的宾语从句又是一个复合句,其中含有一个以 if 引导的条件状语从句,and 连接的两个并列的宾语从句。

为了能够做好长句结构的分析,掌握好专业英语中关于复合句、非谓语动词的知识是非常必要的。在接下来的几课中,将介绍一下关于复合句、非谓语动词的知识。然后我们再结合实例来介绍长句结构分析的方法(见 Lesson 17,物理学专业英语中的长句分析(二))。

9.2 专业英语阅读

9.2.1 Thermal Equilibrium and Temperature(热平衡及温度)

Temperature is familiar to us as the measure of the hotness or coldness of objects or of our surroundings.

Our sense of touch can usually tell us if an object is hot or cold. Early in childhood we learn that to make a cold object warmer, we place it in contact with a hot object and to make a hot object cooler, we place it in contact with a cold object.

When an object is heated or cooled, some of its physical properties change. Most solids and liquids expand when they are heated. A gas, if its pressure is kept constant, will also expand when it is heated, or, if its volume is kept constant, its pressure will rise. If an electrical conductor is heated, its electrical resistance changes. A physical property that changes with temperature is called a **thermometric property.** A change in a thermometric property indicates a change in the temperature of the object.

Suppose that we place a warm copper bar in close contact with a cold iron bar so that the copper bar cools and the iron bar warms. We say that the two bars are in **thermal contact**. The copper bar contracts slightly as it cools, and the iron bar expands slightly as it warms. Eventually this process stops and the lengths of the bars remain constant. The two bars are then in **thermal equilibrium** with each other.

Suppose instead that we place the warm copper bar in a cool running stream. The bar cools until it stops contracting, at the point at which the bar and the water are in thermal equilibrium. Next we place a cold iron bar in the stream on the side opposite the copper bar. The iron bar will warm until it and the water are also in thermal equilibrium. If we remove the bars and place them in thermal contact with each other, we find that their lengths do not change. They are in thermal equilibrium with each other. Though it is common sense, there is no logical way to deduce this fact, which is called the **zeroth law of thermodynamics** (Fig. 9-1):

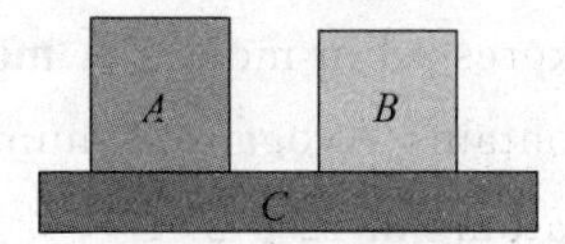

Fig. 9-1 If objects A and B are separately in thermal equilibrium with a third object, C, then A and B are in thermal equilibrium with each other

If two objects are in thermal equilibrium with a third, then they are in thermal equilibrium with each other.

Two objects are defined to have the same temperature if they are in thermal equilibrium with each other.

9.2.2 The Ideal-Gas Law(理想气体定律)

The properties of gases at low densities allow the definition of the ideal-gas temperature scale. If we compress such a gas while keeping its temperature constant, the pressure increases. Similarly, if a gas expands at constant temperature, its pressure decreases. To a good approximation, the product of the pressure and volume of a low-density gas is constant at a constant temperature. This result was discovered experimentally by Robert Boyle (1627—1691), and is known as **Boyle's law**:

$$PV = \text{constant}(\textbf{constant temperature})$$

A more general law exists that reproduces Boyle's law as a special case. As we know, the absolute temperature of a low-density gas is proportional to its pressure at constant volume. In addition—a result discovered experimentally by Jacques Charles (1746—1823) and Joseph Gay-Lussac (1778—1850)—the absolute temperature of a low-density gas is proportional to its volume at constant pressure. We can combine these two results by stating

$$PV = CT \tag{9-1}$$

where C is a constant of proportionality. We can see that this constant is proportional to the amount of gas by considering the following. Suppose that we have two containers with identical volumes, each holding the same amount of the same kind of gas at the same temperature and pressure. If we consider the two containers as one system, we have twice the amount of gas at twice the volume, but at the same temperature and pressure. We have thus doubled the quantity $PV/T = C$ by doubling the amount of gas. We can therefore write C as a constant k times the number of molecules in the gas N

$$C = k\,N$$

Eq. (9-1) then becomes

$$PV = N\,kT \tag{9-2}$$

The constant k is called **Boltzmann's constant.** It is found experimentally to have the same value for any kind of gas

$$k = 1.381 \times 10^{-23}\ \text{J/K} = 8.617 \times 10^{-5}\ \text{eV/K} \tag{9-3}$$

An amount of gas is often expressed in moles. A **mole** (mol) of any substance is the amount of that substance that contains Avogadro's number N_A of atoms or molecules, defined as the number of carbon atoms in 12 g of C^{12}

$$N_A = 6.022 \times 10^{23} \tag{9-4}$$

If we have ν moles of a substance, then the number of molecules is

$$N = \nu N_A \tag{9-5}$$

Eq. (9-2) is then

$$PV = \nu N_A kT = \nu RT \tag{9-6}$$

where $R = N_A k$ is called the **universal gas constant.** Its value, which is the same for all gases, is

$$R = N_A k = 8.314\ \text{J/(mol} \cdot \text{K)} = 0.082\,06\ \text{L} \cdot \text{atm/(mol} \cdot \text{K)} \tag{9-7}$$

An **ideal gas** is defined as one for which $PV/(\nu T)$ is constant for all pressures. The pressure, volume, and temperature of an ideal gas are related by

$$PV = \nu RT \tag{9-8}$$

Eq. (9-8), which relates the variables P, V, and T, is known as the **ideal-gas law**, and is an example of an **equation of state.** It describes the properties of real gases with low densities (and therefore low pressures). At higher densities, corrections must be made to this equation. For any gas at any density, there is an equation of state relating P, V, and T for a given amount of gas. Thus the state of a given amount of gas is determined by any two of the three state variables P, V, and T.

The temperature 0℃ and the pressure of 1 atm are often referred to as **standard conditions.** Under standard conditions, a mol of an ideal gas occupies a volume of 22.4 L.

For a fixed amount of gas, we can see from Eq. (9-8) that the quantity PV/T is constant. Using the subscripts 1 for the initial values and 2 for the final values, we have

$$\frac{P_2 V_2}{T_2} = \frac{P_1 V_1}{T_1} \tag{9-9}$$

Glossary

temperature	温度	thermal equilibrium	热平衡
hotness	热	electrical conductor	电导体
coldness	冷	thermometric property	热力学特性
the zeroth law of thermodynamics	热力学第零定律	thermal contact	热接触
ideal-gas temperature scale	理想气体温标	Boyle's law	波意耳定律
universal gas constant	普适气体常量	constant volume	等体
Boltzmann's constant	玻尔兹曼常量	equation of state	状态方程
mole	摩尔	state variable	状态参量
Avogadro's number	阿伏伽德罗常量	standard condition	标准条件
carbon atom	碳原子	subscript	下标

9.3 专业英语常用表达法-9 热学和波动常用实验仪器

refrigerator 电冰箱

refrigerating box 冷藏箱

digital thermometer 数字温度计

refrigerating fluid 冷冻液、制冷剂

heater 加热器

thermopile 温差电堆、热电堆

heat converter 热交换器

heat exchanger 热交换器

hot-water bag 热水袋

thermos, thermos bottle, vacuum bottle 热水瓶

thermometer 温度计

wet and dry thermometer 干湿温度计

thermal converter 热转换器

empty box barometer 空盒气压计

thermostat 恒温箱、温控器

magnetic barrel calorimeter 磁桶量热器

thermionic tube 热离子管

thermomotor 热力机、热气机、热发动机

calorimeter 热量表、量热器、热量计

boron thermopile 硼温差电堆

sonometer 弦音计、振动频率计

tuning fork 音叉

resonance tuning forks 共振音叉

specific heat capacity converter 比热容转换器

varying-temperature viscosity coefficient tester 变温黏滞系数实验仪

single-junction vacuum thermopile 单接头真空温差电堆

sublimation and condensation of iodine demonstrator 碘的升华与凝华演示器

solid shrink force demonstrator 固体缩力演示器

interconversion of mechanical energy and heat energy demonstrator 机械能热能互变演示器

metal linear expansion demonstrator 金属线膨胀演示器

refrigerated centrifuge 冷冻离心分离机

solidification and melt of naphthalene experiment device 萘的熔解凝固实验器

experimental device for gas law 气体定律实验器

gas law demonstrator 气体定律演示器
heat conduction demonstrator 热传导演示器
thermoelectric couple, thermocouple, thermoelectric pair 热电偶、温差电偶
heat flux density converter 热流密度转换器
thermistor thermometer 热敏温度计、热敏电阻温度计
demonstrator of sound propagation 声传播演示器
quick freezing refrigeration device 速冻制冷装置
standing waves on a string instrument 弦线上驻波实验仪
harmonic wave analysis experimental instrument 谐波分析实验仪
liquid specific heat detector 液体比热测定仪
vacuum flask/dewar flask 真空瓶/杜瓦瓶
vacuum thermopile 真空热电堆、真空温差电堆
refrigeration compressor 制冷压缩机、冷冻机
thermal coefficient meters 导热系数测定仪
meter of metal specific heat capacity 金属比热容测定仪
meter of specific heat ratio of gas 气体比热容比测定仪
thermotics experiment facility 热学实验仪
multi-function constant-temperature controller 多功能恒温控制仪

LESSON 10

10.1 物理学专业英语中的主语从句

前面已经提到过,英语的主从复合句一般分为三大类型:名词性从句、形容词性从句和副词性从句。

名词性从句在整个复合句中起名词作用。主语、宾语、表语和同位语等各种从句统称为名词性从句,在句中起名词作用,充当主语、宾语、表语和同位语。

引导名词性从句的连接词可分为三类:

连接词:that, whether, if (不充当从句的任何成分)

连接代词:what, whatever, who, whoever, whom, whose, which

连接副词:when, where, how, why

主语从句是在复合句中充当主语的从句。

10.1.1 主语从句的引导词

主语从句通常由连词 that 和 whether,连接代词或连接副词,以及关系代词 what 引导。

1) that 引导

That each atom has a tiny but massive nucleus that contains protons and neutrons is a general phenomenon. 每个原子都有一个包含质子和中子的小而重的原子核,是个普遍现象。

That two objects carrying the same type of charge repel each other is certain. 带有同

种电荷的两个物体互相排斥，这是确定无疑的。

That she became a physicist may have been due to her father's influence. 她成为物理学家可能是受她父亲的影响。

There seems little doubt *that radar technology is a permanent and important aspect of research and development in electronics*.（That 引导的从句作实际主语）似乎没有疑问，在电子学的研究和发展中，雷达技术是一个永久并且重要的方面。

2）whether 引导

Whether the experiment of dropping bodies of different weight from the leaning tower of Pisa was performed before a multitude, as some accounts have it, was not known. 我们不知道比萨斜塔落体实验是否像某些资料记载那样在很多人目睹下进行。

Whether the plan of carrying out the experiment is feasible remains to be proved. 这一实验计划是否可行还有待证实。

Whether the discharge will do us harm remains to be seen. 放电是否有害还要看一看。

Whether the experimental result would support the theory was a problem. 实验结果是否会支持这个理论还是一个问题。

Whether a length is exactly a meter is a question no experiment can decide. 一段长度是否精确地是 1 m，没有实验能够决定。

3）连接代词引导

Whoever discovered the phenomena did not make any difference. 到底是谁发现这个现象的，没有什么分别。

Whichever of the particles gets out first does not matter. 无论哪个粒子先出来都没有关系。

Who should be responsible for the environmental degradation is still unknown. 谁对环境恶化负责还不清楚。

4）连接副词引导

When magnetism was recognized is not quite clear. 磁现象是什么时候被发现的，不是十分清楚。

How the experiment was done was a mystery. 这个实验怎样做的是一个谜。

How this happened is not clear to anyone. 这件事怎样发生的，谁也不清楚。

Where Coulomb confirmed the inverse-square law for the electrostatic force remains to be proved. 库仑到底是在哪里确定静电力的平方反比定律还有待证实。

5）关系代词 what 引导

What we need is a precision tool. 我们需要的是精密工具。

What the physicists were waiting for at that time was a solar eclipse. 那时，物理学家们等待的就是一次日食。

What make the lab more impressive are the modern instruments. 为实验室增色的是现代化的仪器设备。

10.1.2 主语从句与形式主语 it

有时为了考虑句子平衡，通常在主语从句的位置使用形式主语 it，而将真正的主语从句移至句末。用 it 作形式主语的 that 从句有以下四种不同的搭配关系：

"It＋be＋形容词＋that 从句"，如：

It is necessary that... 有必要……

It is natural that... 很自然……

It is strange that... 奇怪的是……

"It＋be＋过去分词＋that 从句"，如：

It is known to all that... 众所周知……

It is reported that... 据报道……

It is said that... 据说……

It has been proved that... 已证实……

"It＋be＋名词＋that 从句"，如：

It is a fact that... 事实是……

It is an honor that... 非常荣幸……

It is common knowledge that... ……是常识

"It＋不及物动词＋that 从句"，如：

It appears that... 似乎……

It seems that... 似乎……

It happened that... 碰巧……

It occurs that... 碰巧……

用 it 作形式主语时分以下三种情况：

1）对于以连词 that 引导的主语从句，通常用形式主语代替主语从句：

For example, in electric circuits it is often assumed *that the relations between voltages and currents are linear*. 例如，在电路中通常假设电压和电流的关系是线性的。

It has been found repeatedly in science *that the discovery that two branches are related leads to each branch helping in the development of the other*. 在科学中经常发现，与两个分支相关的发现会导致两方相互促进发展。

It is quite clear *that the whole project is doomed to failure*. 很清楚，整个计划注定要失败。

It's a pity *that he didn't find any useful results in the experiment*. 很遗憾他在实验中没找到什么有用的结果。

It is important that *the experiment should be carried out in vacuum*. 实验必须在真空内进行。

2）对于以连接代词（副词）引导的主语从句，可以使用形式主语代替主语从句，也可直接在句首使用主语从句：

Whether they would support us was a problem. 他们是否会支持我们还是一个问题。

It was a problem *whether they would support us*. 他们是否会支持我们还是一个

问题。

It remains to be seen *whether the material will do us harm or good*. 这材料对我们是利是害，还得看看再说。

3）对关系代词型 what 引导的主语从句，通常直接将主语从句放在句首。如：

What we need is an inertial frame of reference. 我们需要的是惯性参考系。

What I want to know is how we apply Newton's laws to solve problems. 我想知道的就是如何运用牛顿定律解决问题。

有时也可借用形式主语。如：

It is clear enough what the results meant. 结果是什么意思很清楚。

4）如果句子是疑问句，则必须用带形式主语 it 的结构：

Is it true *that the electric field of the source particles is present* whether or not we introduce a test particle into the field? 不论检验电荷是否放入，场源电荷产生的电场都是存在的，是真的吗？

How is it *that the two particles attract each other*? 两个粒子怎么会互相吸引呢？

10.1.3 连词 that 的省略问题

引导主语从句的连词 that 有时可省，有时不能省，其原则是：若 that 引导的主语从句直接位于句首，则 that 不能省略；若 that 引导的主语从句位于句末，而在句首使用了形式主语 it，则 that 可以省略。

That you failed to use Ampere Circuital Theorem correctly was a pity. 很遗憾你未能正确使用安培环路定理。（that 不可省）

It was a pity (that) *you failed to use Ampere Circuital Theorem correctly*。很遗憾你未能正确使用安培环路定理。（that 可省）

10.1.4 主语从句不可位于句首的五种情况

(1) if 引导的主语从句不可居于复合句的句首。

(2) It is said (reported)… 结构中的主语从句不可提前。

(3) It happens…, It occurs… 结构中的主语从句不可提前。

(4) It doesn't matter how/whether… 结构中的主语从句不可提前。

(5) 含主语从句的复合句是疑问句时，主语从句不可提前。

10.1.5 What 与 that 在引导主语从句时的区别

What 引导主语从句时在从句中充当句子成分，如主语、宾语、表语，不能省略。That 不作任何成分，而只在语法上起连接作用，并且在句首使用了形式主语 it 时 that 可以省略。

10.2 专业英语阅读

10.2.1 Heat and the First Law of Thermodynamics(热及热力学第一定律)

Heat is energy that is being transferred from one system to another because of a difference in temperature. In the seventeenth century, Galileo, Newton, and other scientists generally supported the theory of the ancient Greek atomists who considered thermal energy to be a manifestation of molecular motion. In the next century, methods were developed for making quantitative measurements of the amount of heat that leaves or enters an object, and it was found that if objects are in thermal contact, the amount of heat that leaves one object equals the amount that enters the other. This discovery led to the caloric theory of heat as a conserved material substance. In this theory, an invisible fluid called "caloric" flowed out of one object and into another and this "caloric" could be neither created nor destroyed.

The caloric theory reigned until the nineteenth century, when it was found that friction between objects could generate an unlimited amount of thermal energy, deposing of the idea that caloric was a substance present in a fixed amount. The modern theory of heat did not emerge until the 1840s, when James Joule (1818—1889) demonstrated that the increase or decrease of a given amount of thermal energy was always accompanied by the decrease or increase of an equivalent quantity of mechanical energy. Thermal energy, therefore, is not itself conserved. Instead, thermal energy is a form of internal energy, and it is energy that is conserved.

We will always talk about energy transfer to or from some specific *system*. The system might be a mechanical device, a biological organism, or a specified quantity of material such as the refrigerant in an air conditioner. A **thermodynamic system** is a system which can interact (and exchange energy) with it surroundings, or environment. A process in which there are changes in the state of thermodynamic system is called a **thermodynamic process.**

In this part, we define heat capacity, and examine how heating a system can cause either a change in its temperature or a change in its phase. We then examine the relationship between heat conduction, work, and internal energy of a system and express the law of conservation of energy for the thermal systems as the first law of thermodynamics. Finally, we shall see how the heat capacity of a system is related to its molecular structure.

We describe the energy relations in any thermodynamic process in terms of the quantity of heat Q added to the system and the work W done by the system. Both Q and W may be positive, negative, or zero. A positive value of Q represents heat flow into the system, with a corresponding input of energy to it; negative Q represents heat flow out of the system. A positive value of W represents work done by the system against it surroundings, such as work done by an expanding gas, and hence corresponds to energy leaving the system. Negative W, such as work done during compression of a gas in which work is done on the gas by its surroundings, represents energy entering the system.

When the law of conservation of energy was first introduced in previous chapter, it was stated that the mechanical energy of a system is conserved in the absence of nonconservative forces, such as friction. That is, the changes in the internal energy of the system were not included in this mechanical model.

The first law of thermodynamics is a generalization of the law of conservation of energy that includes possible changes in internal energy.

It is a universally valid law that can be applied to all kinds of processes. Furthermore, it provides us with a connection between the microscopic and macroscopic worlds.

We have seen that energy can be transferred between a system and its surroundings in two ways. One is work done by (or on) the system. This mode of energy exchange results in measurable changes in the macroscopic variables of the system, such as the pressure, temperature, and volume of gas. The other is heat transfer, which takes place at the microscopic level.

To put these ideas on a more quantitative basis, suppose a thermodynamics system undergoes a change from an initial state to a final state in which Q units of heat are absorbed (or removed) and W is the work done by (or on) the system. For example, the system may be a gas whose pressure and volume change from P_i, V_i to P_f, V_f. If the quantity $Q-W$ is measured for various paths connecting the initial and final equilibrium states (that is, for various process), one finds that $Q-W$ is the same for all paths connecting the initial and final states. We conclude that the quantity $Q-W$ is determined completely by the initial and final states of the system, and we call the quantity $Q-W$ *the change in the internal energy of the system*. Although Q and W both depend on the path, the quantity $Q-W$, that is, *the change in internal energy is independent of the path*. If we represent the internal energy function by the letter E, then the change in internal energy, $\Delta E = E_f - E_i$, can be expressed as

$$\Delta E = E_f - E_i = Q - W \tag{10-1}$$

where all quantities must have the same energy units. Eq. (10-1) is known as the **first law of thermodynamics.** When it is used in this form, we must note that Q is positive when heat enters the system and W is positive when work is done by the system.

10.2.2 Kinds of Thermodynamic Processes(热力学过程的种类)

In this section we describe four specific kinds of thermodynamic processes that occur often in practical situations. These can be summarized briefly as "no heat transfer" or adiabatic, "constant volume" or isochoric, "constant pressure" or isobaric, and "constant temperature" or isothermal. For some of these we can use a simplified form of the first law of thermodynamics.

Adiabatic Process(绝热过程)

An adiabatic process is defined as one with no heat transfer into or out of a system, $Q=0$. We can prevent heat flow either by surrounding the system with thermally insulating material or by carrying out the process so quickly that there is not enough time for appreciable heat flow. From the first law we find that for every adiabatic process

$$E_2 - E_1 = \Delta E = -W \quad \text{(adiabatic process)} \tag{10-2}$$

When a system expands adiabatically, W is positive (the system does work on its surrounding), so ΔE is negative and the internal energy decreases. When a system is compressed adiabatically, W is negative (work is done on the system by its surroundings) and E increases. In many (but not all) systems an increase of internal energy is accompanied by a rise in temperature.

The compression stroke in an internal-combustion engine is an approximately adiabatic process. The temperature rises as the air-fuel mixture in the cylinder is compressed. The expansion of the burned fuel during the power stroke is also an approximately adiabatic expansion with a drop in temperature.

Isochoric Process(等体过程)

An isochoric process is a constant-volume process. When the volume of a thermodynamic system is constant, it does no work on its surroundings. Then $W=0$, and

$$E_2 - E_1 = \Delta E = Q \quad \text{(isochoric process)} \tag{10-3}$$

In an isochoric process, all the energy added as heat remains in the system as an increase in internal energy. Heating a gas in a closed constant-volume container is an example of an isochoric process. (Note that there are types of work that do not involve a volume change. For example, we can do work on a fluid by stirring it.)

Isobaric Process(等压过程)

An isobaric process is a constant-pressure process. In general, none of the three quantities ΔE, Q, and W is zero in an isobaric process, but calculating W is easy nonetheless.

$$W = P(V_2 - V_1) \quad \text{(isobaric process)} \tag{10-4}$$

Isothermal process(等温过程)

An isothermal process is a constant-temperature process. For a process to be isothermal, any heat flow into or out of the system must occur slowly enough that thermal equilibrium is maintained. In general, none of the quantities ΔE, Q, or W is zero in an isothermal process.

In some special cases the internal energy of a system depends only on its temperature, not on its pressure or volume. The most familiar system having this special property is an ideal gas. For such system, if the temperature is constant, the internal energy is also constant; $\Delta E=0$, and $Q=W$, that is, any energy entering the system as heat Q must leave it again as work W done by the system. For most systems other than ideal gases the internal energy depends on pressure as well as temperature, so E may vary even when T is constant.

Glossary

atomist	原子学家	thermal energy	热能
manifestation	表现形式	molecular motion	分子运动
thermal contact	热接触	internal energy	内能
thermodynamic system	热力学系统	thermodynamic process	热力学过程
caloric	*a*. 热的;*n*. 热(质)	heat capacity	热容量
phase	相	heat conduction	热传导
initial state	初态	final state	末态
adiabatic	绝热的	isochoric	等体积的
isobaric	等压强的	isothermal	等温的
thermally insulating material	绝热材料	adiabatically	绝热地
compression stroke	压缩冲程	internal-combustion engine	内燃机
power stroke	做功冲程		

10.3 专业英语常用表达法-10 精度

1. 精度的表达方法

accuracy control 精确控制

accuracy grade 准确度等级

accuracy index 准确度(精确度)指数

accuracy requirements 准确度要求

high accuracy 高精度

percentage of accuracy 准确度百分数

order of accuracy 准确度

overall accuracy 总精度、总准确度

standard of accuracy 准确度标准

to an accuracy of... 精度达……、准确到……

with accuracy 正确地、准确地

within accuracy of... 精度达……、准确到……

with an accuracy of... 精度达……、以……的精度

within the accuracy of... 在……的精度范围内

the accuracy with/to which... ……的精度

accurate to... 精度达……、精确到……

accurate (to) within... 精度在……的范围之内

be accurate to dimension 符合加工尺寸

be accurate to/within plus or minus five per cent 精确到(精度在)±5%以内

例如:

With care a micrometer will give *an accuracy of* better than 1 part in thousand. 使用

得法时，测微计能给出高于0.1%的准确度。

It is impossible to say *with any accuracy* how many are affected. 无论如何也说不准受影响的有多少。

The volume measuring *accuracy* of this system is ±0.25%. 系统容积的测量精度为±0.25%。

Readings can be obtained to *an accuracy of* one micron. 获得的读数精度可达1 μm。

The discrepancy is *within the accuracy of* the analysis. 误差在分析精度范围之内。

The *accuracy with which* scientists can use laser can also be of help in making different kinds of measurements. 科学家使用激光所能达到的精度对进行其他各种测量也可能有所帮助。

It depends on *the degree of accuracy to which* the parts must be made. 这取决于制造零件所应达到的精度。

All input data are *accurate to* three significant figures. 所有输入的数据精确到3位有效数字。

This apparatus is *accurate within* microseconds. 这台装置的精确度在数微秒之内。

These parameters are known with *an accuracy of* 5%. 这些参数的已知精确度为5%。

The components can be positioned *with an accuracy* better than 0.010 in. 这些部件固定的精确度可高于0.010 in。

2. 用 The function(或 purpose) is + 动词不定式来表达功能

It has several functions and probably *the main one is to hide the structure of the instrument*. 它有好几个作用，主要的可能是为了把这台仪器的内部结构遮盖起来。

At the end we have a clear glass, *its function there is to give you a view of the circuit inside*. 在末端用的是透明玻璃，它的作用是使你看得见里边的电路。

3. 其他表达用途、目的和功能的词语

be designed + 动词不定式(或 for 短语)

help + 动词不定式

have the function of

serve the purpose of

The building *is designed* primarily *for students* rather than teachers. 这座建筑是供学生而不是供教师使用的。

The cast glass *helps to express* the structure of the lab. 压铸玻璃有助于显露出实验室的结构。

It also *serves the purpose of* giving access over the roof for maintenance and for cleaning the glass. 还可以利用它到实验室顶部去进行维修和清洁玻璃。

LESSON 11

11.1 物理学专业英语中的宾语从句

宾语从句就是在复合句中作宾语的名词性从句，通常放在主句谓语动词(及物动词)或介词之后。例如：

For a situation like this, in which the current is discontinuous in space, Ampere's law is not valid. In following, we will see *how Maxwell was able to modify Ampere's law so that it holds for all currents*. 对于像空间中电流不连续的情况，安培定律就不成立了。下面我们将看到麦克斯韦是如何修正安培定律从而使其对所有电流都适用。

This law describes *how electric field lines diverge from a positive charge and converge on a negative charge*. Its experimental basis is Coulomb's law. 这个定律说明了电场线如何发散于正电荷、聚集于负电荷，其实验基础是库仑定律。

11.1.1 宾语从句的引导词

宾语从句通常由连词 that 和 whether/if，连接代词或连接副词，以及关系代词 what 引导。

1) that 引导

It is a common fallacy to suppose *that mathematics is important for physics only because it is a useful tool for making computations*. 人们通常错误地认为数学对物理很重要仅仅因为它是一种有用的计算工具。

Stated in words, Ohm's law says *that the steady current through any portion of an electric circuit equals the potential difference across that portion of the circuit divided by the resistance of that portion of the circuit*. 欧姆定律用语言表述为：电路中任一部分的电流强度等于这部分的电势差除以这部分的电阻。

Archimedes principle states *that a body wholly or partially immersed in a fluid is buoyed up by a force equal to the weight of the fluid displaced by it*. 阿基米德原理表述为：部分或全部浸在水中的物体受到的浮力等于物体排开液体的重量。

That 引导的从句常跟在下列形容词后作宾语：afraid, anxious, aware, certain, confident, convinced, determined, glad, proud, surprised, worried, sorry, thankful, ashamed, disappointed, annoyed, pleased, hurt, satisfied, content 等。例如：

I am afraid *that the shock waves will be very strong in this case*. 在这种情况下，恐怕冲击波会很强。

2) whether/if 引导

Before making this case a general rule, test it by experiment two or three times and see *if the experiment produces the same effect*. 在把这一情况定为常规之前，先做两三次实验看看是否产生同一效果。

In order to determine *whether a system is performing properly and ultimately to control the system performance*, the engineer must know what the system is doing at any instant of time. 为了确定系统是否准确运行以及全面控制系统性能，工程人员必须随时了解系统在做什么。

I wonder *if the voltage is high enough*. 我不知道电压是否够高。

3）连接代词引导

Put them together and see *which is larger*. 把它们加起来看看哪个更大。

We'll do *whatever we can* to measure the magnetic force between the two parallel current-carrying wires. 我们将尽我们所能来测量这两个平行电流间的相互作用磁力。

I don't know *who/whom you mean*. 我不知道你指谁。

You can take *whichever you like*. 你爱拿哪个就拿哪个吧。

Few people knew *what the relativity meant at that time*. 那时很少有人明白相对论。

4）连接副词引导

We often fail to realize *how little we know about a thing* until we attempt to simulate it on a computer. 在我们试图用计算机模拟一件事之前，我们通常意识不到我们对它的了解是多么少。

Our success depends upon *how well we can cooperate with one another*. 我们的成功取决于彼此间的良好合作。

The sum, difference and product of continuous functions are also continuous. The quotient of continuous functions is continuous except *where the denominator vanishes*. 连续函数的和、差、积仍然连续，连续函数的商在分母不为零时也连续。

This essay aims to explore *why quantum was put forward at that time*. 本文将探讨为什么量子的概念会在那个时候提出。

I'd like to know *when the electrification will be finished*. 我很想知道充电什么时候会结束。

5）关系代词 what 引导

In order to determine whether a system is performing properly and ultimately to control the system performance, the experimenter must know *what the system is doing at any instant of time*. 为了确定系统在运行中是否准确全面地控制系统性能，实验人员必须随时了解系统在做什么。

First let's introduce *what the Atwood's machine is like*. 首先让我们介绍一下阿特伍德机。

【注意】 有时介词后可接一个宾语从句(但介词后通常不接 that 和 if 引导的宾语从句)：

From *what Maxwell stated*, the displacement current was reasonable. 根据麦克斯韦所说，位移电流是合理的。

有极个别介词(如 but, except)可接 that 引导的宾语从句：

The experimenters need do nothing about it except *that they should turn the power on*. 实验人员只需要通上电源，其他的什么都不需要做。

11.1.2 宾语从句与形式宾语 it

当宾语从句后跟有宾语补足语时，特别是在带复合宾语的句子中，通常在宾语从句处使用形式宾语 it,而将真正的宾语从句移至句末。

I think *it* best *that you should explain the Guass's law in detail now*. 我认为你最好现在把这个高斯定理仔细解释一下。

He hasn't made *it* known *when he is going to get started with thermodynamics*. 他还没宣布他何时开始讲热力学。

She found *it* difficult *how she could answer the question about Maxwell's Equations*. 她发现回答这个关于麦克斯韦方程组的问题很困难。

The director made *it* a rule *that every graduate student should make at least one presentation in the discussion*. 导师规定每个研究生在讨论时至少发一次言。

11.1.3 连词 that 的省略问题

引导宾语从句的连词 that 通常可以省略。

Einstein thought (*that*) *the quantum theory was unreasonable*. 爱因斯坦认为量子理论难以置信。

She said (*that*) *the equivalent capacitance of a series combination of capacitors would be less than any individual capacitance in the combination*. 她说过串联电容器的总电容会小于其中任何一个电容器的电容。

I promise you (*that*) *I will be there when they maintain the equipment*. 我答应你在他们维护仪器时我会去照看的。

I hoped (*that*) *I would/should succeed in the experiment*. 我曾希望我会在实验中成功。

He thinks (*that*) *they will fix the equipment on time*. 他想他们会按时安装好设备的。

He thought (*that*) *they would set up the equipment in time*. 他本想他们会按时安装好设备的。

I expect (*that*) *the experiment's result will turn out perfect*. 我料想实验的结果将会很完美。

I suggested (*that*) *they should handle the device carefully*. 我建议他们小心操纵设备。

【注意】 有时为了强调,that 引导的宾语从句可位于句首,此时 that 不可省略。

That the electric appliance should be grounded I know. 这电器应该接地,我是知道的。

11.1.4 宾语从句与否定转移

当动词 think, believe, suppose, expect, imagine 后接一个表示否定意义的宾语从句时,其否定通常转移到主语。

I *don't* suppose *that the data they presented is true*. 我认为他们提供的数据不是真的(而不是 I suppose *that the data they presented is not true*.)。

I *don't* think *we need waste much time on the maintenance*. 我想我们不必在仪器维护上面花太多时间(而不是 I think we need not waste much time on the maintenance.)。

11.1.5 不能省略 that 的宾语从句

引导宾语从句的 that 在下列情况下一般不省略。

1）宾语从句前有插入语。如：

We hope, *on the contrary*, *that* he can find the kinetic equation for the simple harmonic motion. 恰恰相反，我们希望他能够找到简谐运动的运动学方程。

2）有间接宾语时。如：

He told *me that* he would apply Huygens Principle to describe the propagation of waves. 他告诉我他要用惠更斯原理来描述波的传播。

3）that 在与之并列的另一个宾语从句之后。如：

The book states (that) the standing waves are very interesting and *that*（不省略）*a standing wave does not transfer energy*. 那本书说驻波很有趣，驻波也不传播能量。

4）在"it(形式宾语)＋补语"之后时。如：

Teachers think *it* necessary *that* physics is taught with multimedia. 教师们认为利用多媒体教授物理学是必要的。

5）that 从句单独回答问题时。如：

—What did he hear? 他听说了什么事？

—*That* Kate had found a new way to prove the theory. (他听说)凯特发现了一个证明这个理论的新方法。

6）在 except 等介词后。如：

The experimenters need do nothing about it except *that* they should turn the power on. 实验人员只需要通上电源，其他的什么都不需要做。

7）位于句首时。如：

That the wave like this is nothing but a standing wave, I believe. 我相信像这样的波就是驻波。

8）在较为正式或不常用的动词(如 reply, object)后。如：

The editor replied *that the paper was accepted*. 编辑回复说文章被接受了。

11.1.6 后边不能直接跟 that 从句的动词

这类动词有 allow, refuse, let, like, cause, force, admire, condemn, celebrate, dislike, love, help, take, forgive 等。这类词后可以用不定式或动名词作宾语，但不可以用 that 引导的宾语从句。

11.1.7 不可用 that 从句作直接宾语的动词

有些动词不可用于"动词＋间接宾语＋that 从句"结构中，常见的有 envy, order, accuse, refuse, impress, forgive, blame, denounce, advise, congratulate 等。例如：

He impressed the expert as an excellent experimenter. (right)

He impressed the expert that he was an excellent experimenter. (wrong)

11.2 专业英语阅读

11.2.1 The Second Law of Thermodynamics(热力学第二定律)

We are often asked to conserve energy. But according to the first law of thermodynamics, energy is always conserved. What then does it mean to conserve energy if the total amount of energy in the universe does not change regardless of what we do? The first law of thermodynamics does not tell the whole story. Energy is always conserved, but some forms of energy are more useful than others. The possibility or impossibility of putting energy to use is the subject of the second law of thermodynamics. For example, it is easy to convert work into thermal energy, but it is impossible to remove energy as heat from a single reservoir and convert it entirely into work with no other changes. This experimental fact is one statement of the second law of thermodynamics.

Kelvin statement of the second law of thermodynamics: *No system can take energy as heat from a single reservoir and convert it entirely into work without additional net changes in the system or its surroundings.*

In this chapter, we will encounter several other formulations of this law.

A common example of the conversion of work into heat is movement with friction. For example, suppose you spend two minutes pushing a block this way and that way along a tabletop in a closed path, leaving the block in its initial. Also, suppose that the block-table system is initially in thermal equilibrium with its surroundings. The work you do on the system is converted into internal energy of the system, and as a result the block-table system becomes warmer. Consequently, the system is no longer in thermal equilibrium with its surroundings. However, the system will transfer energy as heat to its surroundings until it returns to thermal equilibrium with those surroundings. Because the final and initial states of the system are the same, the first law of thermodynamics dictates that the energy transferred to the environment as heat equals the work done by you on the system. The reverse process never occurs—a block and table that are warm will never spontaneously cool by converting their internal energy into work that causes the block to push your hand around the table. Yet such an amazing occurrence would not violate the first law of thermodynamics or any other physical laws we have encountered so far. It does, however, violate the second law of thermodynamics. Thus, there is a lack of symmetry in the roles played by heat and work that is not evident from the first law. This lack of symmetry is related to the fact that some processes are irreversible.

There are many other irreversible processes, seemingly quite different from one another, but all related to the second law. For example, heat conduction is an irreversible process. If we place a hot body in contact with a cold body, heat will flow from the hot body to the cold body until they are at the same temperature. However, the reverse does not occur. Two bodies in contact at the same temperature remain at the same temperature; heat does not flow from one to the other leaving one colder and the other warmer. This experimental fact gives us a second statement of the second law of thermodynamics.

Clausius statement of the second law of thermodynamics: *A process whose only net result is to transfer energy as heat from a cooler object to a hotter one is impossible.*

We will show in this chapter that the Kelvin and Clausius statements of the second law are equivalent.

11.2.2 Equivalence of the Kelvin and Clausius Statements(开尔文表述与克劳修斯表述的等价性)

The heat-engine and refrigerator statements (or the Kelvin and Clausius statements, respectively) of the second law of thermodynamics seem quite different, but they are actually equivalent. We can prove this by showing that if either statement is assumed to be false, then the other must also be false. We'll use a numerical example to show that if the heat-engine statement is false, then the refrigerator statement is false.

Fig. 11-1(a) shows an ordinary refrigerator that uses 50 J of work to remove 100 J of energy as heat from a cold reservoir and rejects 150 J of energy as heat to a hot reservoir. Suppose the heat-engine statement of the second law were not true. Then a "perfect" heat engine could remove energy from the hot reservoir and convert it completely into work with 100 percent efficiency. We could use this perfect heat engine to remove 50 J of energy from the hot reservoir and do 50 J of work (Fig. 11-1 (b)) on the ordinary refrigerator. Then, the combination of the perfect heat engine and the ordinary refrigerator would be a perfect refrigerator, transferring 100 J of energy as heat from the cold reservoir to the hot reservoir without requiring any work, as illustrated in Fig. 11-1 (c). This violates the refrigerator statement of the second law. Thus, if the heat-engine statement is false, the refrigerator statement is also false. Similarly, if a perfect refrigerator existed, it could be used in conjunction with an ordinary heat engine to construct a perfect heat engine. Thus, if the refrigerator statement is false, the heat-engine statement is also false. It then

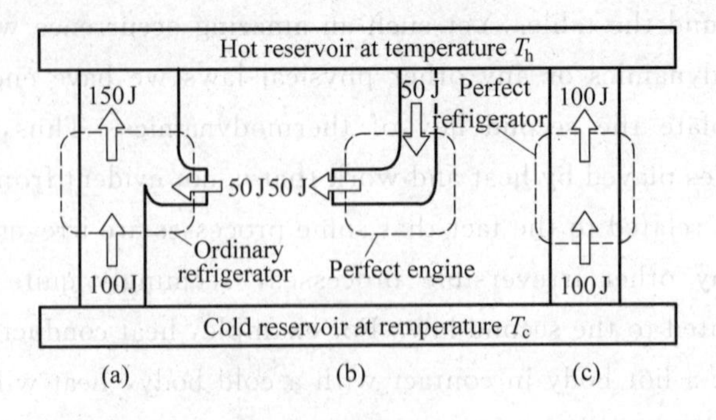

Fig. 11-1

follows that if one statement is true, the other is also true. Therefore, the heat engine statement and the refrigerator statement are equivalent.

11.2.3 Irreversibility and Disorder(不可逆性与无序性)

There are many irreversible processes that cannot be described by the heat-engine or refrigerator statements of the second law, such as a glass falling to the floor and breaking or a balloon popping. However, all irreversible processes have one thing in common—the system plus its surroundings moves toward a less ordered state.

When real (irreversible) processes occur, the degree of disorder of chaos in the system increases. For example, consider the isothermal expansion of an ideal gas in a container with a movable piston. As the gas absorbs heat and gradually expands, it maintains a constant temperature by doing work (pushing on the piston). After the expansion, the gas occupies a greater volume than it did originally. The gas molecules become more disordered in that they are not as localized as they were originally. Left by itself, the gas will not become ordered again by giving up its thermal energy to a reservoir. Thus, the flow of heat is in the direction that increases the amount of disorder. Ordered energy is converted into disordered energy.

Suppose a box containing a gas of mass M at a temperature T is moving along a frictionless table with a velocity v_{cm}(Fig. 11-2(a)). The total kinetic energy of the gas has two components: that associated with the movement of the center of mass $\frac{1}{2}Mv_{cm}^2$, and the energy of the motion of its molecules relative to its center of mass. The center of mass energy $\frac{1}{2}Mv_{cm}^2$ is ordered mechanical energy that could be converted entirely into work. (For example, if a weight were attached to the moving box by a string passing over a pulley, this energy could be used to lift the weight.) The relative energy is the internal thermal energy of the gas, which is related to its temperature T. It is random, non-ordered energy that cannot be converted entirely into work. Now, suppose that the box hits a fixed wall and stops (Fig. 11-2 (b)). This inelastic collision is clearly an irreversible process. The ordered mechanical energy of the gas is converted into random internal energy and the temperature of the gas rises. The gas still has the same total energy, but now all of that energy is associated with the random motion of the gas molecules about the center of mass of the gas, which is now at rest. Thus, the gas has become less ordered (more disordered), and has lost some of its ability to do work.

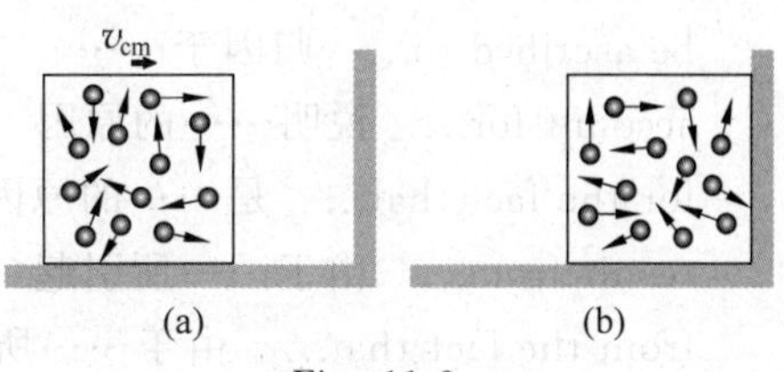

Fig. 11-2

Glossary

second law of thermodynamics	热力学第二定律	reservoir	热源、热库
Kelvin statement	开尔文表述	Clausius statement	克劳修斯表述
irreversible process	不可逆过程	heat-engine	热机
disorder	无序性	chaos	混乱
movable piston	活塞	inelastic collision	非弹性碰撞
random motion	无规则运动		

11.3 专业英语常用表达法-11 原因 理由

1. 常用词语结构

-ed/-ing 短语或介词短语

because of 由于……、因为……

by virtue of 由于……、因为……

in view of 由于……、因为……

on account of 由于……、因为……

due to/owing to 由于……

for... 由于……、因为……

from... 由于……

because/since/as... 因为……

on account of the fact that... 因为……、由于……(这一事实)

owing to the fact that... 因为……、由于……(这一事实)

seeing that... 鉴于……、由于……(的缘故)

now that... 既然……、由于……

in that... 因为……、原因在于……

inasmuch as 因为……、由于……

insofar as... 因为……、由于……

for +分句, 因为……(不用于句首)

arise from... 由于……而引起

be attributed to... 归因于……

be ascribed to... 归因于……

account for... 说明……的原因

for the fact that... 是……的原因、说明……的原因

result from... 由于……所引起

from the fact that... 由于……所引起、由于……产生

in view of the fact that... 因为……、考虑到……(这一事实)

due to the fact that... 因为……、由于……(这一事实)

2. 常用句型

The reason + why 分句 + is + that 分句　（……之所以）……的原因是……

This is/explains + why 分句　这就是……的原因、这说明了为什么……

acceleration *due to* gravity　重力加速度

The radial or normal acceleration is *due to* the time rate of change in direction of the velocity vector. 法向加速度的产生是由于速度矢量的方向随时间的变化率。

Being negative, the electrons are attracted to the anode. 由于电子带负电，故被吸向阳极。

Confined in a boiler, the steam from boiling water develops great pressure on the walls that confine it. 由于被密封在锅炉内，沸水所形成的蒸汽对封闭蒸汽的锅炉壁产生巨大的压力。

For many years, our standard of time was based on astronomical observations of the earth's rotation. *Because of/on account of/owing to/due to* the variations in the earth's rotation, in 1967 the 13th General Conference on Weights and Measures adopted an atomic standard for time. 过去很多年，我们的时间标准都是基于对地球转动的天文观察。由于地球转动的变化，1967 年国际计量大会为时间采用了一种新的原子标准。

The microwaves in home ovens are tuned to the natural frequency of vibration of water molecules. The water molecules in food resonate with the oscillating electric field and absorb large amounts of energy, *accounting for* the rapid cooking times that make microwave ovens so convenient. 家用微波炉的微波波长已经预先设置为水分子的自然振动频率。微波炉之所以方便快捷就是因为食物中的水分子与电磁场共振并吸收了大量的热量所致。

Light travels a long way in a year, *for* its speed is 300 000 kilometers a second. 光在 1 年中能走很远的距离，因为光速是每秒 300 000 km。

The difference between one element and another is *due to the fact that* they have different structures of atoms. 一种元素之所以不同于另一种元素是因为它们具有不同的原子结构。

This equation $\boldsymbol{r}=\boldsymbol{r}_0+\boldsymbol{v}_0 t+\frac{1}{2}\boldsymbol{a}t^2$ says that the displacement vector $\boldsymbol{r}-\boldsymbol{r}_0$ is the vector sum of a displacement $\boldsymbol{v}_0 t$, *arising from* the initial velocity of the particle, and a displacement $\frac{1}{2}\boldsymbol{a}t^2$, *resulting from* the uniform acceleration of the particle. 方程 $\boldsymbol{r}=\boldsymbol{r}_0+\boldsymbol{v}_0 t+\frac{1}{2}\boldsymbol{a}t^2$ 说明：位移矢量 $\boldsymbol{r}-\boldsymbol{r}_0$ 是位移 $\boldsymbol{v}_0 t$（由于粒子的初速度而产生）和位移 $\frac{1}{2}\boldsymbol{a}t^2$（由于匀加速度而产生）的矢量和。

This difference *results from* the fact that there are different numbers of neutrons in the two nuclei. 这种区别是由这两个核中的中子数不同引起的。

The reason why hard water does not lather readily with soap *is that* it contains calcium. 硬水之所以不容易起肥皂泡的原因是它含有钙。

This explains why gases are so easily compressed. 这说明了为什么气体那么容易被压缩。

LESSON 12

12.1 物理学专业英语中的表语从句和同位语从句

12.1.1 表语从句

名词性从句在 be 等系动词后作表语时被称为表语从句,可接表语从句的连系动词有 be,look,seem,sound,appear 等。

1)从属连词 whether,as,as if/though 引导的表语从句

The question that has interested scientists from ancient times is *whether the atom can or cannot be split up*. 原子能否再分割的问题从远古就令科学家感兴趣。

The question remains *whether the experiment will be of any help*. 问题还是这个实验能否有所帮助。

The instrument looked just as *it had looked ten years before*. 这仪器看起来还与十年前一样。

It sounds as if *the argument is reasonable*. 听起来好像论证得有道理。

2)because,why 引导的表语从句

Since the weight depends on g, it varies with geographic location. This is *because g decreases with increasing distance from the center of the earth*. 因为重量与重力加速度 g 有关,所以重量会随地理位置不同而变化。这是因为距离地心越远,g 的数值会越小。

Try opening the door by pushing at different distances from the hinges. The largest effect occurs when the force is applied farthest from the hinges. That's *why doorknobs are put there*! 试着在与铰链距离不同处开门或推门,当力作用在与铰链距离最远处时效果最明显。这就是门把手安装在这里的原因。

【注意】 "That is because..."句型中从属连词 because 引导的名词性从句在此作表语,这也是个常用句型,意为"这就是为什么……/因为……"。"That is because..."与"That is why..."之间的不同在于"That is because..."强调原因或理由,"That is why..."则强调由于各种原因所造成的后果。"That is why..."也是常用句型,意为"这就是……的原因/因此……",该句型通常用于针对前面已经说明过的原因进行总结。"That is why..."与"That is the reason why..."同义,只不过从语法结构上讲,"That is the reason why..."中 why 引导的是一个定语从句,将其中的 the reason 去掉则与"That is why..."结构一样。

what 引导的主语从句表示结果,或名词 reason 作主语而后面的表语从句表示原因时,要用 that 引导,不宜用 because。

What causes the weight to change from position to position is *that the magnitude of g varies with geographic location*. 重量之所以随着位置变化,其原因在于重力加速度 g 的大小会随着地理位置发生变化。

The *reason* for capacitance's increase by a factor ε_r is *that the electric field between the plates of a capacitor is weakened by the dielectric*. 电容增长 ε_r 倍的原因是电容器极板间的电场被电介质削弱了。

3) 连接代词 who, whom, whose, what, which, whoever, whatever, whichever, 和连接副词 where, when, how, why 引导的表语从句

This is *what we call a sinusoidal wave* because the shape in Fig. 20-14 is that of a sinusoid. 这就是所谓的正弦波,因为在图 20-14 中,其形状就是一个正弦曲线。

The problem is *who we can get to carry out the experiment*. 问题是我们能找谁去做实验。

The question is *how we can determine the direction of each particle in the medium*. 问题是该如何确定媒质中每个粒子的(运动)方向。

The question is *who (m) the scholars should trust*. 问题是学者们应当相信谁。

What the researchers want to know is *which project they should adopt*. 研究者们想知道的是他们应采用哪个方案。

That's *why the block can stay rest on the incline*. 这就是木块能够静止在斜面上的缘故。

That's *where Aristotle was wrong*. 这就是亚里士多德的错误所在。

4) 从属连词 that 引导的表语从句

that 作为表语从句的引导词在表语从句中不充当句子成分,一般不能省略。

Galileo's greatest glory was *that in 1609 he was the first person to turn the newly invented telescope on the heavens* to prove that the planets revolve around the sun rather than around the Earth. 伽利略最光辉的业绩在于他在 1609 年第一个把新发明的望远镜对准天空,以证实行星是围绕太阳旋转而不是围绕地球的。

For example, in electric circuits the usual assumption is *that the relations between voltages and currents are linear*. 例如,在电路中通常假设电压和电流的关系是线性的。

5) 使用虚拟语气的表语从句

当句子的主语是:advice, demand, decision, desire, idea, insistence, instruction, order, plan, proposal, recommendation, request, requirement, suggestion, wish 等表示建议、劝告、命令含义的名词时,表语从句谓语动词须用虚拟语气,即:"should+动词原形",should 可省略。如:

My advice is *that you should remember that* electric potential is a scalar quantity when handling problems involving it. 我的建议是你应该在计算与电势有关的问题时记住电势是个标量。

My suggestion *is that students (should) draw a free-body diagram showing all external forces acting on the object* when applying Newton's laws. 我的建议是学生应该在应用牛顿定律时画一个受力分析图,标明物体所受的所有外力。

12.1.2 同位语从句

在复合句中充当同位语的名词性从句称为同位语从句。

一般情况下同位语从句跟在某些名词的后面,用以说明该名词所表达的具体内容。例如:

There comes the news *that China launched its first manned spaceship on Oct* 15,2003. 消息传来,中国于 2003 年 10 月 15 日首次成功发射了载人飞船。

以下名词常用于以上句型：advice, announcement, argument, belief, claim, conclusion, decision, evidence, explanation, fact, feeling, hope, idea, impression, information, knowledge, message, news, opinion, order, probability, promise, proposal, remark, reply, report, saying, statement, suggestion, thought, treat, warning, wish, word.

1）同位语从句的引导词

引导同位语从句的词语通常有连词 that,whether,连接代词和连接副词等。

（1）由 that 引导

The general theory of relativity was proposed by Einstein in 1916 after nine years of grappling with the problem of formulating a theory of gravitation in agreement with the space-time symmetry of special relativity and with the experimental observation known since Galileo *that all bodies, regardless of their mass, follow the same trajectory in a gravitational field*.

(That 引导的从句作 experimental observation 的同位语)为了提出一种引力理论能符合狭义相对论的时空对称要求,同时也与自从伽利略以来就知道的实验观测一致(即一切物体在引力场中运动都遵循相同的轨迹,与质量无关),爱因斯坦通过对此问题长达 9 年的研究,于 1916 年提出了广义相对论。

The scientists have come to the conclusion *that it was unwise to do that*. 科学家得出的结论是：这样做是不明智的。

He referred to Copernicus' statement *that the earth moves round the sun*. 他提到了哥白尼关于地球绕太阳转的说法。

They were all very much worried over the fact *that the telescope is out of order*. 对于望远镜出故障这件事,他们都很焦虑。

（2）由 whether 引导

The question *whether the atom can or cannot be split up* is known to have interested scientists from ancient times. 原子能否再分割的问题从远古就令科学家感兴趣。

The question *whether it is right or wrong* depends on the result. 这个是对还是错要看结果。

We are investigating the question *whether the machine works well*. 我们在调查这台机器是否运行良好。

【注意 1】 whether 可引导同位语从句,但 if 不能引导同位语从句。

【注意 2】 在名词 doubt(怀疑)后的同位语从句用 whether 连接;在 no doubt“不怀疑”之后的同位语从句用 that 连接。例如：

We have some doubt *whether they can complete the experiment on time*. 我们怀疑他们是否能准时完成实验。

There is no doubt *that the electric field between the plates of the capacitor is weakened by the dielectric*. 无疑,电容器极板间的电场被电介质削弱了。

(3) 由连接代词 what 引导

Have you any idea *what time the radiation will be sheltered*? 你知道什么时候辐射会被屏蔽起来吗?

I have no idea *what the equipartition theorem is about*. 我不知道均分定理是什么内容。

(4) 由连接副词 where, when, why, who, whom, whose, how 引导

His forecast *when the hurricane would come* is found entirely successful. 他对飓风何时到来的预报是完全成功的。

The reason *why the nucleus of an atom can stay together* is that the strong nuclear force holds them together. 原子内的核子之所以能够聚集在一起是由于强核力的作用。

It is a question *how he measured the angular momentum of the block*. 问题是他如何测量木块的角动量。

The question *who should do the work* requires consideration. 谁该做这项工作,这个问题需要考虑。

We haven't yet settled the question *where we are going to find suitable capacitors*. 我们还没有确定到什么地方去找合适的电容器。

2) 分隔式同位语从句

有时同位语从句可以不紧跟在它所说明的名词后,而被别的词语隔开,这在语法上叫作分隔式同位语从句。例如:

In a simulation of search and track, decisions will be made *whether we search for new targets or track existing detections on the basis of targets presently being tracked*. 在对搜索和跟踪进行仿真时,要决定去搜索新目标,还是去跟踪已有的被跟踪的目标。(decisions 和 *whether* 引导的从句是同位语。)

The rumour spreads *that the key lab building would be pulled down in a year*. 传言这个重点实验室大楼要在一年后拆掉。

Report has it *that the first capacitor was the Leyden jar invented at the University of Leyden*. 据说历史上第一个电容器是莱顿瓶,那是在莱顿大学发明的。

The thought came to the experimenters *that maybe they could try to store a large amount of charge in a bottle of water*. 实验人员想到或许可以在一瓶水中储存大量电荷。

3) 同位语从句与虚拟语气

在 suggestion, advice, request, order, demand, wish, suggestion, resolution 等意为“建议、命令、要求”的名词后,同位语从句中的谓语动词通常用“should +动词原形”的虚拟语气结构,句中的 should 可以省略。例如:

They were faced with the demand *that this regulation (should) be abolished in the lab*. 他们被要求废除这个实验室的规定。

They expressed the wish *that they (should) cooperate in the field*. 他们表示希望在此领域合作。

There was a suggestion *that Brown should be dropped from the research team*. 建议布朗应该离开研究队伍。

Our teacher gave us some advice on *how we (should) use the demonstrative instrument*. 老师给我们提出了一些如何使用演示仪器的建议。

The head gave the order that *all these houses* (*should*) *be pulled down in three weeks*. 主任下令三个星期内所有这些房子都要拆掉。

12.2 专业英语阅读

12.2.1 Waves(波动)

Waves transport energy and momentum through space without transporting matter. As a water wave moves across a pond, for example, the molecules of water oscillate up and down, but do not cross the pond with the wave. Energy and momentum are transported by the wave, but matter is not. A rowboat will bob up and down on the waves but will not be moved by them across the pond. Water waves, waves on a stretched guitar string, and sound waves all involve oscillation.

The concept of wave is abstract. When we observe a water wave, what we see is a rearrangement of water's surface. Without the water, there would be no wave. In the case of this or any other mechanical wave, what we interpret as a wave corresponds to the disturbance of medium. Therefore, we can consider *a mechanical wave to be the propagation of a disturbance in a medium*. The world is full of other kinds of waves, including sound waves, waves on strings, seismic waves, radio waves, and X-rays. Most waves can be placed in one of two categories. **Mechanical waves** are waves that disturb and propagate through a medium; the ripple in the water due to the pebble and a sound wave, for which air is the medium, are examples of mechanical waves.

In the introduction, we alluded the essence of wave motion—the transfer of a disturbance through space without the accompanying transfer of matter. The propagation of the disturbance also represents a transfer of energy—thus, we can view waves as means of energy transfer. Two entries that depend on waves: mechanical waves and electromagnetic radiation. These are to be contrasted with another entry—matter transfer—in which the energy transfer is accompanied by a movement of matter through space.

All mechanical waves require:

(1) Some source of disturbance.

(2) A medium that can be disturbed.

(3) Some physical mechanism through which particles of the medium can interact with one another.

All waves carry energy, but the amount of energy transmitted through a medium and the mechanism responsible for the energy transport differ from case to case.

Electromagnetic waves are a special class of waves that do not require a medium in order to propagate; it is the varying electromagnetic field that propagates through space. Light waves and radio waves are two familiar examples.

Here are some common properties of waves:

(1) Waves propagate with a definite speed through the medium. This is called **wave speed**.

(2) Waves transport energy and momentum through space from one region to another without transporting matter. The medium itself does not travel through space.

(3) Waves can reflect, refract, interfere, and diffract.

12.2.2 Traveling Waves(行波)

When a medium is disturbed, energy is imparted to it. Suppose that energy is added to a material mechanically, such as by a blow or by compression (in the case of a gas). This sets some of the particles vibrating. Because the particles are linked by intermolecular forces, the oscillation of each particle affects that of its neighbors. The added energy propagates, or spreads, by means of interactions between the particles of the medium. An analogy for this process is shown in Fig. 12-1, where the "particles" are dominoes. As each domino falls, it topples the one next to it. Thus energy is transferred from domino to domino, and the disturbance propagates through the medium.

Fig. 12-1 Energy transfer

In this case, there is no restoring force between the dominoes, so they do not oscillate as do particles in a continuous material medium. Therefore the disturbance moves in space, but it does not repeat itself in time at any one location.

Similarly, if the end of a stretched rope is given a quick shake, the disturbance transfers energy from the hand to the rope, as illustrated in Fig. 12-2. The forces acting between the rope "particles" cause them to move in response to the motion of the hand and a wave pulse travels down the rope. Each "particle" goes up and then back down as the pulse passes by. This motion and that of the wave pulse propagation can be observed by tying pieces of ribbon onto the rope (at x_1 and x_2 in Fig. 12-2). As the disturbance pass point x_1, the ribbon rises and falls, as do the rope "particles". Later, the same occurs for the ribbon at x_2, which indicates that the disturbance energy is propagating or traveling along the rope. This kind of propagation with disturbance is called **traveling waves.** It comes from the meaning of "travel".

Waves may be divided into two types based on the direction of the particles' oscillations relative to the wave velocity.

A **transverse wave** is one for which the particle motion is perpendicular to the direction of the wave velocity. As an experiment, take a string and stretch it taut across your room. If you quickly move your hand up and down once, perpendicular to the axis of the rope, the disturbance you generate propagates along the rope, see Fig. 12-3. But in this case the disturbance (the jiggling—i. e. , the motion of the particles that make up the rope) is perpendicular to the direction the wave disturbance propagates.

Move your hand up and down several times quickly, and you produce multiple oscillations that propagate along the rope. Waves on a rope or string are transverse waves. But the elements do not go with the wave.

In a **longitudinal wave**, the particle oscillation is parallel to the direction of the wave velocity. A longitudinal wave may be produced in a stretched spring by moving the coils back and forth along the spring axis (Fig. 12-4). Alternating pulses of compressions and relaxations move along the spring. A longitudinal wave is sometimes called a **compressional wave**. Longitudinal waves travel by alternately compressing and rarefying (expanding) the medium, and its individual particles undergo back-and-forth motions around their equilibrium positions, but the medium itself does not travel through space.

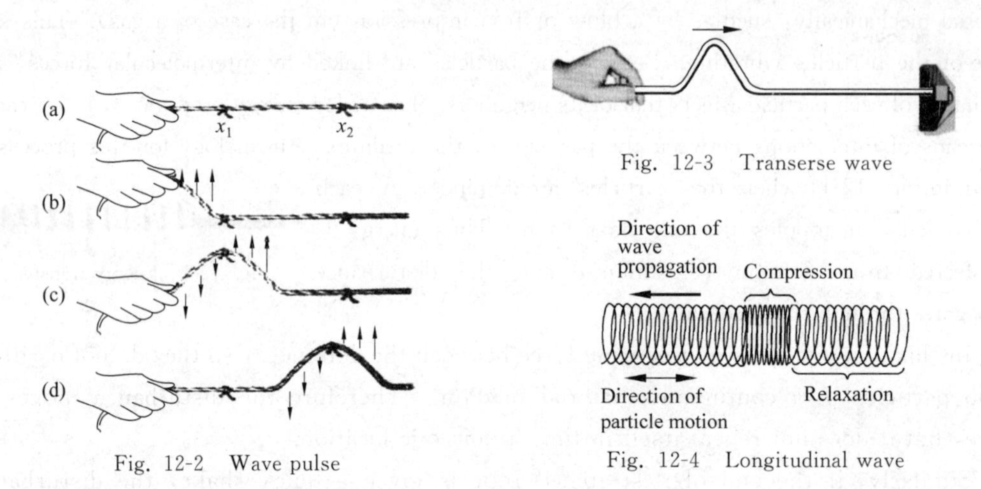

Fig. 12-2 Wave pulse

Fig. 12-3 Transerse wave

Fig. 12-4 Longitudinal wave

Sound waves in air are another example of longitudinal waves—the molecules of a gas, liquid, or solid, through which sound travels oscillate (move back and forth) along the line of propagation, alternately compressing and rarefying (expanding) the medium.

Glossary

disturbance	扰动	propagate	*v*. 传播
propagation	*n*. 传播	medium	媒质、媒介
sound waves	声波	waves on strings	弦上波
seismic waves	地震波	radio waves	无线电波
X-rays	X 射线	mechanical waves	机械波
transmit	传输、发射、传播	electromagnetic waves	电磁波
reflect	反射	refract	折射
interfere	干涉	diffract	衍射
wave speed	波速	traveling waves	行波
intermolecular	分子间的	restoring force	回复力
domino	多米诺牌	wave pulse	波脉冲
compression	压缩	relaxation	张弛

transverse waves	横波	rarefy	使稀疏
longitudinal waves	纵波	compressional wave	压缩波
equilibrium position	平衡位置		

12.3 专业英语常用表达法-12 实验 试验

1. 各种实验与试验

activation experiment 激活试验
acute experiment 急性实验
aerodynamic experiment 空气动力试验
aircraft reactor experiment 航空反应堆实验
analogue experiment 模拟实验
blank experiment 空转试验
bread board experiment 模拟板试验
breakthrough experiment (离子交换)穿透、流穿实验
cascade experiment 串级、级联实验
check experiment 检查试验
choice experiment 选择实验
chronic experiment 慢性实验
cold gas blanket experiment 冷气体包层实验
comparative experiment 比较试验
competitive experiment 竞赛实验
complication/compound experiment 复合实验
computer-aided experiment 计算机辅助实验
confirmatory experiment 验证实验
contrast experiment 对比实验
control experiment 对照实验
controllable extensional experiment 可控延伸试验
controlled experiment 对照实验
controlled computational experiment 受控的计算试验
cookbook laboratory experiment 按作业指南进行的实验室实验
critical experiment 临界、极限试验
cross-over experiment 交叉实验
defect experiment 缺损实验
designed experiment 设计实验
destructive experiment 破坏试验
direct-current experiment 直流电试验
dishpan experiment 转盘实验

drag experiment　阻力试验
exponential experiment　指数实验
factorial experiment　析因实验
Faraday ice bucket experiment　法拉第冰桶试验
fast exponential experiment　快中子指数实验
field experiment　野外/现场试验
Franck-Hertz experiment　弗兰克-赫兹实验
full-scale experiment　全尺寸实验
Gedanken experiment　思想实验
ground experiment　地面试验
group experiment　分组试验
high frequency experiment　高频实验
ice-pail experiment　冰桶实验
inclining experiment　倾斜试验
isolation experiment　隔离试验
laboratory experiment　实验室实验
large-scale experiment　大型试验,工业规模试验
lysimetric experiment　渗漏试验
model experiment　模型试验，模拟实验
molten salt reactor experiment　熔盐反应堆实验
natural experiment　自然实验
non-controllable extensional experiment　非可控延伸实验
numerical experiment　数值实验
open-syphon experiment　敞口虹吸实验
opposite experiment　对比实验
orientative experiment　定向试验
parity experiment　宇称实验
pedagogical experiment　教学实验
perfusion experiment　灌流实验
photo elastic experiment　光(测)弹性实验
photoelectric mixing experiment　光电混频实验
polarization experiment　极化实验
population experiment　群体试验
porous plug experiment　多孔塞实验
quanlitative experiment　定性实验
quantitative experiment　定量实验
random experiment　随机试验

recoil experiment　反冲实验
reconstruction experiment　重组实验
routine experiment　例行试验
sensitivity experiments　灵敏性实验
simulated experiment　模拟实验
structure experiment　结构试验
synchronous altitude spin stabilized experiment　同步高度自旋稳定实验
tentative experiment　初步试验
tracer experiment　示踪/显迹实验
transition experiment　转换实验
transmission experiment　透射实验
triple-jet experiment　三(向)射流实验
two-part experiment　二部实验
two-stage experiment　两阶段实验
Wiener experiment　维纳实验
wind tunnel experiment　风洞试验

2. 与实验相关的词汇

experiment，experimentation　实验
make (do/perform/try/carry out) an experiment on (upon/in/with) …　做实验
experimenter　实验者
laboratory assistant，lab assistant，laboratory technician　实验员
experiment station/establishment　实验站

3. 实验理论常用词汇

measure，measurement　测量
true value　真值
uncertainty　不确定度
theoretical value　理论值
measured value　测量值
series of observations　测量列
arithmetic mean　算术平均值
systematic error　系统误差
random error　随机误差
indication error　示值误差
relative error　相对误差
standard deviation　标准偏差
root-sum-square　方和根
fiducial/confidence probability　置信概率

fiducial/confidence interval　置信区间
significant figures　有效数字
unit conversion　单位换算
scientific notation　科学计数法
round off　舍入
half adjust　四舍五入
round up　进位舍入
round down　不进位舍入
zero correction　零点修正
data processing　数据处理
schedule method　列表法
graphing method　作图法
termwise subtraction　逐差法
least square method　最小二乘法
coordinate paper　坐标纸
equal/unequal observations　等精度/非等精度测量
direct/indirect measurement（observation）　直接/间接测量
directly/indirectly measured quantity　直接/间接测量量
Type A/Type B uncertainty　A 类/B 类不确定度

Unit Four

LESSON 13

13.1 物理学专业英语中的副词性从句

副词性从句也称为状语从句，主要用来修饰主句或者主句的谓语。大致分为九大类，分别表示时间、地点、原因、目的、结果、条件、让步、比较和方式，作什么样的状语就叫什么类型的状语从句。例如：条件状语从句就是用一个句子来作条件状语。

状语从句的起点，由状语连词标明。如果状语从句在全句的前部，则从句的结束处往往用逗号标明。在容易判别时，也可以不加标点符号。

13.1.1 时间状语从句

时间状语从句常用引导词有：when，as，while，as soon as，before，after，since，till，until，once 等。特殊引导词有：the minute，the moment，the second，every time，each time，by the time，at the time，at the moment，the day，the instant，immediately，directly，no sooner... than，hardly... when，scarcely... when 等。时间状语从句一般放在句首或句尾。特别需要注意的是，在时间状语从句中不使用将来时，而是用现在时来表示将来时。如：

The body possesses a definite store of potential energy *while it is in the elevated position*. 把物体举高时，它具有一定的势能。

When the usual road does not lead to success, one should not be content with this determination of impossibility, but should bestir oneself to find a new and more promising route. Physical thought, as such, has no end. 当寻常的方法无法成功时，(一个人)不应该就此认定不可能，而应该打起精神去寻求新的、更有希望的方法。这样的物理思想是永无止境的。

Once the overall system concept is defined, our choice becomes increasingly limited *as we proceed from general concepts to actual design details*. 一旦全系统概念定义后，当我们从一般概念进入设计细节时，我们的选择变得越来越有限了。

When everyone is fully aware of the severity of fresh water scarcity and takes effective measures, I am convinced that we will tackle this problem constructively. 当每个人充分意识到淡水短缺的严重性，并采取有效措施时，我相信我们一定能妥善解决好这个问题。

Each time a call comes through, the signal is sent automatically from the earth station to the new digital exchange. 每当电话一打进来,从地面接收站收到的信号就会自动传到新的数字式电话交换台。

By the time the lorry arrives in the storage park, the gantry will be ready to unload the container and put it in the appropriate place. 当卡车到达货场时,吊车会等在那里把集装箱卸下,并把它堆放在合适的地方。

The engineer wants to be there, on the drill floor *at the moment the bit comes up*, to keep a close eye on things. 当钻头从海底退回到水面时,工程师希望能到现场钻台旁,这样他就能密切注视发生的一切情况。

【注意】 while 也可以作并列连词,表转折的关系,相当于 but,译为"然而"。例如:

The relationship between the transforms of source and response function is simple algebraic relationship, *while* the relationship between corresponding time functions is a differential equation. 源变换和响应函数之间的关系只是简单的代数关系,而对应时间函数间的关系却是微分关系。

13.1.2 地点状语从句

地点状语从句通常由 where, wherever 等引导。

It is well known that all objects, when dropped, will fall toward the earth with nearly constant acceleration. In the idealized case, *where air resistance is neglected*, such motion is referred to as free fall. 众所周知,所有物体被释放下落时几乎匀加速度落向地面。在理想情况下,在空气阻力被忽略的情况下,这种运动被称为自由下落。

13.1.3 原因状语从句

原因状语从句常用引导词有: because, since, as,特殊引导词有: seeing that, now that, in that, considering that, given that, in as much as, in so much as,一般放在句首或句尾。如:

Because $\boldsymbol{a}$ *is a constant*, its components a_x and a_y are also constants. 因为 $\boldsymbol{a}$ 是常量,所以其分量 a_x 和 a_y 也是常量。

We call the force a linear restoring force *because it is proportional to the displacement from the equilibrium position and always directed toward the equilibrium position, opposite the displacement*. 我们把这个力称为线性回复力,因为它正比于离开平衡位置的距离并且总是指向平衡位置,与位移的方向相反。

Angular velocity has units of rad/s, or s^{-1}, *since radians are not dimensional*. 角速度的单位是 rad/s 或 s^{-1},因为弧度没有量纲。

Now that we have covered some of the tools of detailed logic design, let us look at the basic system concepts. 既然我们已经谈过了一些详细逻辑设计的工具,就让我们看看基础系统概念。

After the expansion, the gas occupies a greater volume than it did originally. The gas molecules become more disordered *in that they are not as localized as they were originally*. 膨胀后,气体占据比原来大的体积。气体分子变得更无序,因为分子已经不再局

限在原来的空间。

13.1.4 目的状语从句

目的状语从句常用引导词有：so that，in order that，特殊引导词有：lest，in case，for fear that，in the hope that，for the purpose that，to the end that 等。如：

These calculations were needed *in order that the behavior of different physical processes being studied by engineers and scientists might be understood*. 为了使工程师和科学家们明白所研究的不同物理过程的表现，这些计算是必需的。

We invent a simplified version of a problem *so that we can solve it simply and reasonably*. 为了把问题解决得简单合理，我们建立这个问题的简化模型。

13.1.5 结果状语从句

结果状语从句常用引导词有：so...that，so that...，such...that，特殊引导词有：such that，to the degree that，to the extent that，to such a degree that 等。结果状语从句一般表示已经发生的事情。如：

Nothing is *so* beautiful *that it does not betray some defect on close inspection*. 没有什么能如此完美，以致严格检查也无缺陷。

A consequence of the definition is that moving a vector *so that it remains parallel to itself* does not change it. 这样定义的结果是，平行移动一个矢量并不改变这个矢量。

13.1.6 条件状语从句

常用引导词有：if，unless，特殊引导词有：as/so long as，only if，providing/provided that，suppose that，in case that，on condition that，whereas 等。

Kinematics is concerned with two basic questions，"Where" and "When". Though the questions are simple，the answers are potentially quite complicated *if we inquire about phenomena outside our ordinary daily experiences*. 运动学探讨两个基本问题：在哪里和在何时。尽管问题简单，可是如果我们探究日常经验以外的现象，答案可能相当复杂。

Unless there is motion，however，there is no work. 除非有运动，否则不会有做功存在。

Whereas negative numbers were created to make the operation of subtraction possible without any exception，subtraction as an independent operation ceased to exist by virtue of that creation. 鉴于有了负数以后，就可以取代所有的减法运算，因而减法作为独立的运算就没有存在的必要了。

13.1.7 让步状语从句

引导让步状语从句的常用引导词有：though，although，even if，even though，特殊引导词有：as(用在让步状语从句中必须要倒装)，while(一般用在句首)，no matter +疑问词，in spite of the fact that，whatever，whoever，wherever，whenever，however，whichever，whether...or...等。

In other words，*although the amplitude of successive harmonics may go up and down as n increases*，it will always tend to zero as *n* increases. 换言之，尽管连续谐波的振

幅可能随着 n 的增加而上下振荡,但是它最后总会随着 n 的增加而趋于零。

The computer remembers an incredible amount of data, but *no matter how large its brain becomes* it will always need help. 计算机会记住大量的数据,但是不论其记忆容量变得多大,它总是需要帮助。

It has the same result *whichever way you carry out the experiment*. 不管你怎么做实验,结果都一样。

Whatever happens everyone must be calm. 不管发生什么情况大家都要镇静。

Much as I respect him, I can't agree to his proposal. 尽管我很尊敬他,我却不同意他的建议。

Young as he is, he is quite experienced in this work. (=Though he is young) 尽管他年轻,对这个工作却经验丰富。

We'll keep on *even if/though the working condition is bad*. 即使工作条件不好,我们也要坚持下去。

You'll have to attend the workshop *whether you're free or busy*. 不管你忙不忙,都要参加这个讨论会。

【注意】 有时 while 也可以引导让步状语从句,但一般要位于句首。例如:

While I like the color, I don't like the shape. 我虽然喜欢那颜色,但不喜欢那形状。

13.1.8 比较状语从句

常用引导词有:as(同级比较), than(不同程度的比较)。特殊引导词有:the more... the more..., just as..., so..., A is to B what /as X is to Y, no... more than, not A so much as B 等。

The physics lab is three times *as big as the chemistry lab*. 物理实验室是化学实验室的 3 倍大。

Food is to men what oil is to machine. 食物之于人,犹如油之于机器。

The more the angle of incidence increases, *the farther the refracted ray diverges from the normal*. 入射角增加的越大,折射线就偏离法线越远。

13.1.9 方式状语从句

常用引导词有:as, just as, as if, how, as though 等。特殊引导词有:the way.

The young man made the experiment *just as the teacher had taught him*. 年青人按照老师所教的做了实验。

Everything went on as usual *as if nothing had happened*. 每件事都一切如常,就像什么都没发生过。

13.2 专业英语阅读

13.2.1 Introduction to Electromagnetics(电磁学引言)

The laws of electricity and magnetism play a central role in the operation of devices such as radios, televisions, electric motors, computers, high-energy particle accelerators, and a host of electronic devices used in medicine. More fundamental, however, is the fact that the interatomic and intermolecular forces responsible for the formation of solids and liquids are electric in origin. Furthermore, such forces as the pushes and pulls between objects in contact and the elastic force in a spring arise from electric forces at the atomic level.

Chinese documents suggest that magnetism was recognized as early as about 2000 B. C. The ancient Greeks observed electric and magnetic phenomena possibly as early as 700 B. C. They found that a piece of amber, when rubbed, attracted pieces of straw or feathers. The existence of magnetic forces was known from observations that pieces of a naturally occurring stone called magnetite were attracted to iron. (The word electric comes from the Greek word for amber, elektron. The word magnetic comes from Magnesia, on the coast of Turkey where magnetite was found.)

In 1600, the Englishman William Gilbert discovered that electrification was not limited to amber but was a general phenomenon. Scientists went on to electrify a variety of objects, including chickens and people! Experiments by Charles Coulomb in 1785 confirmed the inverse-square law for the electrostatic force.

It was not until the early part of the 19th century that scientists established that electricity and magnetism are related phenomena. In 1820, Hans Oersted discovered that a compass needle, which is magnetic, is deflected when placed near an electric current. In 1831, Michael Faraday in England and, almost simultaneously, Joseph Henry in the United States showed that, when a wire loop is moved near a magnet (or, equivalently, when a magnet is moved near a wire loop), an electric current is observed in the wire. In 1873, James Clerk Maxwell used these observations and other experimental facts as a basis for formulating the laws of electromagnetism as we know them today. Shortly thereafter (around 1888), Heinrich Hertz verified Maxwell's predictions by producing electromagnetic waves in the laboratory. This achievement was followed by such practical developments as radio and television.

Maxwell's contributions to the science of electromagnetism were especially significant because the laws he formulated are basic to all forms of electromagnetic phenomena. His work is comparable in importance to Newton's discovery of the laws of motion and the theory of gravitation.

13.2.2 Charge(电荷)

Suppose we rub a plastic rod with fur and suspend it from a string so that it is free to rotate. Now we bring a second similarly rubbed plastic rod near it. The rods repel each other. We get the same results if we use two glass rods that have been rubbed with silk. But a plastic rod rubbed with fur and a glass rod rubbed with silk attract each other. If we repeat the experiment with various materials, we find that all objects fall into one of just two groups—those like the plastic rod rubbed with fur and those like the glass rod rubbed with silk. When materials behave this way, they are said to have become **electrically charged.**

Experiments demonstrate that there are two kinds of **charge** in nature, given the names **positive** and **negative** by Benjamin Franklin (1706—1790), these names have remained to this day. *Two objects that carry the same type of charge repel each other, and two objects that carry opposite charges attract each other.*

13.2.3 Coulomb's Law(库仑定律)

Electric forces between charged objects were measured quantitatively by Charles Coulomb (1736—1806) using a torsion balance of his own invention. In Coulomb's experiment, the charged spheres were much smaller than the distance between them so that the charges could be treated as **point charges.** Coulomb used the method of charging by induction to produce equally charged spheres. For example, beginning with charge q_0 on each sphere, he could reduce the charge to $q_0/2$ by temporarily grounding one sphere to discharge it and then placing the two spheres in contact. The results of the experiments of Coulomb and others are summarized in **Coulomb's law.**

The force exerted by one point charge on another acts along the line between the charges. It varies inversely as the square of the distance separating the charges and is proportional to the product of the charges. The force is repulsive if the charges have the same sign and attractive if the charges have opposite signs.

The magnitude of the electric force exerted by a charge q_1 on another charge q_2 is thus given as

$$F_{21} = k\frac{|q_1 q_2|}{r_{21}^2} \tag{13-1}$$

where r_{21} is the distance between q_1 and q_2, k is an experimentally determined constant called the **Coulomb constant**, which has the value

$$k = 8.9880 \times 10^9\ \mathrm{N \cdot m^2/C^2} \approx 9.0 \times 10^9\ \mathrm{N \cdot m^2/C^2}$$

Coulomb's law is given in vector equation by

$$\boldsymbol{F}_{21} = k\frac{q_1 q_2}{r_{21}^2}\boldsymbol{e}_{r21} \tag{13-2}$$

where $\boldsymbol{e}_{r21}$ is a unit vector pointing from q_1 to q_2 and $\boldsymbol{F}_{21}$ is the force exerted by q_1 on q_2. If q_1 and q_2 have the same sign, the force on q_2 is in the same direction as $\boldsymbol{e}_{r21}$, and is directed away from q_1 (Fig. 13-1). If q_1 and q_2 are of opposite signs, the force on q_2 is in the direction opposite to $\boldsymbol{e}_{r21}$, and is directed toward q_1. From Newton's third law, we see that the electric force exerted by q_2 on q_1 is equal in

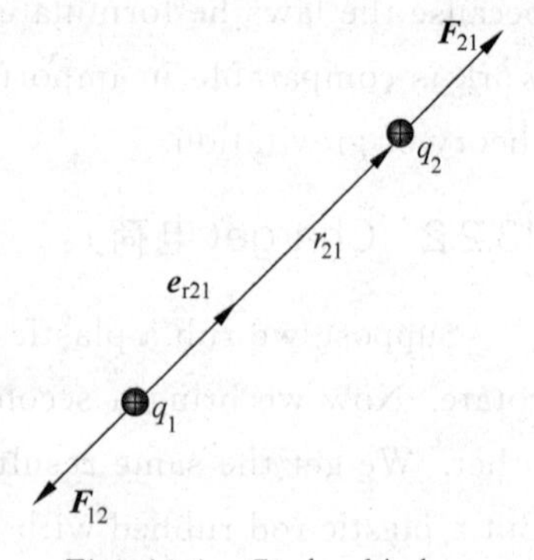

Fig. 13-1 Coulomb's law

magnitude to the force exerted by q_1 on q_2 and in the opposite direction; that is, $\boldsymbol{F}_{21} = -\boldsymbol{F}_{12}$. The Coulomb constant k is also written as

$$k = \frac{1}{4\pi\varepsilon_0}$$

where ε_0, known as **the permittivity of free space**, has the value

$$\varepsilon_0 = 8.85 \times 10^{-12}\ \mathrm{C}^2/\mathrm{N} \cdot \mathrm{m}^2$$

Therefore, in free space the expression of Coulomb's law is also written as

$$\boldsymbol{F}_{21} = \frac{q_1 q_2}{4\pi\varepsilon_0 r_{21}^2}\boldsymbol{e}_{r21} \tag{13-3}$$

Glossary

electricity	电	magnetism	磁
high-energy particle accelerator	高能粒子加速器	electronic device	电子设备
		intermolecular	分子间的
interatomic	原子间的	rub	摩擦
amber	琥珀	electrification	充电、带电、起电
magnetite	磁铁矿	inverse-square law	平方反比定律
electromagnetism	电磁学	electromagnetic waves	电磁波
magnet	磁铁	plastic rod	塑料棒
charge	电荷	repel	排斥
fur	毛皮	suspend	悬挂
attract	吸引	opposite charges	相反电荷
electrically charged	带电的	torsion balance	扭秤
Coulomb's Law	库仑定律	point charge	点电荷
charged sphere	带电球	proportional to	正比于
ground	*v.* 接地	square	平方
discharge	放电	repulsive	排斥的
product	乘积	Coulomb constant	库仑常数
attractive	吸引的	permittivity of free space	真空介电常数
opposite signs	相反符号		

13.3 专业英语常用表达法-13 电磁学常用实验仪器 1

AC bridge experiment facility 交流电桥实验仪

AC / DC resistance box 交直流电阻箱

AC millivoltmeter 交流毫伏表

adjustable capacitor 可变电容器

adjustable inductance 可调电感

adjustable resistor 可调电阻器

air core inductance　空心电感线圈

air dielectric capacitor　空气(介质)电容器

alkaline accumulator　碱性蓄电池

ammeter/amperemeter/amperometer　安培计、电流计

analogue standard resistance　模拟标准电阻

4 and 1/2-bit DC digital voltmeter/amperemeter　4位半直流数字电压表/电流表

bridge comprehensive experiment facility　电桥综合实验仪

bridge　电桥

by-pass resistor　分路电阻器、旁路电阻器

capacitor　电容器

classical AC bridge　经典交流电桥

combined inductance box　组合电感箱

commutator modulator　换向调制器

commutator motor　整流式电动机

commutator rectifier　换向整流器

commutator　换向器、转换开关

core inductance　铁芯电感

DC bridge　直流电桥

DC double bridge　直流双臂电桥

DC galvanometer　直流检流计

DC regulated power supply　直流稳压电源

DC resistance bridge　直流电阻电桥

DC single-and-double bridge　直流单双臂电桥

DC single bridge　直流单臂电桥

DC voltage divider　直流分压器

decimal capacitance box　十进式电容箱

decimal inductance box　十进式电感箱

digital micro-current meter　数字式微电流计

direct current supply　直流电源

dry accumulator　干式蓄电池

electric accumulator　蓄电池

electrical inductance　电感

electroscope　验电器

emergency supply　应急电源

experiment facility of Helmholtz coil magnetic field　亥姆霍兹线圈磁场实验仪

experiment facility of horizontal component of earth magnetic field　地磁场水平分量实验仪

four-terminal resistor　四端电阻器
Hall Effect combined experiment facility　霍尔效应组合实验仪
heat-variable resistor　热变电阻
high potential DC potentiometer　高电势直流电位差计
hysteresis loop experiment facility　磁滞回线实验仪
Kelvin bridge　凯文电桥
low-potential DC potentiometer　低电势直流电位差计
multimeter，universal meter　万用表
non-linear circuit chaotic experiment facility　非线性电路混沌实验仪
packaged double bridge　组装式双臂电桥
photoresistor，photosensitive resistor　光敏电阻
portable DC potentiometer　便携式直流电位差计
portable universal bridge　便携式万能电桥
power supply　电源
precise DC resistance box　精密直流电阻箱
regulated power supply　稳定电源、稳压电源
resistance box　电阻箱
resistor　电阻器
RLC experiment facility　RLC 实验仪
rotary resistance box　旋转式电阻箱
spherical capacitor　球形电容器
standard capacitance box　标准电容箱
standard cell，normal cell　标准电池
standard inductance box　标准电感箱
standard resistance　标准电阻
student-purpose potentiometer　学生式电位差计
switch　开关
thermal resistor，thermistor　热敏电阻
thin-film resistor　薄膜电阻器
tuning inductance　调谐电感
unbalanced bridge　非平衡电桥
variable resistor　可变电阻器
voltage dependent resistor　压敏电阻
volt-ampere characteristic tester　伏安特性实验仪
voltmeter　伏特计
Wheatstone bridge　惠斯顿电桥，单臂电桥
wire-wound resistor　线绕电阻(器)

LESSON 14

14.1 物理学专业英语中的形容词性从句

在复合句中,具有形容词功能、修饰一个名词、词组或代词的从句叫作形容词性从句或定语从句;被定语从句所修饰的词或词组叫先行词;引导定语从句的词叫关联词。定语从句通常位于先行词之后,由关系代词或关系副词引导。

定语从句有限制性定语从句和非限制性定语从句两种。

14.1.1 限制性定语从句

限制性定语从句对先行词起修饰作用,紧接先行词之后,无逗号,若省去,原句意思不完整。

1. 由关系代词 who,whom,whose,that,which 引导的定语从句

引导定语从句的关系代词有 who,whom,whose,which,that,as 等。who,whom,whose 用于指人,whose 有时也可指物,相当于 of which;which 用于指物;that 既可指人也可指物,但只用于限制性定语从句中。关系代词除了引导定语从句,替代先行词外,还在从句中担任主语、宾语、定语等。

(1) 由 that 引导的定语从句

一般情况下 that 可指人或物,可以代替 who,whom 和 which,在从句中作主语或谓语动词的宾语,但不能作介词的宾语。如:

Whether a length is exactly a meter is a question *that no experiment can decide*. 一段长度是否是精确的一米,没有实验能够决定。

Switching circuits are circuits *that perform logical functions*. 开关电路就是执行逻辑函数的电路。

There is hardly a task ***that** we perform **that** does not involve the generation, transmission or reaction to information in some form*. 我们执行的任务几乎没有不包括以某种形式产生、传送、反应信息的(本句中有两个 that 引导的定语从句)。

【注意 1】 当先行词是 all,anything,everything,something,nothing 等不定代词,或先行词前有序数词或 last,any,few,much,some,no,the only,the very 以及形容词最高级修饰时,只能用关系代词 that 引导从句。

This is the first laser *that was introduced from abroad*. 这是第一台从国外引进的激光器。

That's all *that I know about the research project*. 关于研究项目我知道的就这些。

Is there anything *that you want to put in the lab*? 你有什么东西要放在实验室吗?

【注意 2】 先行词既有人又有物时,关系代词只用 that,不用 which。例如:

The scientist and his achievements *that you told me about* are admired by us all. 我们所有人都钦佩你告诉我的那位科学家和他所取得的成就。

【注意 3】 关系代词在限定性定语从句中作表语并带有类比含义时,通常只用 that,

不用 which。例如：

He is no longer the scholar *that he was*. 他不再是过去的那位学者了。

The key lab is no longer the lab *that it used to be*. 这个重点实验室不再是以前的那个实验室了。

【注意 4】 句中其他位置已出现 which，为避免重复，不用 which 而用 that 引导限定性定语从句。例如：

Which is the lab *that we paid a visit last year*? 我们去年参观的是哪个实验室？

(2) 由 who，whom 和 whose 引导的定语从句

who 在从句中作主语，whom 在从句中作宾语，whose 在从句中作定语。如：

Madame Curie is one of the world's greatest women scientists *who will always be remembered as the discoverer of radium*. 居里夫人是世界上最伟大的女科学家之一，作为镭的发现者她被永远铭记。

He is the experimenter *whom you wanted to find*. 他就是你想要找的那个实验人员。

The cross product of two vectors $\boldsymbol{A}$ and $\boldsymbol{B}$ is defined to be a vector $\boldsymbol{C}=\boldsymbol{A}\times\boldsymbol{B}$ *whose magnitude equals the area of the parallelogram formed by the two vectors*. 两个矢量 $\boldsymbol{A}$ 和 $\boldsymbol{B}$ 的矢积定义为 $\boldsymbol{C}=\boldsymbol{A}\times\boldsymbol{B}$，其大小等于两个矢量形成的平行四边形的面积。

For instance, we may visualize a system as an electrical circuit *whose input is a voltage signal and whose output is a current reading*. 比如，我们可以把系统设想为一种电路，它的输入是电压信号，它的输出是电流示数。

(3) 由 which 引导的定语从句

which 指物，在从句中作主语或谓语动词的宾语。如：

A theorem is a statement *which reasoning shows to be true*. 定理是经推导得到的正确论断。

For example, in a mechanical system an ideal model may be a mass *which is concentrated at a point* such that the numerical value of the mass is independent of forces, velocities or acceleration. 例如，力学中的理想模型可以是集中于一点的质量，其数值与力、速度和加速度无关。

【注意 1】 在介词提前到关系代词之前形成“介词＋关系代词”结构来修饰表示事物的先行词时关系代词必须用 which。例如：

A curve is a line *of which no part is straight* and which changes direction without angles. 曲线就是没有直的部分、转向时又没有角的线。

We regard a system as a structure *into which something (matter, energy, or information) may be put at certain times* and which itself puts out something at certain times. 我们把系统看作这样一种结构，有时可以对它输入(比如物质、能量或信息)，有时它自己也输出。

An inertial frame of reference is one *in which an object, subject to no force, moves with constant velocity*. 在一个参考系中，如果一个不受力的物体做匀速运动，那么这个参考系就是惯性参考系。

如果介词不放在修饰事物的限定性定语从句的句首，which 就可换为 that，例如：

This is the question *which/that we've had so much discussion about*. = This is the question *about which we've had so much discussion*. 这就是我们已经多次讨论过的问题。

【注意 2】 先行词为“those＋表示事物的复数名词”时，关系代词通常只用 which 而不用 that。例如：

Experimenters should keep in mind those regulations *which keep them off dangers*. 实验人员应牢记那些保护自己免受危险的规则。

【注意 3】 关系代词在定语从句中作宾语时，关系代词常省去。如：

The device (*which/that*) *I handled just now* has been repaired twice. 我刚才使用的仪器已经维修了两次。

In this chapter, we shall take a closer look at the structure of a few particularly interesting atoms and molecules in order to see how they are fitted to the use (*which/that*) *man makes of them*. 在本章中，我们将更仔细地研究一下几个特别有趣的原子和分子的结构，看一看怎样使它们便于人类利用。

(4) as 引导的定语从句

as 引导的定语从句主要用于“such...as”及“the same...as”的结构中，代替先行词是人或物的名词。as 引导非限制性定语从句时，代替整个主句，从句可位于主句之前、之后或中间。

These are not such problems *as can be easily solved*. (as 代替先行词 problems) 这是些不容易解决的问题。

2. 由关系副词 where，when 和 why 引导的定语从句

引导定语从句的关系副词 where 在从句中作地点状语，when 在从句中作时间状语，why 在从句中作原因状语。意义上相当于一个“介词＋which”的结构。如：

Benjamin Franklin would never forget the moment *when* (＝*at which*) *he was knocked out by the charge*. 富兰克林永远忘不了被电击昏的那个时刻。

Ever since then, Einstein has never been back to the town *where* (＝*in which*) *he was born*. 从那以后，爱因斯坦再也没回过他出生的小镇。

I don't know the reason *why* (＝*for which*) *the capacitor can store charge*. 我不知道电容器能够储电的原因。

14.1.2 非限制性定语从句

非限制性定语从句既可修饰先行词，也可修饰整个主句，起补充说明作用，与主句之间有逗号隔开，若省去，原句意思不受影响。不可用 that 引导非限制性定语从句。关系词不可省略。

This arises because of interference with the reflected waves, *which have the same wavelength, amplitude, and speed*. 这种现象的发生是由于与反射波的干涉，反射波具有相同的波长、振幅和速度。

Moving around the nucleus are extremely tiny particles, called electrons, *which revolve around the nucleus in much the same way as the planets do around the sun*. 围绕着原子核运动的是一些极其微小的粒子，称为电子，这些电子围绕着原子核旋转，正像行星围绕着太阳旋转一样。

For a situation like this, *in which the current is discontinuous in space*, Ampere's law is not valid. 在这样的情况下，电流是不连续的，安培定律不适用。

【注意】 as 与 which 引导的定语从句

两者均可引导非限制性定语从句，有时可以互换，但下列情况多用 as。

(1) 关系代词引导的定语从句居句首时。

As we all know, oceans cover more than 70% of the earth. 我们都知道，海洋覆盖地球表面的 70%以上。

As is known to everybody, the moon travels round the earth once every month. 众所周知，月球每个月绕地球运行一周。

(2) 当与 such 或 the same 连用时，一般用 as。

Such physics books *as you tell me* are interesting. 你告诉我的物理书很有趣。

Different statements of the second law of thermodynamic have the same meanings *as you can find*. 你会发现，关于热力学第二定律的不同表述具有等价性。

(3) 当从句和主句语义一致时，用 as，反之则用 which。

She failed again, *as was expected*.

She failed again, *which was unexpected*.

(4) as 在从句中作主语时，后面常接行为动词的被动语态，如 be known, be said, be reported 等，如从句中行为动词是主动语态，一般要用 which 作主语。

No single company or group can control what happens on the Internet, *as is mentioned above*，如上所述，没有哪个公司和组织能够控制互联网上的事。(as 代替主语)

The model was useful, *which made us happy*. 这个模型很有用，令我们很高兴。

(5) as 与 which 的差别还表现在以下方面：

(a) This cable is like a snake, *as /which everybody can see*. = As everybody can see, this cable is like a snake. 任何人都看得出来，这电线像条蛇。

(b) Einstein's general relativity turned out to be correct, *which made the physicists very excited*. 爱因斯坦的广义相对论结果是正确的，这使得物理学家们很兴奋。

这两个例句中，as 和 which 所代表的都是整个主句所表示的内容。但有两点不同之处：

① 在形式上 as 引导的非限制性定语从句可位于主句的后面，也可位于主句的前面；而 which 引导的非限制性定语从句只能位于主句的后面，不能位于主句的前面。

② 在意义上，as 引导的定语从句和主句的关系一般为一致关系，常译为“正如……”，“就像……”，而 which 引导的定语从句和主句的关系是因果关系，或 which 引导的定语从句是对主句的评论。因此，在意思通顺的情况下，which 可代替 as，而 as 许多时候不能代替 which，如例句(b)。再如：

He was late again, *as/which we had expected*. = As we had expected, he was late again. 正如我们所料，他又迟到了。/他又迟到了这是我们早就料到的。

The engineer uses a lot of oil to lubricate the machine, *which makes it very dirty*. 工程师用很多油来润滑机器，结果把它搞得很脏。(which 为主动语态的主语，不用 as)

Many people ever tried hard to invent machines violating the second law of thermodynamic, *which was impossible*. 许多人曾努力想发明违背热力学第二定律的机

器，但这是不可能的。（不用 as）

He maintains the machine every month, *which has done a lot of good to its performance*. 他每月保养机器，这对它的性能很有好处。（which 为主动语态的主语，不用 as）

As has already been pointed out, the concept of idealized models is extremely important in all physical science and technology. 正如已经指出的那样，理想模型的概念在所有物理科学和技术中是极其重要的。（置于句首，不用 which）

14.2 专业英语阅读

14.2.1 Gauss's Law（高斯定理）

Gauss's law, which relates the electric field on a closed surface to the net charge within the surface, is of fundamental importance in the study of electrostatic fields. In words, *Gauss's law states that in the electrostatic field in vacuum the net electric flux through any closed surface is equal to the net charge inside the surface divided by* ε_0.

In principle, Gauss's law is valid for all surfaces and all charge distributions. In practice, however, it can be used to calculate the electric field only for charge distributions that have high degrees of symmetry. For static charge distributions, Gauss's law and Coulomb's law are equivalent. However, Gauss's law is more general in that it is always valid whether or not the charges are static.

14.2.2 Conductors in Electrostatic Equilibrium（导体的静电平衡）

A good electrical conductor contains charges (electrons) that are not bound to any atom and are free to move about within the material. When no motion of charge occurs within the conductor, the conductor is in **electrostatic equilibrium**. In this situation, every charge in the conductor is a particle in equilibrium, experiencing zero net force. As we shall see, an isolated conductor (one that is insulated from ground) in electrostatic equilibrium has the following properties:

The electric field is zero everywhere inside the conductor.

If the isolated conductor carries a net charge, the net charge resides entirely on its surface.

The electric field just outside the charged conductor is perpendicular to the conductor surface and has a magnitude σ/ε_0, where σ is the surface charge density at that point.

Every point on the surface of a charged conductor in electrostatic equilibrium is at the same electric potential, that is, the conductor is an equipotential volume and its surface is an equipotential surface.

On an irregularly shaped conductor, the surface charge density σ is highest at locations where the radius of curvature of the surface is smallest.

14.2.3 Capacitance and Capacitors（电容和电容器）

When charge is placed on an isolated conductor, the potential of the conductor increases. The ratio of the charge to the potential is called the **capacitance** of the conductor. A useful device for storing charge and energy is the **capacitor**, which consists of two conductors, closely spaced but insulated from each other. When attached to a source of potential difference such as a battery, the conductors carry equal and opposite charges. The ratio of the magnitude of the charge on either conductor to the potential difference between the conductors is the capacitance of the capacitor. Capacitors have many uses. The flash attachment for your camera uses a capacitor to store the energy needed to provide the sudden flash of light. Capacitors are also used in the tuning circuits of devices such as radios, televisions, and cellular phones, allowing them to operate at specific frequencies.

Any two conductors separated by an insulator are said to form a capacitor. In most cases of practical interest the conductors have charges of equal magnitude and opposite sign. A capacitor is usually charged by transferring a charge Q from one conductor to the other, and then leaving one of the conductors with a charge $+Q$ and the other with a charge $-Q$. The electric field in the region between the conductors is proportional to the magnitude of this charge, and it follows that the potential difference U between the conductors is also proportional to the charge magnitude Q. The capacitance of a capacitor is defined to be Q/U.

Capacitors find many applications in electrical circuits. Capacitors are used for tuning radio circuits and for "smoothing" the rectified current delivered by a power supply. A capacitor is used to eliminate sparking when a circuit containing inductance is suddenly opened. The ignition system of every automobile engine contains a capacitor to eliminate sparking of the "points" when they open and close. The efficiency of alternating current power transmission can often be increased by the use of large capacitors.

Glossary

Gauss's law	高斯定理	closed surface	闭合面
electrostatic field	静电场	vacuum	真空
electric flux	电通量	charge distribution	电荷分布
degrees of symmetry	对称度	equivalent	等价的
in principle	原则上	in practice	实际上
conductor	导体	electrostatic equilibrium	静电平衡
equilibrium	平衡	isolated conductor	孤立导体
insulate	*v.* 绝缘	perpendicular to	垂直于
charged conductor	带电导体	electric potential	电势
equipotential	等电势的	radius of curvature	曲率半径
capacitance	电容	capacitor	电容器
flash	闪光灯	attachment	配件
tuning circuit	调谐电路	cellular phone	移动电话
electrical circuit	电路	sparking	火花

ignition system　　（汽车的）点火系统　　alternating current　　交流
power transmission　　电力传输

14.3　专业英语常用表达法-14　图的表示法

1. figure 通常指几何图形或图案；此外书中的插图不管是什么图都可以用 figure 表示。例如：

Figure 3 shows an electrics motor being used to lift a heavy object. 图 3 所示是正在起吊重物的一台电动机。

Figure 5 is the schematic diagram of a d/a converter. 图 5 是数/模转换器的示意图。

2. diagram 是专业英语中用得比较广泛的一个词，用于一般的图形、图表。例如：

schematic diagram　示意图，原理图

a diagram of the apparatus　装置简图

a diagram of a flask showing its outline　一只长颈烧瓶的轮廓简图

rough out a diagram　草拟一张图表

This simplified block diagram is valuable. 这个简化了的框图很有用。

3. graph 一般用于指曲线图、标绘图，也可指图解。例如：

relating to a graph　与图表有关

presented by a graph　用图表表达

The graph of this equation has been drawn above. 此方程所对应的曲线图已在上面画出。

The graph showed the increase in instruments between the years 2005 and 2009. 这张曲线图所示是 2005—2009 年仪器增长的情况。

4. graphics 表示（作单数用）制图法，制图学，图表算法，图形。例如：

specification of tables of graphics　对图表的说明

5. plot 往往可以与 graph 互换。例如：

The plot of I versus V is a straight line. I 随 V 的变化曲线是一条直线。

6. view 一般用于机械图或各种结构图中的各向视图，也用来指航海中使用的“对景图”（实物图片）。例如：

The top and cross-sectional views of the memory structure are shown below. 这种存储器结构的俯视图和截面图如下面所示。

A front view shows only the width and height of an object. 前视图（正面图）只给出一物体的宽度和高度。

7. profile 一般指零（部）件的外形轮廓图，剖面形状图，profile 也可指各种分布图。例如：

stratigraphic profile　地层纵断面图（剖面图）

density profile　密度分布图

pressure profile　压力分布图

8. pattern 用于指图形、图案、花纹等。例如：

diffraction/interference pattern　衍射/干涉图样

A mask with 1000 tiny individual patterns is placed over the chip and exposed to light. 将带有1000个微小分立图样的掩膜放在晶片上并曝光。

9. drawing 主要用于各种工程图纸及有关工程设计的附图、插图、工程图。例如：

perspective drawing 透视图

projection drawing 投影图

10. chart 多用于表示航线图和某种方向的图。例如：

nautical chart 航海图

flight chart 航空图

make a chart of 绘制……的图表

a chart representing numerical relationships 代表数字关系的图表

11. map 指地图、天体图、布局图以及一些专用图形。例如：

seismic map 地震图

weather map 气象图

radar map 雷达图

12. sketch 一般指粗略勾画的示意图，草图，图样。例如：

a sketch of the apparatus used for distillation 蒸馏装置的草图

13. layout 着重于被描绘对象的总体概貌、外形轮廓、总体安排和设计。例如：

Among the disadvantages are great layout difficulties. 缺点之一是(电路)排图困难。

14. line 一般仅用于线条描绘的轮廓和外形图，如船体型线图。例如：

The engineers are making the designing of the hull lines. 工程师们正在设计船体型线图。

15. scheme 指草图、示意图。例如：

wiring scheme 配线图，接线图，布线图

16. illustration 插图，图解。例如：

The colored illustrations in that book are most attractive. 那本书中的彩色插图非常吸引人。

17. image 指像，图像。例如：

Project the image on a screen. 把图像投射到屏幕上。

18. outline 指轮廓图，略图。

19. draft 含有底稿的意思，用得较少。

20. delineation 只用于草图、略图、示意图，用得较少。

LESSON 15

15.1 物理学专业英语中的非谓语动词(一)动词不定式和动名词

仔细分析各种物理专业英语文献，我们会发现其中的很多句子结构并不复杂，都只有一套主谓结构。前面的分析曾经提到科技英语的句子比较长，这种的现象看似矛盾，原因之一是非谓语动词的大量使用。

动词不定式、分词(现在分词，过去分词)和动名词在句中均不能作谓语用，所以统称为

非谓语动词。现代英语将现在分词和动名词合为一大类叫作 *v.* ＋ing 形式。这些动词的形式不能在句中单独作谓语用，因而没有语法主语，但可以有逻辑主语。由于没有语法主语，也就不受人称和数的限定，由于不是谓语，也就没有时态和语态，但这些词仍能表示动作和状态，所以仍有确定的形式，用来表示相对于其他动词的时间关系。由于具有逻辑主语，因此也有表示主动、被动的形式，同时也有自己的宾语和状语，一起构成非谓语动词短语（动词不定式短语、分词短语、动名词短语）。非谓语动词在专业英语语法中占有特殊且重要的地位。非谓语动词形式多样，应用广泛，在句中起着重要的作用。

15.1.1 动词不定式

动词不定式由"to＋动词原形"构成。动词不定式虽然不能作谓语动词用，但仍保留动词的特征，它可以带有自己的宾语或状语而构成动词不定式短语，如 to locate a point in space，to be quite important。

1. 动词不定式的基本用法

动词不定式能起名词、形容词和副词的作用，可在句中作主语、表语、宾语补足语、定语和状语用。

（1）作主语

To say that a distance is 25 is meaningless. The magnitude of the physical quantity must include both a number and a unit. 说一段距离是 25 是没有意义的。物理量的大小必须包括数值和单位两个部分。

动词不定式作主语时，一般可用 it 作形式主语，而将作主语的动词不定式置于句末。如：

It is convenient *to use Equation*(4-5) *in its component form*. 采用方程(4-5)的分量形式是比较方便的。

（2）作表语

If we wish to locate a point in space, the simplest way is *to describe* how far the point is from some other known point, and in what direction. 如果我们想在空间定位一个点，最简单的方法就是描述这个点距离某个已知的点有多远以及在哪个方向上。

（3）作宾语

① 作动词的宾语，如：

A car starts *to move* from rest with speed increasing uniformly along a circular path. 一辆车沿着圆周从静止开始以匀加速度运动。

This relation, known as Hooke's law, turns out *to be* quite important. 这个关系，就是胡克定律，是非常重要的。

② 作形容词的宾语，如：

Space and time, by themselves, are concepts that are difficult (or perhaps impossible) *to define* in terms of anything simpler. 时间和空间，就其本身而言，是非常难以（或许也不可能）用更简单的东西来定义的概念。

③ 动词不定式一般不作介词的宾语，但动词不定式之前如有疑问词时，就可作介词的宾语。如：

Can you give us some advice on what *to do* next? 接下来怎么办，能给我们点建议吗？

(4) 作宾语补足语

Every coordinate system involves a particular point, the origin (O), and specified directions or axes that enable points in space *to be described* relative to the origin and the axes. 每个坐标系包括一个特定点(原点 O)和一些规定的方向(坐标轴)，从而使空间中的点可以相对于原点和坐标轴加以描述。

We assume space *to be continuously uniform and isotropic*. 我们假设空间是连续均匀和各向同性的。

【注意】 如果句中的谓语动词为 see, hear, watch, notice, have, make, let 等，作宾语补足语的动词不定式须将 to 省去。如：

The students watch the object *fall* freely to the ground. 学生观察物体自由下落到地面。

(5) 动词不定式在句中作宾语，如带有宾语补足语时，须先用 it 作形式宾语，而将该动词不定式后置。如：

I don't think *it* right *to handle the apparatus that way*. 我认为那样操作仪器不对。

(6) 作定语

动词不定式作定语时，须位于被其修饰的名词或代词之后。如：

In this case, the particle has maximum tendency *to rotate* about the origin. 在这种情况下，粒子具有绕原点旋转的最大趋势。

(7) 作状语

动词不定式可以作下列状语：

① 目的状语

To measure the distance between two points, we need a standard unit, such as a meter. 为了测量两点间的距离，我们需要一个标准单位，例如米这个单位。

为了强调不定式表示目的的作用，可在不定式前加 in order to 或 so as to(以便或为了)，但应注意 in order to 位于句首或句中均可，而 so as to 不能位于句首。如：

In order to make an idealized model of a system, we have to overlook quite a few minor effects to concentrate on the most important features of the system. 为了给系统做一个理想模型，我们不得不忽略很多不重要的因素，而把注意力集中在系统的重要特征上。

【注意】 将表示目的的不定式置于句首，也可起到强调目的的作用。

② 结果状语

We need to use some judgment and creativity to construct a model that simplifies a problem enough *to make it manageable*, yet keeps its essential features. 为了构建一个模型，我们需要发挥一下判断力和创造力，把问题简化到既能得到有效解决，又能保留必要的特征。

The electric force binds atoms and molecules in compounds *to form ordinary matter*. 电力把原子和分子约束在化合物内，形成了物质。

③ too＋形容词或副词＋动词不定式，表示"太……以至于不能……"的结果。如：

In physics, a model is a simplified version of a physical system that would be *too complicated to analyze* in full detail. 物理学中的模型是那些难以详尽分析的物理系统的

简化描述。

2. 复合结构不定式

由 for+名词(或代词宾格)+ 动词不定式即构成复合结构的动词不定式。其中 for 本身无意义。for 后面的名词或代词是不定式的逻辑主语,这种不定式在句中可作主语、表语、宾语、定语或状语。如:

Note that the net external force need not be zero *for angular momentum to be conserved*. 注意:角动量守恒合外力不必为零。

It is very important *for us to get everything ready* for the experiment. 我们为实验准备好一切是很重要的。

当作表语用的形容词表示不定式的逻辑主语的性质或特征时,就用介词 of 而不用 for 引出不定式的逻辑主语,这些形容词一般有 good, nice, kind, wise, silly, stupid, foolish, right, wrong, careless, impolite 等。如:

It is very kind *of you to measure the temperature of the oil* for me every day. 你真好,每天都为我测量油温。

3. 疑问词+动词不定式

疑问代词和疑问副词后可加动词不定式构成不定式短语,在句中可作主语、表语或宾语。如:

How to prevent the temperature from getting too high in this heat engine is a problem. 如何在热机中防止高温是一个问题。

4. 动词不定式的否定式

动词不定式的否定式是由 not +动词不定式构成。如:

It's wrong of you *not to apply Ampere's law for a situation like this*. 在这种情况下你不应用安培定律是不正确的。

5. 动词不定式的时态所表示的时间关系

(1) 一般式:动词不定式一般式所表示的动作是和谓语动词所表示的动作同时发生,但在多数情况下,是在谓语动词所表示的动作之后发生。如:

We decided *to demonstrate a standing wave* with a rope with one end fixed on the wall. 我们决定利用一根绳子,把其一端固定在墙上来演示驻波(其后)。

The assistants often watch the professor *operate the spectrograph*. 助教经常观察教授操作摄谱仪(同时)。

(2) 完成式:动词不定式完成式所表示的动作发生在谓语动词所表示的动作之前。如:

I am sorry *to have kept you waiting* outside the lab. 对不起,让你在实验室外面等着。

(3) 进行式:动词不定式进行式所表示的动作正在进行中,而且与谓语动词所表示的动作同时发生。如:

She happened *to be repairing a multimeter* in the room when I came in. 我进来时她正在修理万用表。

6. **动词不定式的被动语态**

如果动词不定式的逻辑主语为这个不定式所表示的动作的承受者时，不定式一般就用被动语态形式。如：

Resistance welding is performed by passing a very heavy current through the pieces of metal *to be joined*. 电阻焊接是让强大电流流过待焊金属来完成的。

15.1.2 动名词

动名词是动词的一种非谓语形式。动名词的构成为：*v*. ＋ing。

动名词具有名词的特点，可在句中作主语、宾语和表语。动名词也具有动词的某些特点，可带有宾语和状语，构成动名词短语。动名词也有主动和被动、一般时和完成时等形式。

1. **动名词及其短语在句中可以作主语、宾语和表语**

例如：

Heating the water changes it into vapor. 把水加热可以使水变为蒸汽（Heating the water 在句中作主语，water 是 heating 的宾语）。

Closing the switch connects the cranking motor to the battery. 合上开关就把启动马达与蓄电池接通了（句中 closing the switch 作主语，switch 为 closing 的宾语）。

Oxygen does not burn, but does support *burning*. 氧不自燃，但能助燃（burning 是动名词，作 support 的宾语）。

This is equivalent to *stating* that it has no tendency to rotate about the origin. 这相当于说物体根本没有绕原点运动的趋势（动名词 stating 为介词 to 的宾语）。

Energy is involved in *doing* work, or in *heating* on object. 能量与做功有关，或使物体变热（动名词 doing 和 heating 均为介词 in 的宾语）。

In the dynamo, mechanical energy is used for *rotating* the armature between the poles of an electromagnet. 在直流发电机中，机械能被用来转动电磁铁两极之间的电枢（动名词 rotating 是介词 for 的宾语）。

One way to fit the pieces of a broken iron bar together is *heating* the iron until it softens and *pounding* it with a hammer. 把一根断成数截的铁棍连接起来的一个方法就是把铁加热至软化，然后用铁锤去锻打（句中 heating 和 pounding 均为表语）。

2. **动名词的被动语态**

当动名词的逻辑主体是动名词的动作对象时，该动名词通常用被动形式。例如：

When a vapor condenses, the energy that it absorbed in *being vaporized* is again released as heat. 当蒸汽冷凝时，它汽化时所吸收的能量又以热的形式释放出来（it 用来代替 vapor，it 是 vaporize 这一动作的承受者，所以要用动名词的被动形式 being vaporized）。

It is assumed that every substance which is capable of *being magnetized* consists of a very large number of molecular magnets. 人们假定，能够被磁化的各种物质都是由大量的分子磁体组成的（句中 which 用来代替 substance，which 是动词 magnetize 这一动作的对

象,所以 magnetize 的动名词应用被动形式)。

3. 动名词的完成式

如果动名词的动作发生在谓语动词之前,则该动名词通常用完成式。例如:

Having discovered the property of the atom, scientists began finding ways of using atomic energy for industrial purposes. 发现了原子的特性之后,科学家们开始寻找将原子能用于工业的途径(discover 发生在谓语动词 began 之前,即先"发现了原子的特性",后"开始寻找"用于工业的途径)。

Nuclear fuel will be used in the reactor after its *having been compressed* into the form of rods. 核燃料被压成棒状之后,可用于反应堆中(句中 having been compressed 是完成式被动语态动名词,作 after 的宾语,其行为发生在谓语动词 will be used 之前,即先"压成棒状",后"用于反应堆")。

4. 动名词的逻辑主体

动名词短语前面可以加一个物主代词(如 their,its 等),名词的所有格(如 teacher's)或名词的普通形式来表示这个动名词的逻辑主体。例如:

The main trouble is *their not having* enough voltage. 主要困难是它们没有足够的电压(句中物主代词 their 是动名词 having 的逻辑主体)。

Physically we understand forces as due to *matter pushing* on other matters. 在物理意义上,我们可以把力理解为一物体对另一物体的推力(句中动名词 pushing 是介词 due to 的宾语,matter 是 pushing 的逻辑主体)。

We know of the *earth behaving* as a large magnet. 我们知道,地球相当于一个大磁体(动名词 behaving 作 of 的宾语,the earth 是 behaving 的逻辑主体)。

By using phototubes, doors can be opened without *anyone's having touched them*. 使用光电管,可以使门无人自开(having touched 是动名词完成式,在句中作介词 without 的宾语,anyone's 是 having touched 的逻辑主体)。

The process of *one substance mixing* with another because of molecular motion is called diffusion. 一物质与另一物质由于分子运动而混合的过程称为扩散(动名词 mixing 是 of 的宾语,one substance 是 mixing 的逻辑主体)。

Magnetism may be induced into an iron bar without *it touching* the magnet. 铁棒无须接触磁铁就可受到磁感应(动名词 touching 是 without 的宾语,it 是 touching 的逻辑主体)。

15.2 专业英语阅读

15.2.1 Nature of Magnetic Force(磁力的本质)

More than 2000 years ago, the Greeks were aware that a certain type of stone (now called magnetite) attracts pieces of iron, and there are written references to the use of magnets for navigation dating from the twelfth century.

In 1269, Pierre de Maricourt discovered that a needle laid at various positions on a spherical natural magnet orients itself along lines that pass through points at opposite ends of the sphere. He called these points the **poles of the magnet**. Subsequently, many experimenters noted that every magnet of whatever shape has two poles, designated the **north and south poles**, where the force exerted by the magnet is strongest. It was also noted that the like poles of two magnets repel each other and the unlike poles attract each other.

Although electric charges and magnetic poles are similar in many respects, there is an important difference: *Magnetic poles always occur in pairs*. When a magnet is broken in half, equal and opposite poles appear at either side of the break point. The result is two magnets, each with a north and a south pole. There has long been speculation as to the existence of an isolated magnetic pole, and in recent years considerable experimental effort has been made to find such an object. Thus far, there is no conclusive evidence that an isolated magnetic pole exists.

Magnetic force is the interaction between moving electric charges. This is the nature of magnetic force.

15.2.2 The Biot-Savart Law(毕奥-萨伐尔定律)

The earliest known sources of magnetism were permanent magnets. One month after Oersted announced his discovery that a compass needle is deflected by an electric current, Jean Baptiste Biot and Felix Savart announced the results of their measurements of the force on a magnet near a long, current-carrying wire and analyzed these results in terms of the magnetic field produced by each element of the current. Andre-Marie Ampere extended these experiments and showed that current elements also experience a force in the presence of a magnetic field and that two currents exert forces on each other.

The magnetic field $\mathrm{d}\boldsymbol{B}$ produced by a current element $I\mathrm{d}\boldsymbol{l}$ is given by

$$\mathrm{d}\boldsymbol{B} = \frac{\mu_0}{4\pi}\frac{I\mathrm{d}\boldsymbol{l}\times\boldsymbol{e}_r}{r^2} \tag{15-1}$$

where $\boldsymbol{e}_r$ is a unit vector that points from the current element $I\mathrm{d}\boldsymbol{l}$ to the field point P_1 (Fig. 15-1) and μ_0 is a constant of proportionality called the **permeability of free space**, which has the value

$$\mu_0 = 4\pi\times10^{-7}\ \mathrm{N/A^2}$$

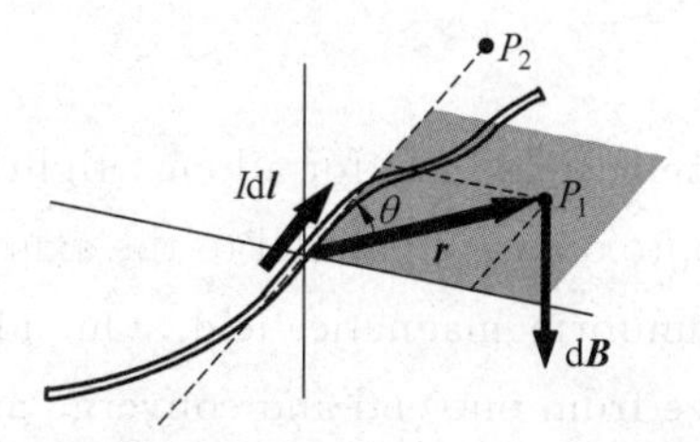

Fig. 15-1 The current element $I\mathrm{d}\boldsymbol{l}$ produces a magnetic field at point P_1 that is perpendicular to both $I\mathrm{d}\boldsymbol{l}$ and $\boldsymbol{r}$. It produces no magnetic field at P_2, which is along the line of $I\mathrm{d}\boldsymbol{l}$

Eq. (15-1), known as the **Biot-Savart law**, was also deduced by Ampere. The source of the magnetic field is a moving charge $q\boldsymbol{v}$ or a current element $I\mathrm{d}\boldsymbol{l}$, just as the charge q is the source of the electrostatic field. The magnetic field decreases with the square of the distance from the moving charge or current element, just as the electric field decreases with the square of the distance from a point charge. However, the directional aspects of the electric and magnetic fields are quite different. Whereas the electric field points in the radial direction $\boldsymbol{e}_r$ from the point charge to the field point (for a positive charge), the magnetic field is perpendicular both to $\boldsymbol{e}_r$ and to the direction of motion $\boldsymbol{v}$ of the charges, which is along the direction of the current element. At a point along the line of a current element, such as point P_2 in Fig. 15-1, the magnetic field due to that element is zero, because $I\mathrm{d}\boldsymbol{l}$ is parallel to the vector $\boldsymbol{e}_r$.

The magnetic field due to the total current in a circuit can be calculated by using the Biot-Savart law to find the field due to each current element and then summing (integrating) over all the current elements in the circuit. This calculation is difficult for all but the simplest circuit geometries.

15.2.3 B due to a Current in a Solenoid(通电螺线管的磁场)

A **solenoid** is a wire tightly wound into a helix of closely spaced turns, as illustrated in Fig. 15-2. It is used to produce a strong, uniform magnetic field in the region surrounded by its loops. Its role in magnetism is analogous to that of the parallel-plate capacitor, which produces a strong, uniform electric field between its plates. The magnetic field of a solenoid is essentially that of a set of N identical current loops placed side by side. Fig. 15-3 shows the magnetic field lines for two such loops.

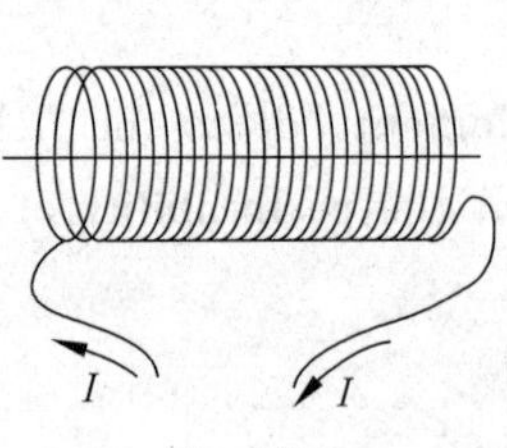

Fig. 15-2

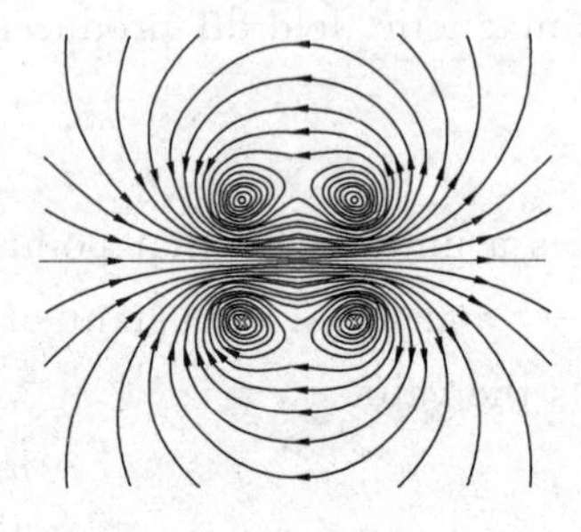

Fig. 15-3

Fig. 15-4 shows the magnetic field lines for a long, tightly wound solenoid. Inside the solenoid, the field lines are approximately parallel to the axis and are closely and uniformly spaced, indicating a strong, uniform magnetic field. Outside the solenoid, the lines are much less dense. They diverge from one end and converge at the other end.

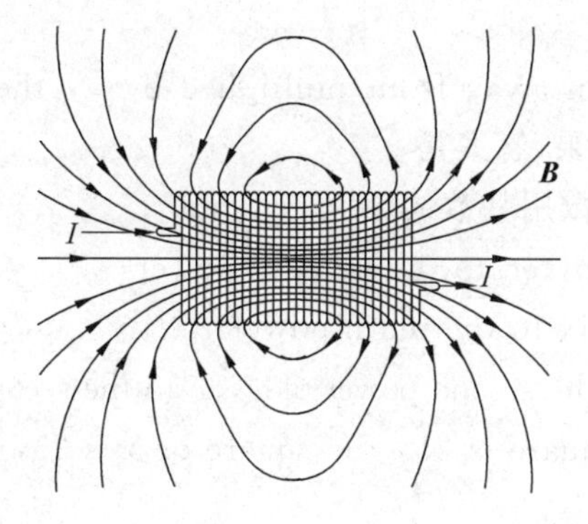

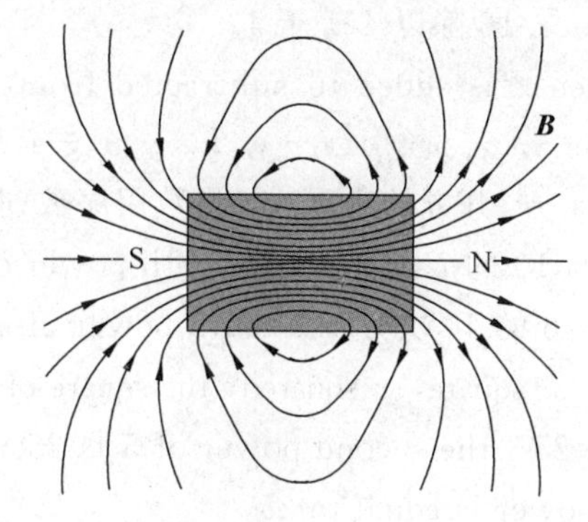

Fig. 15-4

Glossary

magnetite	磁铁矿石	magnet	磁铁
pole of the magnet	磁铁的极	magnetic pole	磁极
interaction	相互作用	Biot-Savart Law	毕奥-萨伐尔定律
permanent magnet	永久磁铁	current-carrying wire	载流导线
current element	电流元	solenoid	螺线管
turn	匝	parallel-plate capacitor	平行板电容器
diverge	发散	converge	聚合
permeability of free space	真空磁导率		
constant of proportionality	比例常数		

15.3 专业英语常用表达法-15 加减乘除 乘方 开方

1. 加用 plus,and 或 add 表示;等于用 is,make,equal 等词表示。

2+3=? How much is two plus three?

2+3=5 Two plus three is five. 或者 Two and three is equal to five. 或者 Two and three makes five. 或者 Two added to three equals five. 或者 If we add two to/and three, we get five. 2 加 3 等于 5。

2. 减用 minus 或 take from 表示。

10−6=? How much is ten minus six?

10−6=4 Ten minus six is four. 或者 Take six from ten and the remainder is four. 或者 Six (taken) from ten is four. 10 减去 6 等于 4。

3. 乘用 time(动词)或 multiply 表示。

3×4=? How much is three times four?

3×4= 12 Three times four is/are twelve. 或者 Multiply three by four, we get twelve. 或者 Three multiplied by four makes twelve. 3 乘以 4 等于 12。

4. 除用 divide 的过去分词形式表示。

16 ÷ 4= ? How much is sixteen divided by four?

16 ÷ 4= 4 Sixteen divided by four is four. 或者 Sixteen divided by four equals/gives/

makes four. 16 除以 4 等于 4。

When x is added to/subtracted from/taken away from/multiplied by y, the result/answer is z. 若 $x+y/x-y/x\times y/x\div y$,则结果/答案是 z。

5. 乘方、开方的表示法：指数用序数词,底数用基数词。

10^7　10 的 7 次方 the seventh power of ten(ten to the seventh power)

6^{10}　6 的 10 次方 the tenth power of six(six to the tenth power)

x^2　x square, x squared, the square of x, the second power of x, x to the second power

$5^2=25$　the second power of 5 is 25, 5 square is 25, the square of 5 is 25, 5 to the second power is equal to 25

y^3　y cube, y cubed, the cube of y, y to the third power, y to the third

y^{-10}　y to the minus tenth (power), y to the power of minus ten

$\sqrt{4}=\pm 2$　the square root of 4 is/equals plus or minus 2

$\sqrt[3]{a}$　the cube root of a

$\sqrt[5]{x^2}$　the fifth root of x square

$\sqrt{518}$　the square root of five hundred and eighteen

$\sqrt[3]{930}$　the cubic root of nine hundred and thirty

$3x=5$　three times x equals 5

$\frac{x^3}{5}=y^2$　x raised to the third power divided by five equals y squared

$x^2+y^2=10$　x squared with y squared equals 10

更多关于数学符号的表示方法请参阅本书附录部分。

LESSON 16

16.1　物理学专业英语中的非谓语动词(二)分词

分词也属于非谓语动词,是动词的一种非谓语形式,分词分为现在分词和过去分词两种。现在分词表示正在进行或主动的动作;过去分词表示被动的或是已完成的动作。例如:

boiling water　沸水,boiled water　开水

developing country　发展中国家,developed country　发达国家

falling leaves　下落的叶子,fallen leaves　落叶

moving body　运动的物体,the moved body　被移动的物体

分词具有动词的某些特征,可以带宾语和状语组成分词短语。在专业英语中,分词短语在句中主要作定语和状语。过去分词在句中作某种成分时,其逻辑主语一般为该分词所表示的动作的承受者。

1. 作定语

单个分词作定语时,一般放在它所修饰的名词前(也可放在名词之后);分词短语作定语时,一般放在它所修饰的名词后;分词在修饰由 some/any/no+thing/body/one 所形成

的不定代词时应后置。

All *moving* bodies have kinetic energy. 所有运动的物体都具有动能。

A *lifted* weight increases in its potential energy. 被举起的重物势能会增加。

The work *done* is the scalar product of the force and the displacement. 力所做的功等于力和位移的标积。

We see that impulse is a vector quantity *having a magnitude* equal to the area under the force-time curve. 我们看到冲量是一个矢量,其大小等于力-时间曲线下的面积。

When a body floats in a fluid the buoyancy is equal to the weight of the fluid *displaced*. 当物体漂浮在液体中时,浮力等于所排开的流体的重量。

A direct current is a current *flowing always in the same direction*. 直流电是一种总是沿同一方向流动的电流。

Above the troposphere is the stratosphere, *followed by the ionosphere* and *capped by the exosphere*. 在对流层上是平流层,接着是电离层,最上层是外逸层。

Heat is the energy *produced by the movement of molecules*. 热是分子运动所产生的能量。

An alloy is a substance *composed of two or more metals fused together*. 合金是由两种或两种以上融合在一起的金属所组成的物质。

2. 作状语

分词短语作状语时,往往具有时间、条件、原因、结果、让步、方式、补充说明等含义。它可以放在句首、句中或句尾,通常用逗号与句子格开。一般来说,时间、原因、条件、让步等分词短语多放在句首;而表结果、伴随的分词短语放在句尾。分词短语作状语时,前面可用 when, while, if, unless, though 等词来加强时间、条件等含义。

Being negative, an electron moves always from negative to positive. 由于电子的负电性,所以总是由负(极)向正(极)运动。(原因)

Being relatively close, the capacitor's two plates produce relatively large electric field in between. 由于两个极板距离相对较近,所以电容器极板间的电场相对很大。(原因)

Considering some common aspects of the performance of the steam engine and automobile engine, we notice that both reject heat. 在研究蒸汽发动机与汽车发动机性能的共同点时,我们注意到两者都排除热量。(时间)

All metals are fairly hard, *compared with nonmetals*. 与非金属相比,金属是相当硬的。(条件)

In a liquid or solid, the molecules are much closer together, *resulting in much more material in a given volume*. 在液体或固体中,分子结合得紧密得多,因而在一定体积内的物质也多得多。(结果)

The particle motion may be nearly circular at the surface but become more elliptical with depth, eventually *becoming longitudinal*. 粒子的运动在表面时接近圆轨道,但是随着深度加深更加椭圆化,最终变为纵向的。(结果)

Utilizing the material obtained by means of astronomical instruments, we learned more about the moon. 由于利用了天文仪器得到的资料,我们了解到了更多有关月球的知

识。(原因)

Molecules vary considerably in size and weight, *ranging from the tiniest micromolecules to the largest macromolecules*. 各种分子的大小和重量都有很大的差别，范围从最小的微观分子直到最大的宏观分子。(补充说明)

Objects begin to move because a force is acting on them, *pushing or pulling them*. 物体开始运动是因为受力的作用，即受到推力或拉力的作用。(补充说明)

When heated, a body will expand. 物体受热时就会膨胀。(时间或条件)

Iron, *when combined with carbon and other elements*, is used most widely in industry. 铁与碳和其他元素结合时，在工业上获得了极为广泛的应用。(时间或条件)

A substance that expands *while melting* has its fusing point slightly raised by an increase of pressure. 熔化时会膨胀的物质，通过加压可使其熔点略有升高。(时间或条件)

The photon, *while not having a material mass*, may be considered as having a radiation mass. 光子虽然没有物质质量，但可以认为它具有辐射质量。(让步)

3. 分词和分词短语作宾语补足语

在英语中，有些及物动词以分词、动词、形容词等作宾语补足语以表达完整的意思。这些动词在专业英语中也经常出现。这类动词有：see, watch, make, have, keep, start, find 等。

If we put wet clothes above the fire, we can see steam *rising from the wet clothes*. 如果把湿的衣服挂在火的上方，就会看见蒸汽从湿衣服上升起。

To keep a body *moving* on the ground, you need apply less force than you did to get it started. 保持一个物体在地面运动比启动该物体所施加的力要小。

A body at rest will remain at rest if there is no enough force to start it *moving*. 如果没有足够的力来启动静止的物体，那么静止的物体将保持静止状态。

We need to keep the temperature high enough to keep the reaction *going*. 为了使反应进行下去，我们需要保持温度足够高。

4. 分词的时态和语态

现在分词有一般式和完成式，有主动语态和被动语态。现在分词被动态所表示的动作与句子中谓语动词所表示的动作同时发生或正在进行。例如：

The Hall effect instrument *being repaired* there is ours. 正在修理的霍尔效应仪是我们的。

Being cooled in the air, the metal hardened. 金属在空气中冷却时就硬化了。

现在分词的完成时态表示的动作发生在句子谓语动词所表示的动作之前。完成时态的分词在句中只作状语用。例如：

Bound charge is deliberately ignored on the right side of this equation, *having been taken into account by the introduction of* ε_r *on the left side*. 方程右侧的束缚电荷被故意略掉，因为已经被考虑在左侧引入的 ε_r 中。

Having defined two units of power, the horse-power and the kilowatt, we may use these in turn to define two new units of work, the horsepower-hour and the kilowatt-hour.

在定义了功率的两种单位(hp,kW)之后,我们就可以转而用这两个单位来定义功的两个新单位(hp·h,kW·h)。

5. 分词的独立结构

分词短语作状语时,通常它的逻辑主语也就是句子中的主语。如:

Electrifying a capacitor, you store electric energy. 给电容器充电时,你就是在储存电能。

分词 electrifying 的逻辑主语就是句中的主语 you。但是,如果分词短语中的逻辑主语与句子中的主语不一致时,分词短语就可以有自己的逻辑主语。这种带逻辑主语的分词短语叫作"分词独立结构"。这种结构在句中只起到状语作用,可以表示时间、原因、条件和附加说明等。

Carbon steel being hard ferromagnetic material, we can use it for memory units. 碳钢作为硬磁性材料,我们可以把它用作记忆元件。(原因)

Soft iron is magnetically soft, *the hysteresis effect being small*. 软铁是软磁性材料,因为磁滞效应小。(原因)

A change of velocity, *the mass being constant*, must mean a change in momentum. 如果质量不变,速度的变化就意味着动量的变化。(条件)

To explain this remarkable behavior, each electron is considered to possess an electric charge, *the charge being a numerical measure of the repulsion experienced between two electrons*. 为了解释这种奇特现象,可认为每一个电子带一个电荷,即电荷成为两个电子之间排斥力的量度。(附加说明)

The world is full of other kinds of waves, *including sound waves, seismic waves, radio waves, and* X-*rays*. 世界上充满其他种类的波,包括声波、地震波、无线电波和 X 射线。(附加说明)

A molecule of hydrogen has only two atoms, *each having one proton and one electron*. 氢分子仅有两个原子,每个原子有一个质子和一个电子。(附加说明)

An object can move uniformly in a straight line, *there being no cause to change that motion*. 如果没有改变物体运动状态的原因,那么物体会做匀速直线运动。(条件)

6. 分词复合结构

某些分词独立结构由介词 with 或 without 引出,形式是:

with(或 without)+名词+分词,这种结构在英语中称为"分词复合结构",在句中可作定语和状语,在专业文章中这种结构经常用来作为补充说明。其中 with 引导的分词结构中有时 with 可省略,分词结构前用逗号分隔。

The ferromagnetic materials can become paramagnetic, with the temperature *being above the Curie temperature*. 温度高于居里点温度时,铁磁质会变成顺磁质。(条件)

The volume of an ideal gas varies inversely proportional to pressure, with temperature *being constant*. 若温度不变,则理想气体的体积与压强成反比。(条件)

Maxwell pointed out that an object may be hot without the motion in it *being visible*. 麦克斯韦指出,即使看不见物体内部的分子运动,物体也可能是热的。(让步)

If we shake the end of the string, the string will take on the shape shown by the curve in the graph, with this shape *remaining the same but moving toward the right*. 如果我们摇动绳的一端,绳就会具有插图所示的曲线的形状,并保持不变的形状向右传播。(附加说明)

If something is moving, *with nothing touching it*, it will go on forever, moving at a uniform speed in a straight line. 运动中的物体,在没有受到任何外力触动时,将永远继续做匀速直线运动。(条件)

Ice, of relative density 0.9, floats in water *with nine-tenth submerged*. 冰的相对密度为 0.9,所以冰浮在水面上时,9/10 淹没在水里。(方式方法)

An atom of oxygen has eight protons and eight neutrons in its nucleus, *with eight electrons circling about that nucleus*. 一个氧原子核内有 8 个质子和 8 个中子,核外有 8 个电子绕着原子核旋转。(附加说明)

7. 非谓语动词短语代替从句

非谓语动词短语常用来代替从句而使句子简洁。例如:

The concept of time *which was considered as an absolute quantity in the Newtonian theory* received as a basically different interpretation in the theory of relativity *which was announced by Einstein in* 1905. 在牛顿理论中被认为是绝对量的时间概念,在爱因斯坦于 1905 年宣布的相对论中,得到了根本不同的解释。

可以用两个分词短语代替两个定语从句。请比较:

The concept of time *considered as an absolute quantity in the Newtonian theory* received as a basically different interpretation in the theory of relativity *announced by Einstein in* 1905.

16.2 专业英语阅读

16.2.1 Ampere's Circuital Law(安培环路定理)

In previous chapter we found that for highly symmetric charge distributions, we could calculate the electric field more easily using Gauss's law than Coulomb's law. A similar situation exists in magnetism. Ampere's law, which relates the tangential component of $\boldsymbol{B}$ summed around a closed curve L to the current I_{inside} that passes through the curve can be used to obtain an expression for the magnetic field in situations that have a high degree of symmetry. In mathematical form, **Ampere's Circuital Law** is

$$\oint_L \boldsymbol{B} \cdot \mathrm{d}\boldsymbol{l} = \mu_0 \sum I_{\text{inside}} \quad (L \text{ is any closed curve}) \tag{16-1}$$

where I_{inside} is the net current that penetrates the area bounded by the curve L. Ampere's law holds for any curve L as long as the currents are continuous, that is, they do not begin or end at any finite point. It is useful in calculating the magnetic field $\boldsymbol{B}$ in situations that have a high degree of symmetry so that the line integral $\oint_L \boldsymbol{B} \cdot \mathrm{d}\boldsymbol{l}$ can be written as the product of B and some distance. Ampere's law and Gauss's law are both of considerable theoretical importance, and both hold whether or not there is symmetry, but if there is no symmetry, neither is useful in calculating electric or magnetic fields.

16.2.2 Magnetic Field due to Varying Electric Field(与变化的电场相联系的磁场)

Ampere's law relates the line integral of the magnetic field around some closed curve L to the current that passes through any area bounded by that curve as Eq. (16-1).

Maxwell recognized a flaw in Ampere's law. Fig. 16-1 shows two different surfaces bounded by the same curve L, which encircles a wire carrying current into a capacitor plate. The current through surface S_1 is I, but there is no current through surface S_2 because the charge stops on the capacitor plate. There is thus ambiguity in the phrase "the current through any surface bounded by the curve." Such a problem always arises when the current is discontinuous.

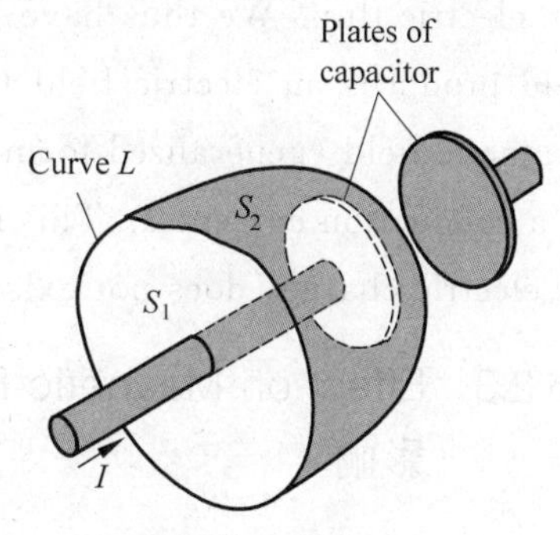

Fig. 16-1

Maxwell showed that the law can be generalized to include all situations if the current I in the equation is replaced by the sum of the conduction current I and another term I_d, called **Maxwell's displacement current**, defined as

$$I_d = \varepsilon_0 \frac{d\phi_e}{dt} \tag{16-2}$$

where ϕ_e is the flux of the electric field through the same surface bounded by the curve L. The generalized form of Ampere's law is then.

$$\oint_L \boldsymbol{B} \cdot d\boldsymbol{l} = \mu_0 (I + I_d) = \mu_0 I + \mu_0 \varepsilon_0 \frac{d\phi_e}{dt} \tag{16-3}$$

We can understand this generalization by considering again. Let us call the sum $I + I_d$ the generalized current. According to the argument just stated, the same generalized current must cross any area bounded by the curve L. Thus, there can be no net generalized current into or out of the closed volume. If there is a net true current I into the volume, there must be an equal net displacement current I_d out of the volume. In the volume in the Fig. 16-1, there is a net conduction current I into the volume that increases the charge within the volume

$$I = \frac{dQ}{dt} \tag{16-4}$$

The flux of the electric field out of the volume is related to the charge by Gauss's law

$$\phi_{net} = \oint_s E_n dS = \frac{1}{\varepsilon_0} Q_{inside} \tag{16-5}$$

The rate of increase of the charge is thus proportional to the rate of increase of the net flux out of the volume

$$\frac{dQ}{dt}=\varepsilon_0\frac{d\phi_{e,net}}{dt}=I_d \tag{16-6}$$

Thus, the net conduction current into the volume equals the net displacement current out of the volume. The generalized current is always continuous.

Maxwell's modification of Ampere's law shows that a changing electric flux produces a magnetic field whose line integral around a curve is proportional to the rate of change of the electric flux. We thus have the interesting reciprocal result that a changing magnetic field produces an electric field (Faraday's law) and a changing electric field produces a magnetic field (generalized form of Ampere's law). Note that there is no magnetic analog of a conduction current I. This is because the magnetic monopole, the magnetic analog of an electric charge, does not exist.

16.2.3 Effect on Magnetic Field Caused by Magnetic Media(磁介质对磁场的影响)

Materials fall into three categories-paramagnetic, diamagnetic, and ferromagnetic, according to the behavior of their magnetic moments in an external magnetic field. **Paramagnetism** arises from the partial alignment of the electron spins (in metals) or of atomic or molecular magnetic moments by an applied magnetic field in the direction of the field. In paramagnetic materials, the magnetic dipoles do not interact strongly with each other and are normally randomly oriented. In the presence of an external magnetic field, the dipoles are partially aligned in the direction of the field, thereby increasing the field. However, in external magnetic fields of ordinary strength at ordinary temperatures, only a very small fraction of the molecules are aligned because thermal motion tends to randomize their orientation. The increase in the total magnetic field is therefore very small. **Ferromagnetism** is much more complicated. Because of a strong interaction between neighboring magnetic dipoles, a high degree of alignment occurs even in weak external magnetic fields, causing a very large increase in the total field. Even when there is no external magnetic field, a ferromagnetic material may have its magnetic dipoles aligned, as in permanent magnets. **Diamagnetism** arises from the orbital magnetic dipole moments induced by an applied magnetic field. These magnetic moments are opposite the direction of the applied magnetic field so they decrease the total magnetic field B. This effect actually occurs in all materials, but because the induced magnetic moments are very small compared to the permanent magnetic moments, diamagnetism is masked by paramagnetic or ferromagnetic effects. Diamagnetism is thus observed only in materials that have no permanent magnetic moments.

Glossary

highly symmetric	高度对称的	tangential component	切向分量
closed curve	闭合曲线	degree of symmetry	对称度
penetrate	穿过	bounded by...	以……为边界
finite point	有限点	line integral	线积分
encircle	环绕	discontinuous	不连续的
conduction current	传导电流	displacement current	位移电流
generalized current	广义电流	magnetic monopole	磁单极
modification	修正	magnetic medium	磁介质
diamagnetic	抗磁的	paramagnetic	顺磁的
ferromagnetic	铁磁的	magnetic moment	磁矩
paramagnetism	顺磁性	partial alignment	部分取向
electron spin	电子自旋	magnetic dipole	磁偶极子
randomly oriented	任意取向	ferromagnetism	铁磁性
diamagnetism	抗磁性	mask	*v*. 掩盖

generalized Ampere's Law　广义安培环路定理
orbital magnetic dipole moment　轨道磁偶极矩
induced magnetic moment　感生磁矩
permanent magnetic moment　固有磁矩

16.3　专业英语常用表达法-16　导致　产生　概括　结论

1. 可以表示"导致、产生、引起、造成"等含义的动词

bechance　发生、产生
cause　引起、使发生、导致
create　引起、造成、是……的原因
elicit　导致、诱出
engender　产生、引起、使出现、发生
entail　引起、使成为必要
excite　唤起、引起、诱发
evoke　唤起、引起
generate　使形成、造成、引起、导致
give　产生、引起、带来
incur　招致、引起、带来
induce　引起、招致、感应
initiate　使出现、使存在
introduce　引导、引入、使开始
invite　招致、引起
lead　促使、致使、引导
originate　发起、引起、导致
produce　产生、引起、导致
provoke　激起、引起、诱发
raise　唤起、激起、引起、招致

短语有：

be at the bottom of　是……的根本原因
be creative of　产生
bring about　带来、引起、导致、产生
bring on　引起、导致、带来
call forth　唤起、使出现
come about　发生、产生
come out of　由……产生、从……出来
conduce to　引起、导致、有助于
draw forth　引起、引出、唤起
give birth to　引起、产生、造成

give occasion to 引起、使发生
give rise to 引起、导致、使发生
issue in 导致、造成
lead to 导致、引起、发生
lead up to 为……的先导、导致
open a door to 导致、使成为可能
open the door for 使成为可能、导致
result in 导致、造成、引起
result from 起因于、由……造成
set off 引起、激起
set up 引起、导致
spark off 导致、引发、诱发
touch of 触发、引起

The increase in space between the molecules of a solid *causes* the expansion. 固体分子之间间隔的扩大导致了膨胀。

Rapid expansion of a gas *leads to* the decrease of the internal energy and the accompanying temperature reduction. 气体的迅速膨胀导致内能减少，温度下降。

The Lenz's law can be used to give the direction of the induced electromotive force (emf) and induced current *resulting from* electromagnetic induction. 楞次定律可以说明感生电动势和感生电流的方向，而二者均是由电磁感应所引起的。

The most common method to "pump" a gas laser, that is, to invert the population in the states that *give rise to* light amplification, is to use a gas discharge. "泵浦"气体激光(也就是对那些能产生光放大的状态进行粒子数反转)的最常见方法就是运用气体放电。

The results show that the ECR plasma source may *generate* stable, high density plasma with low electron temperature. 实验证明，ECR 等离子体源能够稳定地产生电子温度较低的高密度等离子体。

Shock waves are *produced* when a body moving at hypersonic speeds compresses the air in front of it. 以高超音速运动的物体压缩它前面的空气时，会产生冲击波。

When reading a thermometer, it is essential to hold the instrument in an upright vertical position otherwise the reading error may be *introduced*. 对温度计读数时，一定要使温度计保持直立的位置，否则会引起读数误差。

2. 可用于表达"概括"含义的词语

above all 首先、最重要的是
all in all 总的来说
all things considered 总的来看
altogether 总之、总的看来
anyway 无论如何、总之
basically 总而言之
briefly＝in brief 总之、简言之
by and large 总的来说、一般地说
first and last 整体来看、总的看来
generally speaking 一般来说
generally 一般地、概括地
given all that 综上所述
in a few words 总之、简单地说
in a nut shell 概括地说、简单来说
in a word 总而言之
in any case 无论如何、总之
in broad outline 概括地说
in fine 简言之，总之、最后
in general 总之、一般来说
in short 总之、简言之
in summary 总的来说、概括起来
in the lump 就整体而言
on the whole 总的来说、大体上
overall 总的来说、总体上
the long and the short of 总之、整体上……
to be brief 简言之

to have a run-through of　归纳起来……
to make a long story short　长话短说
to put it briefly　简言之
to sum up　总而言之、总之
to summarize　简言之、总之
to take/taking it all in all　总的来说

3. 可用于引出"结论"的词语

accordingly　因此、从而、相应地、于是
as/so far as I know　据我所知
as a result　作为结果、因此、结果是
as has been mentioned　正如所提到的
as is known (to all)　众所周知
as we all know　众所周知
as we have seen　正如我们所见
at last　最后
clearly　很明显、显然
consequently　所以、因此
final consideration　最后要考虑的
finally　最后
given all that　综上所述
hence　所以、因此
in closing　最后、在结束前
in conclusion　最后、总之
in consequence　因此、结果
in the end　最后
in the final analysis　归根结底
lastly　最后
so　所以、因此
taking all into consideration　全面考虑
the point is　重点是
therefore　所以、因此
thus　因此、于是
to close by　最后、以……来结束
to conclude　最后、总而言之
to conclude with 以……来结束
to draw a conclusion　得出结论

Unit Five

LESSON 17

17.1 物理学专业英语中的长句分析(二)

在前几课中,介绍了关于复合句、从句、非谓语动词等知识,对这些知识的良好掌握是进行长句结构分析的基础。现在我们再结合实例来介绍一下长句结构分析的方法。

1. 物理学专业英语中长句的特点

由于物理专业内容的复杂性、综合性、多样性等因素,专业英语中长句用得较多,看起来较为复杂。特别是复合句式,在科技标准、专利规范、原理说明中尤为多见。有时,一个句子长达数行、数十行,包含几十个乃至上百个单词,或包含若干个从句和非谓语动词短语。它们互相制约、互相联系、互相依附,以至于形成从句中有短语、短语中带从句的复杂语言现象。

值得强调的是,在专业英语中多用一些过渡连词或短语来实现句子、段落和上下文的有效衔接,使陈述内容连贯一致,成为整体,或形成规范合理、紧凑连贯的语篇,最终达到准确表达内容的目的。

下面是专业英语中长句的几个特点:①修饰语多,特别是后置定语很长;②并列的成分多;③语言结构层次多;④指代关系复杂;⑤习惯搭配和成语经常出现。

2. 长句的分析方法

在分析长句时,不要因为句子太长而产生畏惧心理。采用我们前面介绍过的方法:抓住主干,添枝加叶(见 Lesson 9,物理学专业英语中的长句分析(一)),一般就会对长句子做出正确分析。无论是多么复杂的句子,它都是由一些基本的成分组成的。因此,首先我们要分析句子的基本成分,如主语、谓语、宾语、状语等,抓住主干部分,然后逐层分析各词意思以及相互间的逻辑、语法关系,再进行翻译。其次,要弄清英语原文的句法结构,找出整个句子的中心内容及其各层意思,然后分析几层意思之间的相互逻辑关系。最后,再按照汉语的特点和表达方式,正确地理解原文的意思。

根据专业英语长句的特点,在分析长句时,采用抓住主干,添枝加叶的方法,一般要进行下面几个步骤:

(1) 抓住主干:即通过形态识别,找出全句的主、谓、宾、表等主干成分,了解其"骨架含义",从整体上把握句子的结构。

(2) 查找分支:找出句中所有的谓语结构、非谓语动词、介词短语和从句的引导词。

(3) 辨别分支：分析从句和短语的功能。对于结构复杂、层次纷纭的主从复合句要抓住引导各种从句的连词，弄清各种从句的性质，如：是否为主语从句、宾语从句、表语从句等，若是状语从句，它是表示时间、原因、结果，还是表示条件等。

(4) 理清分支：分析词、短语和从句之间的相互关系，理解这些成分之间的逻辑关系和修饰关系，如：定语从句所修饰的先行词是哪一个等。

(5) 找寻叶子：注意插入语等其他成分，分析句子中是否有固定词组或固定搭配。

这种方法从分析英语句子结构入手，化繁为简，化长为短，化难为易，进而理解英语长句、难句的意思，从而使在英语长句、难句在理解、翻译过程中所遇到的困难得以解决。

下面就应用此方法对一些专业长句子进行分析：

例 1：Copernicus by announcing his heliocentric world system, may be said to have spoken the prologue preceding the rise of the curtain on modern physics revealing among the first actors, Galileo and Kepler. 哥白尼，由于宣布了他的太阳中心世界系统，可以认为已经为现代物理学的升幕作了个开场白，升幕后，最早出现的一批演员中有伽利略和开普勒。

句子主干是：Copernicus may be said to have spoken the prologue. 可以认为哥白尼已经作了个开场白。

逐个加上修饰语：

(1) by announcing his heliocentric world system(哥白尼)由于宣布了他的太阳中心世界系统(介词短语作方式状语)。

(2) preceding the rise of the curtain 在升幕之前(分词短语作状语)。

(3) on modern physics 为谁升幕？“为现代物理学”(介词短语作定语)。

(4) revealing among the first actors, Galileo and Kepler 升幕后，就“显露出了最早的一批演员(指现代物理学家)，其中有伽利略和开普勒”(分词短语作状语)。

例 2：The general theory of relativity was proposed by Einstein in 1916 after nine years of grappling with the problem of formulating a theory of gravitation in agreement with the space-time symmetry of special relativity and with the experimental observation known since Galileo that all bodies, regardless of their mass, follow the same trajectory in a gravitational field.

为了提出一种引力理论能符合狭义相对论的时空对称要求，同时也与自从伽利略以来就知道的实验观测一致(即，一切物体在引力场中的运动都遵循相同的轨迹，与质量无关)，爱因斯坦通过9年时间对此问题进行研究，于1916年提出了广义相对论。

句子主干是：The general theory of relativity was proposed by Einstein. 广义相对论是爱因斯坦提出来的。

(1) after nine years of grappling“在九年搏斗之后提出的……”(从 after 开始到句尾一个长长的介词短语作状语，修饰 proposed)。

(2) with the problem of formulating a theory of gravitation 搏斗的对象是“形成一个引力理论的问题”。

(3) in agreement with the space-time symmetry of special relativity 这个理论“符合狭义相对论的时空对称要求”。

(4) and with the experimental observation known since Galileo“还要符合自从伽利略以来就知道的实验观测”(划线的两个 with 是并列的成分，in agreement with 是固定搭配，在此引导定语修饰 a theory of gravitation)。

(5) that all bodies, regardless of their mass, follow the same trajectory in a gravitational field. 这个实验观察就是“一切物体在引力场中的运动都遵循相同的轨迹，与质量无关”(that 引导的从句作的 the experimental observation 的同位语)。

例 3：Stated in words, Ohm's law says that the steady current through any portion of an electric circuit equals the potential difference across that portion of the circuit divided by the resistance of that portion of the circuit. 欧姆定律用文字表述为：电路中任一部分的稳恒电流强度等于这部分的电势差除以这部分的电阻。

句子主干是：Ohm's law says that...欧姆定律说的是……(从 that 开始到句尾是 that 引导的从句作宾语)。

宾语从句的主干是：the steady current equals the potential difference(主谓宾)。

(1) Stated in words 用文字表述(分词短语作状语)。

(2) the steady current through any portion of an electric circuit 电路中任一部分的稳恒电流强度(through 引导介词词组作定语修饰从句主语 steady current)。

(3) the potential difference across that portion of the circuit 电路中这部分的电势差(across 引导介词词组作定语修饰从句宾语 potential difference)。

(4) divided by the resistance of that portion of the circuit 除以这部分的电阻(分词短语作定语修饰从句宾语 potential difference)。

例 4：The great challenge in the study of these novel processes is not so much how to solve a particular pattern by means of a particular method, as how to find new methods to furnish solutions to whole class of problems. 在研究这些新过程中的大难关主要不在于如何用某一个方法去解决某一个问题，而是在于如何去找到能使整个一类问题得到解决的新方法。

本句虽然很长，但却是只有一套主系表结构的简单句。句子主干是：The great challenge is not so much how to solve a particular pattern as how to find new methods. 大的难题主要不是如何解决某一特殊典型而是如何去找到新的方法。

可以更简化为 The challenge is not so much A as B 要克服的困难主要不是 A 而是 B。not so much A as B 是一个固定词组，表示“不是 A 而是 B”。

(1) in the study of these novel processes 在研究这些新过程中(介词词组作定语修饰语)。

(2) how to solve a particular pattern by means of a particular method 如何用某一个方法去解决某一个问题(不定式短语作表语，by means of a particular method 作状语修饰 solve)。

(3) how to find new methods 如何找到新的方法(不定式短语作表语)。

(4) to furnish solutions to whole class of problems 能使整个一类问题得到解决(不定式短语作目的状语)。

例 5：In order to present a sufficiently complete picture we feel it is necessary to connect theory, on the one hand, with a variety of applications from radar, sonar, speech, music, and medical signal processing, and on the other hand, with the digital component technology that is the main driving force for progress in this field as well as the overall field of computer design. 为了呈现一幅充实的全景图,我们认为有必要把理论,一方面,与雷达、声呐、话音、音乐及医学信号处理的多种应用相结合,另一方面,与数字分量技术相结合,这种技术是本领域及计算机设计全部领域的主要推动力。

句子主干是:We feel it is necessary. 我们觉得那有必要。

(1) In order to present a sufficiently complete picture 为了呈现一幅充实的全景图(目的状语)。

(2) 从 it is necessary 到句尾的从句是整个句子的宾语,前面省略的引导词 that。

(3) 从 to connect theory 到句尾是包含从句的 to 动词不定式短语,是宾语从句的实际主语,而前面的 it 是形式主语。

(4) 两个较长的以 with 引入的状语短语是平行的,由 on the one hand, on the other hand 引导,connect...with...是固定搭配。

(5) that is the main driving force for progress in this field as well as the overall field of computer design 是本领域及其计算机设计全部领域的主要推动力(该定语从句,修饰 technology,两个 field 是平行的,由 as well as 连接)。

(6) for progress in this field as well as the overall field of computer design(for 引导介词词组作定语修饰 force)。

例 6: While numerous basic advances were made during this period its most important characteristic is the development and the firm establishment of the experimental method as a means of scientific inquiry, as is well illustrated by Galileo's famous experiment of dropping two bodies of unequal weight from the leaning tower of Pisa, thereby proving by experiment the incorrectness of the assertion of Aristotle that the heavier body would fall more rapidly, an assertion which had been believed implicitly for nearly two thousand years. 这一时期取得非常多的基础进展,其最重要的特征是实验方法作为一种科学探索的方法得到发展和奠定,正如伽利略的著名比萨斜塔落体实验所展示的那样,通过实验证明了亚里士多德的、已经被盲目信奉了近两千年的、关于较重物体落得更快的论断的错误性。

句子主干是:its most important characteristic is the development and the firm establishment of the experimental method 其最重要的特征是实验方法得到发展和奠定。

(1) While numerous basic advances were made during this period 这一时期取得非常多的基础进展(状语从句)。

(2) as a means of scientific inquiry 作为一种科学探索的方法(定语,修饰 experimental method)。

(3) as is well illustrated by Galileo's famous experiment of dropping two bodies of unequal weight from the leaning tower of Pisa 正如伽利略的著名比萨斜塔落体实验所展示的那样(定语从句,修饰主句)。

(4) 从 thereby 到句尾,分词短语作状语。

(5) that the heavier body would fall more rapidly 较重物体落得更快(assertion of

Aristotle 的同位语从句）。

（6）which had been believed implicitly for nearly two thousand years（定语从句，修饰 an assertion）。

例 7：The physicist rightly dreads precise argument, since an argument that is convincing only if its precise loses all its force if the assumptions on which it is based are slightly changed, whereas an argument that is convincing though imprecise may well be stable under small perturbations of its underlying assumptions. 物理学家恐惧于那些精确无误的论断不无道理，因为某种只有在精确无误的条件下才令人置信的论断，一旦它赖以建立其上的假设稍有变化，便会失去它一部分的作用；而与此相反，一个尽管并不精确无误但却令人置信的论断，在其基本假设稍受干扰的情况下，仍然有可能是站得住脚的。

句子主干是：The physicist dreads precise argument. 物理学家恐惧于那些精确无误的论断。

此句中的主句和从句加在一起共有八个：

（1）由 since 引导的是原因状语从句。

（2）由 whereas 连接了两个长句子。

（3）whereas 后面的句子中的主语 an argument 之后又跟了一个定语从句 that is convincing though imprecise，修饰 argument。

（4）since 从句中主语 an argument 由 that 引导的定语从句修饰。

（5）定语从句中的 only if it is precise 是条件状语从句。

（6）since 从句中的 if the assumptions on which it is based are slightly changed 是条件状语从句。

（7）定语从句 on which it is based 修饰主语 assumptions。

例 8：Thus, for instance, it may come as a shock to mathematicians to learn that the Schrödinger equation for the hydrogen atom is not a literally correct description of this atom, but only an approximation to a somewhat more correct equation taking account of spin, magnetic dipole, and relativistic effects; and that this corrected equation is itself only an imperfect approximation to an infinite set of quantum field-theoretical equations. 因此，举例来说，对数学家而言，了解到下述情形可能会令其惊愕不已，即薛定谔的氢原子方程式并非是对该原子做出的一种决然正确的描述，而仅仅是个近似值，趋近于一个在某种程度上更为正确的将自旋、磁性偶极子以及相对论效应考虑在内的方程式；而这个得以纠正的方程式就其本身而言也只是一个不完美的近似值，是趋近于无穷无尽的一整套量子场论方程式。

句子主干是：it may come as a shock to mathematicians 下述情形可能会令数学家惊愕不已。to learn that...; and that...是并列成分作句子的主语，两个 that 引导的都是 learn 的宾语从句。

例 9：Whether the experiment of dropping bodies of different weight from the leaning tower of Pisa was performed before a multitude, as some accounts have it; or whether the bodies were of lead or stone and what their weight were we do not know. 我们不知道比萨斜塔落体实验是否像某些记载那样在很多人目睹下进行，也不知道落体是铅制的还是石头

的、重量是多少。

句子主干是：we do not know whether...，or whether...两个 whether 引导的都是宾语从句。

掌握好专业英语中关于长句结构分析的方法在专业英语的学习中是非常重要的，它直接影响到读者对专业文献的阅读、理解和翻译。具有良好长句结构分析水平的读者，在专业文献的阅读、理解和相关翻译工作甚至专业写作中也会得心应手、游刃有余。

在接下来关于专业文献的阅读和翻译的部分中，关于长句结构分析的方法还会经常出现，读者也应该把提高长句结构分析水平一直贯穿在自己的专业英语学习过程当中。

17.2 专业英语阅读

17.2.1 Faraday's Law of Electromagnetic Induction（法拉第电磁感应定律）

In the early 1830s，Michael Faraday in England and Joseph Henry in America independently discovered that a changing magnetic field induces a current in a wire（Fig. 17-1）. The two experiments shown in Fig. 17-1 have one thing in common. In both cases，an emf is induced in a circuit when the magnetic flux through the circuit changes with time. The emfs and currents caused by changing magnetic fields are called **induced emfs** and **induced currents.** The process itself is referred to as **magnetic induction.**

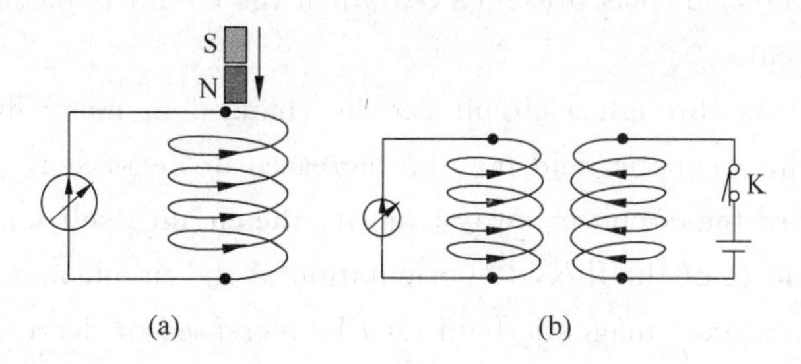

Fig. 17-1

When you pull the plug of an electric cord from its socket，you sometimes observe a small spark. Before the cord is disconnected，it carries a current，which produces a magnetic field encircling the current. When the cord is disconnected，the current abruptly ceases and the magnetic field around it collapses. The changing magnetic field produces an emf that tries to maintain the original current，resulting in a spark across the plug. Once the magnetic field reaches zero it is no longer changing，and the emf is zero.

Changing magnetic fields can result from changing currents or from moving magnets. The ends of a coil are attached to a galvanometer and a strong magnet is moved toward or away from the coil. The momentary deflection shown by the galvanometer during the motion indicates that there is an induced electric current in the coil-galvanometer circuit. A current is also induced if the coil is moved toward or away from the magnet，or if the coil is rotated in a fixed magnetic field. A coil rotating in a magnetic field is the basic element of a generator, which converts mechanical energy into electrical energy.

All of the various methods of magnetic induction can be summarized by a single relation known as **Faraday's law of induction**, which relates the induced emf in a circuit to the change in magnetic flux through the circuit.

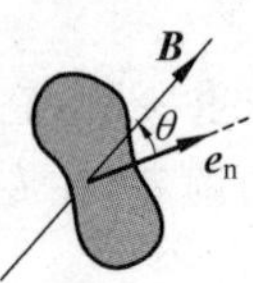

Fig. 17-2

The flux of a magnetic field through a surface is defined similarly to the flux of an electric field. Let dS be an element of area on the surface and $\boldsymbol{e}_n$ be the unit vector perpendicular to the element (Fig. 17-2). The magnetic flux ϕ_m is then defined to be

$$\phi_m = \int_S \boldsymbol{B} \cdot \boldsymbol{e}_n \mathrm{d}S = \int_S B_n \mathrm{d}S \tag{17-1}$$

The unit of magnetic flux is that of magnetic field times area, tesla-meter squared, which is called a **weber** (Wb)

$$1 \text{ Wb} = 1 \text{ T} \cdot \text{m}^2$$

Since B is proportional to the number of field lines per unit area, the magnetic flux is proportional to the number of lines through the area.

Experiments by Faraday, Henry, and others showed that if the magnetic flux through an area bounded by a circuit is changed by any means, an emf equal in magnitude to the rate of change of the flux is induced in the circuit. We usually detect the emf by observing a current in the circuit, but it is present even when the circuit is incomplete (not closed) and there is no current.

The magnetic flux through a circuit can be changed in many different ways. The current producing the magnetic field may be increased or decreased, permanent magnets may be moved toward the circuit or away from it, the circuit itself may be moved toward or away from the source of the flux, the orientation of the circuit may be changed, or the area of the circuit in a fixed magnetic field may be increased or decreased. In every case, an emf is induced in the circuit that is equal in magnitude to the rate of change of the magnetic flux.

Fig. 17-3 shows a single loop of wire in a magnetic field. If the flux through the loop is changing, an emf is induced in the loop.

$$\varepsilon = -\frac{\mathrm{d}\phi_m}{\mathrm{d}t} \tag{17-2}$$

This result is known as **Faraday's law of induction** which states *the emf induced in a circuit is directly proportional to the time rate of change of magnetic flux through the circuit*.

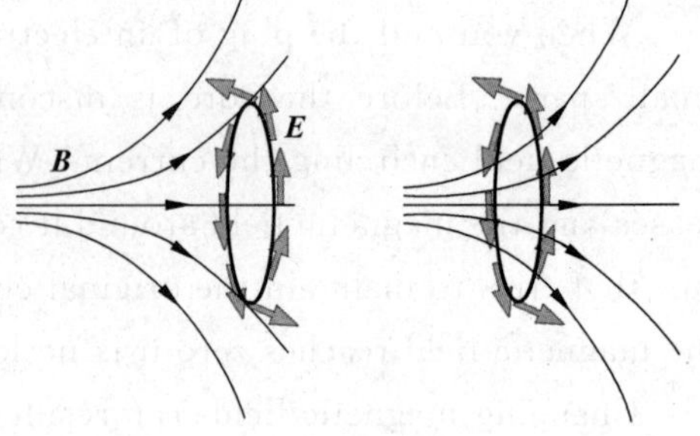

Fig. 17-3

The negative sign in Faraday's law has to do with the direction of the induced emf, which we will discuss shortly.

17.2.2 Lenz's Law(楞次定律)

The negative sign in Faraday's law has to do with the direction of the induced emf, which can be found from a general physical principle known as **Lenz's law**:

The polarity of the induced emf is such that it tends to produce a current that will create a magnetic flux to oppose the change in magnetic flux through the loop.

Note that we didn't specify just what kind of change causes the induced emf and current. We purposefully left the statement vague to cover a variety of conditions, which we will now illustrate.

Fig. 17-4 shows a bar magnet moving toward a loop that has a resistance R. Since $\boldsymbol{B}$ from the bar magnet is to the right, out of the north pole of the magnet, the movement of the magnet toward the loop tends to increase the flux through the loop to the right. (The magnetic field at the loop is stronger when the magnet is closer.) The induced current in the loop produces a magnetic field of its own. This induced current is in the direction shown, so the magnetic flux it produces is opposite that of the magnet. The induced magnetic field tends to decrease the flux through the loop. If the magnet were moved away from the loop, which would decrease the flux through the loop due to the magnet, the induced current would be in the opposite direction from that in Fig. 17-4. In that case, the current would produce a magnetic field to the right, which would tend to increase the flux through the loop. As we might expect, moving the loop toward or away from the magnet has the same effect as moving the magnet. Only the relative motion is important.

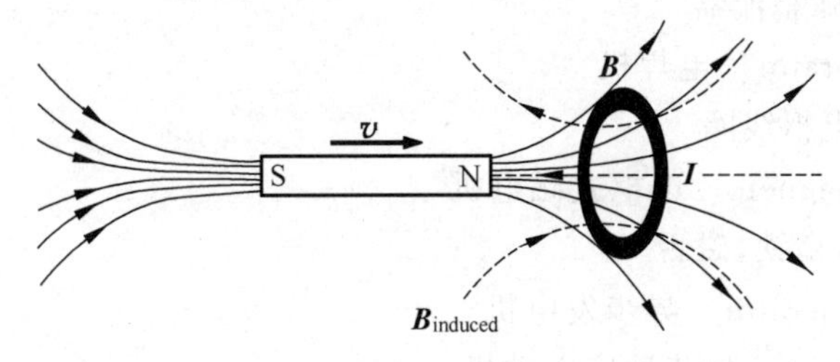

Fig. 17-4

Lenz's law is required by the law of conservation of energy. If the current in the loop in Fig. 17-4 were opposite the direction shown, the induced magnetic moment of the loop would attract the magnet when it is moving toward the loop and cause it to accelerate toward the loop. If we begin with the magnet a great distance from the loop and give it a very slight push toward the loop, the force due to the induced current would be toward the loop, which would increase the velocity of the magnet. As the speed of the magnet increases, the rate of change of the flux would increase, thereby increasing the induced current. This would further increase the force on the magnet. Hence, the kinetic energy of the magnet and the rate at which Joule heat is produced in the loop (I^2R) would both increase with no source of energy. This would violate the law of conservation of energy.

Glossary

electromagnetic induction	电磁感应	emf (electromotive force)	电动势
magnetic flux	磁通量	induced emf	感生电动势
induced current	感应电流	magnetic induction	磁感应
plug	插头	socket	插座
encircle	环绕	cord	软线，电源线
galvanometer	检流计	deflection	偏转
coil	线圈	generator	发电机
Lenz's Law	楞次定律	polarity	极性

17.3 专业英语常用表达法-17 电磁学常用实验仪器 2

air cooled motor 气冷马达/电动机
alternating accumulator 交流蓄电池
alternating current generator 交流发电机
alternating current meter 交流电流计
alternating current motor 交流电动机
alternating current rectifier 交流整流器
alternative switch 转换开关
alternator 交流发电机
arc lamp 弧光灯
arc rectifier 电弧整流器
arc welding generator 电焊机
armature coil 电枢线圈
armature type magneto 电枢式磁电机
armature 电枢，衔铁，磁舌
asynchronous generator 异步发电机
asynchronous motor 异步感应电动机
beam power amplifier 电子束功率放大器
cadmium nickel accumulator 镉镍蓄电池
cathode ray accelerator 阴极射线加速器
commutator ammeter 整流式安培计
conductivity meter 电导仪
copper-zinc accumulator 铜锌蓄电池
crest ammeter 巅值/峰值安培计
current amplifier 电流放大器
DC ammeter 直流电流计
DC amplifier 直流放大器
differential ammeter 差动安培计

diode　二极管
dust accumulator　集尘器
Edison accumulator　爱迪生蓄电池
electric arc lamp　电弧灯
electrodynamic ammeter　电动安培计
electromagnetic ammeter　电磁安培计
electron accelerator　电子加速器
electron microscope　电子显微镜
electrostatic actuator　静电激发器
feedback amplifier　回馈/反馈放大器
feedback relay amplifier　回馈继电放大器
helical accelerator　螺旋加速器
high-frequency ammeter　高频安培计
hydraulic accumulator　水压蓄电池
induction alternator　感应交流发电机
iron-nickel accumulator　铁镍蓄电池
optical ammeter　光学电流计
thermal expansion ammeter　热胀安培计
thermocouple ammeter　热电偶安培计
transistor　晶体管

LESSON 18

18.1　物理学专业英语文章的阅读与信息获得

当代科学技术发展迅速，物理学的发展也是日新月异，每天都会有大量的专业英语文章、科技新闻及其他专业资料发表在各类报刊和网络上，新的物理专业英语书籍也不断涌现，要想把这些英语专业文献资料都及时翻译成中文是不可能的。如果想要及时掌握最新的专业发展信息，把握住物理学及其相关领域的最新发展脉搏，最好的解决办法是直接阅读英语文献资料，获得第一手信息。

专业英语阅读活动的主要目的是从读物中获取所需要的信息，在阅读中培养从所读材料中获取信息的能力和技巧。阅读能力是一种综合性的能力，在阅读中既要了解语言点，更要分析读物的内容和作者的立场观点；既要求理解具体的事实，也要求理解抽象的概念；既要求理解文章的字面含义，也要求理解内容的深层次含义；既要求理解文中描述的具体细节，也要求理解语段或语篇之间的逻辑关系；既要分析字面意义，更要分析字里行间所隐含的意思。有时还必须根据整体意思，联系各个部分的内容，进行意义上的逻辑推理，才能理解真正的含义，进而达到对专业内容的理解。此外，专业资料中还常常包含若干辅助阅读部分（如插图、图形、图表等），在这些部分中仍然含有很多标记性词汇、符号、算式、数值等，都有助于对阅读内容的理解。只要有较好的英语基础，再掌握英语阅读的技巧，做到这一点并非难事。

18.1.1 浏览

浏览的主要目的是了解阅读内容的概貌。

在正式阅读前进行浏览,可以大概了解所需要的信息,决定该材料是否有读的必要,有时也能够激发阅读兴趣和欲望。

浏览的内容主要有:书名/标题和作者,因为书名或标题本身就点出了读物的中心和主题,作者信息也可以作为是否阅读的参考。书名或标题下的副标题或用特型字印刷的引子比标题更详细地提供了有关主要内容的信息,也是浏览的内容。浏览内容提要、摘要、关键词、目录、总结可以了解读物的主要内容。浏览前言、结束语或后记可以了解背景知识、作者意图、使用方法等。读文内小标题可以了解整体结构。浏览文中插图、表格、图片可以了解资料与自己所研究内容的相关程度。

18.1.2 略读

略读找出中心思想。中心思想是作者在文章中要表达的主要内容,是贯穿全文的核心。作者在文章中会努力通过各种支持材料和论述来阐明中心议题。因此,在进行专业资料的阅读时,首先要明确阅读时的主要目标,那就是寻找中心思想,明确 What is the main idea (subject) of this article/passage? What does this article/passage mainly (primarily) concerned? 在阅读中应自始至终把注意力集中在猎取所需信息上,努力抓住读物的中心内容、把握主要思想,这对于全文内容理解具有重要意义。抓住中心思想的方法是:阅读开头部分以了解主题、背景、意图;找出主要段落的主题句以了解读物的主要内容和整体结构;再读结尾部分以了解结论、用途、意义。"抓住中心"的指导思想要贯穿阅读全过程,始终保持清醒的头脑,不至于在攻克难词、难句、难概念的拼搏中迷失大方向,同时也能加深对读物的印象。

在寻找中心思想的过程中,要注意以下几个方面:

(1) 重视标题

物理学专业文献的标题和文学作品、故事、新闻的标题有显著的不同。专业文献的标题总是以最恰当、最简明词语的逻辑组合反映出文章最重要的内容。标题具有简明、准确、醒目的特点,也就是实话实说,用最少的词语反映出最为确切的论文内容。读者在阅读和检索文献时看懂标题也就知道了文献的最主要的内容。

(2) 细读摘要

国际上正式出版并对外发行的英文学术期刊都要求刊中的调查报告、实验方法或技术性文章、综述性文章、研究论文等在正文中附上英文摘要。摘要一般包括如下内容:陈述研究目的和研究内容及需要解决的问题;简要介绍研究所采用的方法和基本步骤;简要描述主要结果、主要结论及论文的价值。

(3) 抓住反映中心思想的句子

采用略读的方法浏览全文,理解文章主要内容。阅读时要注意抓住反映中心思想的句子,它们在文章中的位置不尽相同。但大多数情况下,文章的开头、结尾及段落的段首句和段尾句特别重要,因为其中往往包含文章的中心议题。

（4）关注结果、结论与讨论部分

在物理学专业文献中，结果部分是对研究过程中所获取的数据的分析和对所观察到的现象的总结，是一篇专业文献的核心。结果部分会引发讨论，导出推理，最后给出结论。讨论内容一般从实验和观察的结果出发，从理论上对其分析、比较、阐述、推论和预测；推论中作者会提出自己的新见解，会着重讨论新发现、新发明和新的启示以及从中得出的结论。所以对这部分的认真阅读也是获得中心思想的捷径。

18.1.3 研读

为了领会和掌握所读材料的精神实质和内容细节以及作者的立场、观点、意图，就必须使用字斟句酌、边阅读边理解边深入思考的阅读方法，这种方法就是研读。研读时要求理解准确、透彻，不但要掌握文献的字面意义，而且还要深入理解作者的意图及文献的价值；不但要理解文献里的信息，而且还要通过分析、比较、联想等思维方法，用已有的知识、经验对这些信息进行理解和分析，把它们置入我们的知识系统，真正消化吸收所读的材料，丰富自己的知识，提高专业水平。研读不等于慢读，要注意培养按意群阅读的习惯，集中注意领会文献的信息，立足于读一遍就能理解全部内容，努力提高阅读速度。

在研读过程中，要尽量做到以下几点。

（1）善于抓住段落的主题句和主要细节

为了对一个段落进行快速理解，从主题句出发，展开阅读是广泛运用并行之有效的手段。绝大多数段落都有一个主题句，而大多数主题句又安排在每一段的开头(有时也把主题句放在段尾或段内其他位置)。只要抓住了主题句就抓住了该段的段意。例如：

The law of charge conservation is one of the important characteristics of electric charge. In certain interactions among elementary particles, charged particles such as electrons are created or annihilated. However, in all these processes, equal amounts of positive and negative charge are produced or destroyed, so the net charge of the universe is unchanged. When two initially neutral objects are charged by being rubbed together, charge is not created in the process. The objects become charged because electrons are transferred from one object to the other. One object gains some amount of negative charge from the electrons transferred to it while the other loses an equal amount of negative charge and hence is left with a positive charge. That is, charge is conserved. The law of conservation of charge is a fundamental law of nature.

第一句“电荷守恒定律是电荷的重要特点之一”为此段的主题句，读完第一句就知道这一段的段意。

在抓住主题句后，接下来的阅读就是要抓住主要细节。除主题句外，段落还必须有把主题句扩展成段落的细节，有时，这些细节比主题句更有价值。必须指出，并不是所有的细节都同等重要，有些细节是主要的，而有些则是说明细节的“细节”，称为次要细节。阅读时要注意区分主要细节和次要细节，注意掌握好重要细节。

为了说明“电荷守恒定律是电荷的重要特点”这个主题，作者列举了两个细节：In certain interactions... 和 When two initially neutral objects are charged by being rubbed together...。这些细节对电荷守恒定律的说明都是非常重要的，对理解主题也非常有帮助。

(2) 记住中心内容和主要细节

记住读过的主要内容对全文的理解大有好处,甚至对改善我们的知识结构、提高今后应用所学知识解决问题的能力都大有好处。方法主要包括:阅读时思想高度集中,努力加深第一印象,只围绕目的有重点地阅读和记忆,阅读过程中不时地停下来回想一下前面的内容,一边阅读一边思考、联想、推导、归纳、总结,把读物中的知识转化为大脑的知识,标记或摘录重点等。

(3) 区分对待事实、观点、推断和猜测

在阅读中,针对事实材料,要考虑材料的出处、来源数据的准确度和精确度,要衡量它们的价值和潜在用途。遇到某些观点时,要注意哪些是作者个人的观点,哪些是专家的证言,还要去分析判断这些观点正确与否,进而决定接受、部分接受、还是否定这些观点。在处理推论和猜测时,一方面要注意衡量这些推论所依据的是事实还是主观臆断,还要考虑所依据的事实是否充分;另一方面要注意衡量这些推论是否符合逻辑。总之,在阅读过程中要注意科学思维、进行正确的推论、做出合理的判断、得出合乎逻辑的结论,力求对内容全面准确地进行理解。

例如,下列的句子表达的就是事实,这些事实都可以得到证实。

In fact, the electric field of the source charge is present whether or not we introduce a test charge into the field. 事实上,不论检验电荷是否放入,场源电荷产生的电场都是存在的。

The case is the more the angle of incidence increases, the farther the refracted ray diverges from the normal. 实际情况是,入射角越大,反射线就偏离法线越远。

Ever since then, Einstein has never been back to the town where he was born. 从那以后,爱因斯坦再也没回过他出生的小镇。

表达事实的句子有时会包含如 in fact, actually, as a matter of fact, the case is, the truth is, the fact is that 等词句。

下列句子表达的是观点。

It sounds as if the argument is reasonable. 听起来好像论证得有道理。

Galileo's greatest glory was that in 1609 he was the first person to turn the newly invented telescope on the heavens to prove that the planets revolve around the sun rather than around the Earth. 伽利略最光辉的业绩在于他在 1609 年第一个把新发明的望远镜对准天空,以证实行星是围绕太阳旋转而不是围绕地球的。

I am afraid that the shock waves will be very strong in this case. 在这种情况下,恐怕冲击波会很强。

I think it best that you should explain the theorem in detail now. 我认为你最好现在把这个定理仔细解释一下。

I don't suppose that it is true. 我认为那不是真的。

表达观点的句子可能包含 as if, seem, sound, think 等表达观点的动词,还有诸如 great, too much, wonderful, boring, should, must 等一些用来表达通过判断所得出的感觉、感受的词汇。

下列句子表达的是推断,斜体部分就是推断的结果。

Since the weight depends on **g**, *it varies with geographic location*. This is because **g** decreases with increasing distance from the center of the earth. 因为重量与重力加速度 **g** 有关,所以重量会随地理位置不同而变化。原因在于距离地心越远,**g** 的数值会越小。

After the expansion, the gas occupies a greater volume than it did originally. *The gas molecules become more disordered* in that they are not as localized as they were originally. 膨胀后,气体占据比原来大的体积。气体分子变得更无序,因为分子已经不再局限在原来的空间。

下列句子表达的是猜测。

From what Maxwell stated, the displacement current was reasonable. 根据麦克斯韦所说,位移电流是合理的。

That the wave like this is nothing but a standing wave, I guess. 我猜测像这样的波就是驻波。

The experiment of dropping bodies of different weight from the leaning tower of Pisa was probably performed before a multitude, according to some accounts. 根据某些记载所述,比萨斜塔落体实验可能是在很多人目睹下进行的。

18.2 专业英语阅读

18.2.1 Maxwell's Equations(麦克斯韦方程组)

Maxwell's equations, first proposed by the great Scottish physicist James Clerk Maxwell, relate the electric and magnetic field vectors $\boldsymbol{E}$ and $\boldsymbol{B}$ to their sources, which are electric charge, currents, and changing fields. These equations summarize the experimental laws of electricity and magnetism—the laws of Coulomb, Gauss, Biot-Savart, Ampere, and Faraday. These experimental laws hold in general except for Ampere's law, which does not apply to discontinuous currents such as those that occur when charging or discharging a capacitor. Maxwell was able to generalize Ampere's law with the invention of the displacement current. He was then able to show that the generalized laws of electricity and magnetism imply the existence of electromagnetic waves.

Maxwell's equations play a role in classical electromagnetism analogous to that of Newton's laws in classical mechanics. What's more, Maxwell's equations are of great theoretical importance. For example, Maxwell showed that these equations can be combined to yield a wave equation for the electric and magnetic field vectors $\boldsymbol{E}$ and $\boldsymbol{B}$. Such electromagnetic waves are caused by accelerating charges, for example, the charges in an alternating current in an antenna. They were first produced in the laboratory by Heinrich Hertz in 1887. Maxwell showed that the speed of electromagnetic waves in free space should be

$$c = 1/\sqrt{\mu_0 \varepsilon_0} \tag{18-1}$$

where ε_0, the permittivity of free space, is the constant appearing in Coulomb's and Gauss's laws and μ_0, the permeability of free space, is the constant appearing in the Biot-Savart law and

Ampere's law. When the measured value of ε_0 and defined value of μ_0 are put into Eq. (18-1), the speed of electromagnetic waves is found to be about 3×10^8 m/s, the same as the measured speed of light. Maxwell noted this "coincidence" with great excitement and correctly surmised that light itself is an electromagnetic wave.

Maxwell's equations in medium are

$$\oint_S \boldsymbol{D}\cdot \mathrm{d}\boldsymbol{S} = q_{\text{free-inside}} \tag{18-2a}$$

$$\oint_S \boldsymbol{B}\cdot \mathrm{d}\boldsymbol{S} = 0 \tag{18-2b}$$

$$\oint_L \boldsymbol{E}\cdot \mathrm{d}\boldsymbol{l} = -\int_S \frac{\partial \boldsymbol{B}}{\partial t}\cdot \mathrm{d}\boldsymbol{S} \tag{18-2c}$$

$$\oint_L \boldsymbol{H}\cdot \mathrm{d}\boldsymbol{l} = \int_S \left(\boldsymbol{j} + \frac{\partial \boldsymbol{D}}{\partial t}\right)\cdot \mathrm{d}\boldsymbol{S} \tag{18-2d}$$

Eq. (18-2a) is Gauss's law. This law describes how electric field lines diverge from a positive charge and converge on a negative charge. Its experimental basis is Coulomb's law.

Eq. (18-2b), sometimes called Gauss's law for magnetism, states that the flux of the magnetic field vector $\boldsymbol{B}$ is zero through any closed surface. This equation describes the experimental observation that magnetic field lines do not diverge from any point in space or converge on any point; that is, it implies that isolated magnetic poles do not exist.

Eq. (18-2c) is Faraday's law; it describes how electric field lines encircle any area through which the magnetic flux is changing, and it relates the electric field vector $\boldsymbol{E}$ to the rate of change of the magnetic field vector $\boldsymbol{B}$.

Eq. (18-2d) is Ampere's law modified to include Maxwell's displacement current. This law describes how the magnetic field lines encircle an area through which a current is passing or through which the electric flux is changing.

18.2.2 Electromagnetic Waves(电磁波)

Fig. 18-1 shows the electric and magnetic field vectors of an electromagnetic wave. The electric and magnetic fields are perpendicular to each other and both are perpendicular to the direction of the propagation of the wave. Electromagnetic waves are thus **transverse waves.** The vectors $\boldsymbol{E}$ and $\boldsymbol{B}$ are in phase and their magnitudes are related by

$$E = cB \tag{18-3}$$

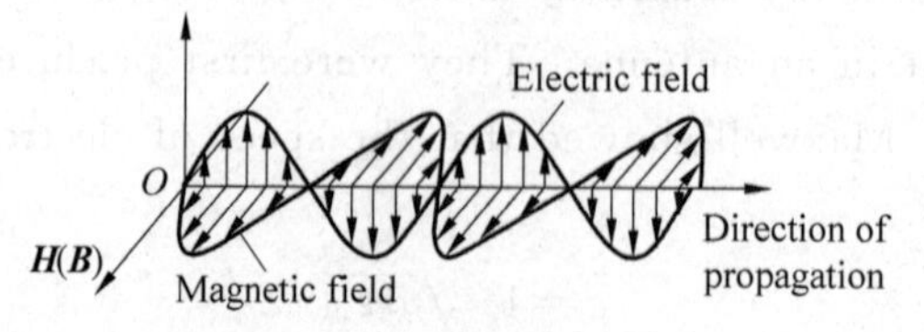

Fig. 18-1

where $c=1/\sqrt{\mu_0\varepsilon_0}$ is the speed of the wave. In general, the direction of propagation of an electromagnetic wave is the direction of the cross product $\boldsymbol{E}\times\boldsymbol{B}$.

Maxwell's equations imply that both $\boldsymbol{E}$ and $\boldsymbol{B}$ obey the wave equation similar to

$$\frac{\partial^2 y(x,t)}{\partial x^2}=\frac{1}{v^2}\frac{\partial^2 y(x,t)}{\partial t^2} \tag{18-4}$$

We consider only free space, in which there are no charges or currents, and we assume that the electric and magnetic fields $\boldsymbol{E}$ and $\boldsymbol{B}$ are functions of time and one space coordinate only, which we will take to be the x coordinate. Such a wave is called a **plane wave**, because field quantities are constant across any plane perpendicular to the x axis. For a plane electromagnetic wave traveling parallel to the x axis, the x components of the fields are zero, so the vectors $\boldsymbol{E}$ and $\boldsymbol{B}$ are perpendicular to the x axis and each obeys the wave equation

$$\frac{\partial^2 \boldsymbol{E}}{\partial x^2}=\frac{1}{c^2}\frac{\partial^2 \boldsymbol{E}}{\partial t^2},\quad \frac{\partial^2 \boldsymbol{B}}{\partial x^2}=\frac{1}{c^2}\frac{\partial^2 \boldsymbol{B}}{\partial t^2} \tag{18-5}$$

where $c=1/\sqrt{\mu_0\varepsilon_0}$ is the speed of the waves, too. Eq. (18-5) are wave equations for $\boldsymbol{E}$ and $\boldsymbol{B}$ respectively.

The Electromagnetic Spectrum(电磁波谱)

The various types of electromagnetic waves—light, radio waves, X-rays, gamma rays, microwave, and others—differ only in wavelength and frequency, which are related to the speed c in the usual way, $f=c/\lambda$. The names corresponding to different parts of the electromagnetic spectrum usually associate with the various frequency and wavelength ranges. These ranges are often not well defined and sometimes overlap.

The human eye is sensitive to electromagnetic radiation with wavelengths from about 400 to 700 nm, the range called **visible light.** The shortest wavelengths in the visible spectrum correspond to violet light and the longest to red light, with all the colors of the rainbow falling between these extremes. Electromagnetic waves with wavelengths just beyond the visible spectrum on the short-wavelength side are called **ultraviolet ray**, and those with wavelengths just beyond the visible spectrum on the long-wavelength side are called **infrared waves.** Heat radiation given off by bodies at ordinary temperatures is in the infrared region of the electromagnetic spectrum. There are no limits on the wavelengths of electromagnetic radiation; that is, all wavelengths are theoretically possible.

Energy Carried by Electromagnetic Waves(电磁波的能量)

Mechanical waves carry energy. Electromagnetic waves also carry energy, and as they propagate through space they can transfer energy to objects placed in their path. The rate of flow of energy in an electromagnetic wave is described by a vector, called the **Poynting Vector**, defined by the expression

$$\boldsymbol{S}=\boldsymbol{E}\times\boldsymbol{H} \tag{18-6}$$

where $\boldsymbol{H}=\frac{\boldsymbol{B}}{\mu_0\mu_r}=\frac{\boldsymbol{B}}{\mu}$, is the magnetic intensity vector. *The magnitude of the Poynting vector represents the rate at which energy flows through a unit surface area perpendicular to the flow and its direction is along the direction of wave propagation*, that is, in the same direction of the wave velocity, **c**. Thus, the Poynting vector represents *power per unit area*. The SI units of the Poynting vector are $J/s \cdot m^2 = W/m^2$.

Glossary

transverse wave	横波	in phase	同相
propagation	传播	wave equation	波方程
plane wave	平面波	wavelength	波长
visible spectrum	可见光谱	infrared wave	红外波
radiation	辐射	ultraviolet ray	紫外线
electromagnetic spectrum	电磁波谱	Poyngting vector	坡印亭矢量

18.3 专业英语常用表达法-18 物理学中常用数学运算

add/addition 加/加法

subtract/subtraction 减/减法

multiply/multiplication 乘/乘法

divide/division 除/除法

addend/augend 加数/被加数

subtrahend/minuend 减数/被减数

multiplier/multiplicand 乘数/被乘数

divisor/dividend 除数/被除数

numerator/denominator 分子/分母

sum/difference/product/quotient 和/差/积/商

perform an operation 进行1次运算

reduce a fraction 约分

reduce fractions to a common denominator 通分

cancel the denominator 消去分母

raise x to the nth power x^n（取 x 的 n 次方）

extract the nth root of x $\sqrt[n]{x}$ （求 x 的 n 次根）

combine like terms 将同类项合并

cancel common terms 消去同类项

establish a formula 列出方程式

find the root to three decimals 求根至3位小数

rearrange the expression 重新排列表达式

bracket 加括号

remove parentheses 去括号

take a factor outside 拆出因子

factor (x^2-a^2)　(x^2-a^2)因子

substitute B into A　将 B 代入 A

substitute A by B/substitute B for A　以 B 代 A

solve the equation for x　解方程式求 x 的值

remove a common factor　移出公因子、提取公因子

round off/half adjust/round　四舍五入

round up/down　(按四舍五入法)调高/调低

differentiate a function implicitly　求一函数的隐微分

graph a function　用图表示出函数

9 and 12 can be cancelled out by 3　9 和 12 可用 3 来通约

dx (an increment of x considered as tending to zero) dee of x/dee x/differential x　x 的增量

$\frac{dy}{dx}$ the differential coefficient of y with respect to x; the first derivative of y with respect to x, differentiate y with respect to x　对 x 求 y 的导数(微分)

$\frac{d^2y}{dx^2}$ the second derivative of y with respect to x　对 x 求 y 的二阶导数(微分)

$\frac{d^ny}{dx^n}$ the nth derivative of y with respect to x　对 x 求 y 的 n 阶导数(微分)

$\frac{\partial y}{\partial u}$ the partial derivative of y with respect to u, where y is a function of u and another variable (or other variables)　对 u 求 y 的偏导数(偏微分)

$\int_a^b$ integral between limits a and b　在区间$[a,b]$之间积分

LESSON 19

19.1　物理学专业英语阅读能力的培养

阅读能力是学习任何语言的基本技能之一。物理学专业英语的交际功能可以通过听、说、读和写四种语言技能来实现,其中说和写的技能属于主动性技能,比较难掌握,而听和读的技能属于被动性技能,相对容易掌握。阅读技能相对来讲是最容易掌握的技能。

对英语阅读技能和技巧的分类有不同的方法,但基本上可分为四种。第一种是精读(intensive reading skill);第二种是泛读(extensive reading skill);第三种是略读(skimming reading skill);第四种是掠读(scanning reading skill)。每种阅读技能和技巧都有其特点和要求。

1. 精读技能

精读在四种阅读技能和技巧中要求最高,对所阅读的文章,要做到逐字细读,不但要了解文章的主题思想和内容,了解文章的文法结构,而且要掌握每个字词的意义和用法,必要时还需要把文章翻译成中文。精读阅读技能是学习专业英语的人、讲授专业英语的人或从

事专业翻译的人员必须要掌握的技能。但对只需要查阅英语科技文献和信息的专业工作者,在多数情况下不需要使用精读技能。是否运用精读技能,不是看文章的长短,而在于是否需要对所阅读的英语资料做深度了解和研究。

下面以一段专业文献为例,简要说明通过精读要达到什么样的标准。

①In announcing that "every particle of matter in the universe attracts every other particle with a force inversely proportional to the square of the distance between the two particles", in showing that the one universal and comparatively simple law governs not only the motion of the planets round the sun and of the satellites round their planets but, probable, also the relative motions of all the heavenly bodies, Newton gave to the world a truth the importance of which in all branches of human thought can hardly be overestimated. ② Of value to science, of course, from microphysics to macrophysics. ③But consider the effect on man's concept of nature and of his relations thereto of realizing, indeed, of having proven to him, for the first time that the physical universe is governed by law, not by caprice; and if the physical universe, why not the biological universe, even the moral universe.

分析:这一段虽然比较长,但只有三个句子。第一句是一个复杂的长复合句,其主句是 Newton gave to the world a truth,In announcing that... heavenly bodies 是该句的状语。the importance of which in all branches of human thought can hardly be overestimated 是定语从句修饰 a truth,the importance of which 实际上相当于 whose importance;这个定语从句中的 in all branches of human thought 是修饰 importance 的。第二句是一个简单句,可以认为是 It is of value to science...的缩写。第三句也是一个并列复合句,其主句是 But consider the effect... why not... the moral universe;on man's concept of nature and of his relations thereto 作定语修饰 effect;of nature and of his relations thereto 是 concept 的定语;of realizing, indeed, of having proven to him, for the first time that the physical universe is governed by law, not by caprice 也是 concept 的定语;that the physical universe is governed by law, not by caprice 是 realizing 的宾语从句;by law, not by caprice 是状语;if the physical universe 是状语从句的省略。其中分号以前的子句 But consider ... caprice 可以简化为 But consider the effect on man's concept of realizing that...;分号以后的子句 and if the physical universe, why not the biological universe, even the moral universe 相当于 if the physical universe is governed by law, why is the biological universe, even the moral universe, not so governed?

只有在了解了句子结构后,才能准确理解文章内容。该段文献的中文意思是:通过宣布"宇宙中的一个质点吸引另外一个质点的力反比于他们距离的平方",表明一个普适的并且相对简单的定律不仅制约着行星绕太阳、卫星绕行星的运动,而且或许还制约着所有天体的相对运动,牛顿给世界一个真相,其重要性(在人类思想各分支中)是怎么过高评价都不过分的。这对于科学是非常有价值的,不论是微观物理学还是宏观物理学。但是,考虑一下对人类的观念造成的影响吧,对自然、对人类与自然的关系、人类第一次认识到(实际上已经获得了证实)物理的宇宙是由定律来制约,而不是由怪想制约的;如果物理的宇宙是这样,为什么生物的世界、道德的世界并非如此呢?

除此之外，也可以对文章的词汇含义、时态等做分析。不过，进行句子分析的目的是为了理解句子所包含的专业知识内容，千万要避免为分析而分析。只要懂得了句子的意思就应停止，不要无限制地分析下去。分析时首先应确定它是简单句还是复合句。如果是简单句，应先找出主语和谓语，然后重点分析其并列成分；如果是主从复合句，首先找出主句，然后仔细地分析各个从句，判定它们的作用和与其他句子的关系。

精读技能不是一朝一夕就能学会的，它需要有比较全面的英语知识和专业背景，需要通过大量的阅读训练才能掌握好。

另外三种阅读技能和技巧，对需要通过英语及时了解大量物理专业信息的人们，更为实用，使用频率高，也比较容易掌握。熟练掌握了这三种英语阅读技能和技巧，对提高精读技能水平很有帮助。

2. 泛读技能

通过阅读大量的专业英语文章，可以扩大知识面、丰富专业词汇量、提高专业英语语感。为了快速提高专业英语水平，要坚持经常阅读一定数量的专业英语文章，一般每周要阅读2～3篇文章。专业人员工作忙，时间少，开始做泛读练习时，可以从500字开始，以后逐渐增加长度，一般控制在2000～3000字为宜。如果能坚持下去，一年以后，专业英语水平一定会有显著提高。

在做泛读练习时，要求虽然不像精读那样严格，但是必须要从头到尾把文章读完，对文章的关键词语和句子要有所理解。为了达到这样的要求，最好根据文章的内容或特点，事先提出一些问题，在阅读之后，回答这些问题，以巩固阅读效果。

下面一些问题可以作为参考。

(1) What is main theme of the article? 文章的主题思想是什么？

(2) How many paragraphs does this article have? 此文有几个段落？

(3) What is the topic sentence of each paragraph? 每段的主题句是什么？

(4) Pick up the sentences you are most interested and analyze them. 挑选出你最感兴趣的句子，并做分析。

(5) List the important and useful words and phrases. 记录下重要的和有用的单词和短语。

除了上面这些问题外，读者还要时刻记着几个问题：文章谈论的是谁？谁来谈论？与谁有关？(who)，讨论什么内容？重点是什么？(what)，什么时间？时效和期限如何？(when)，在哪里？(where)，文章写作的原因？为什么？(why)，哪一类？(which)，怎样进行？采取哪些方法和技术？(how)。总之，阅读时要保持头脑清晰，有针对性地阅读，做到随记要点。

3. 略读技巧

略读是一种快速阅读技巧，在查询大量英语文献时非常有用。略读是快速阅读文章，目的是了解文章主题思想。略读时不需要逐字去读，而是寻找文章内关键词语、主题句，从而了解文章的主题思想。在这个基础上，决定是否选取此资料，是否需要进一步精读。现以下面的短文为例，说明如何运用略读技巧，找出此文的主题思想。

Rutherford's Model

A student of Thomson, *Ernest Rutherford, did an experiment that radically changed his teacher's model of the atom.* A key tool in Rutherford's work was the newly discovered radioactivity, the spontaneous emission of fast-moving atomic particles from certain elements. Rutherford had previously shown that one of these radioactive products is a helium atom without its electrons. These particles have two units of positive charge and are known as *alpha particles*.

Rutherford bombarded materials with alpha particles and observed how the alpha particles recoiled from collisions with the atoms in the material. This procedure was much like determining the shape of an object hidden under a table by rolling rubber balls at it. Rutherford used the alpha particles as his probe because they were known to be about the same size as atoms. (One doesn't probe the structure of snowflakes with a sledgehammer.) The target needed to be as thin as possible to keep the number of collisions by a single alpha particle as small as possible, ideally only one. Rutherford chose gold because it could be made into a very thin foil.

The alpha particles are about 4 times the mass of hydrogen atoms and 7300 times the mass of electrons. Thomson's model predicted that the deflection of alpha particles by each atom (and hence by the foil) would be extremely small. The positive material of the atom was too spread out to be a very effective scatterer. Consider the following analogy. Use 100 kilograms of newspapers to make a wall one sheet thick. Glue ping-pong balls on the wall to represent electrons. If you were to throw bowling balls to the wall, you would clearly expect them to go straight through with no noticeable deflection.

Initially, Rutherford's results matched Thomson's prediction. However, when Rutherford suggested that one of his students look for alpha particles in the backward directions, he was shocked with the results. *A small number of the alpha particles were actually scattered in the backward direction*! Rutherford later described his feelings by saying, "It was almost as incredible as if you fired a 19-inch shell at a piece of tissue paper and it came back and hit you!"

On the basis of his experiment, *Rutherford proposed a new model in which the positive charge was not spread out but was concentrated in a very, very tiny spot at the center of the atom, which he called the nucleus.* Returning to our analogy, imagine what would happen if all the newspapers were crushed into a small region of space: Most of the bowling balls would miss the paper and pass straight through, but occasionally a ball would score of direct hit on the 100-kilogram ball of newspaper and recoil.

These experiments gave him an upper limit on the size of the gold nucleus. The nucleus has a diameter approximately 100,000 times smaller than its atom. This means that atoms are almost entirely empty space.

Rutherford's "solar system" model led to a new explanation for the emission of light from atoms. The orbiting electrons are accelerating, and accelerating charges radiate electromagnetic waves.

这篇短文约500字。用略读技巧,应该在3～5分钟内读完全文。文中用斜体字表示的,是阅读时应该注意的句子或词语。此文主题是介绍Rutherford(卢瑟福)如何通过自己的实验否定了他的老师Thomson(汤姆森)提出的关于原子的模型,最后提出了关于原子的"太阳系"模型。了解了短文的主题思想,就等于有了一条线索,能帮助读者更好地理解文章。略读是在大量英语文献中查找和选择所需要的资料时非常有用的阅读技巧。

4. 掠读技巧

掠读也是一种快速阅读技巧,与略读不同的是,它用来查询文章中的特定信息。人们在日常生活和工作中,都在自觉或不自觉地使用掠读技巧。例如,看文献的目录时,查找特定主题的文章;查字典时,查阅所需要的词汇;看文章时,查找特定的信息等。下面是一个国际会议的征稿通知,比如说对有关Antennas(天线)方面的内容感兴趣,使用掠读的办法快速找出相关主题。掠读时,是找特定信息,所以只需要查看有Antennas一词的主题,其他完全不需要看,下面的内容应该在一分钟内就读完,迅速找到所有含Antennas一词的主题。

Call for Papers

ISAPE2008 offers a forum for antennas, propagation, and electromagnetic theory researchers to exchange ideas and experiences in the latest development and designs. You are invited to submit papers in all areas of antennas, propagation, electromagnetic theory, computational electromagnetics, and EMI.

Relevant topics of interest include the following:

Active and Integrated Antennas√	Non-linear electromagnetics
Adaptive and Smart Antennas√	Optical Technology in Antennas√
Antennas for Wireless Communications√	Radio Meteorology and Radio Astronomy
Aperture Antennas and Feeds√	Random Media and Rough Surfaces
Bio-electromagetics	Remote Sensing
Earth-Space and Terrestrial Propagation	SAR Polarimetry and Interferometry
EM fields in complex media	Scattering, Diffraction, and RCS
Geo-electromagnetics	Seismo-Electromagnetics
High-Frequency Techniques	Small Antennas√
Inverse Problems and Imaging	Time Domain Techniques
Ionospheric Propagation	Transient EM fields
Measurement Techniques	Tunnel Propagation
Millimeter and Optical Wave Propagation	Waveguiding Structures
Mobile and Indoor Propagation	Wire and Slot Antennas√

查找结果,一共有七条有关Antennas的相关主题,画有√ 的即是相关内容。

略读和掠读技巧对阅读的速度是有要求的,要越快越好。在训练这两种阅读技能时,最好用表计时,来督促自己不断提高阅读速度。

在了解了专业英语四种基本阅读技能和技巧后,实际学习中需要做的就是要通过练习,熟练掌握这些阅读技能。掌握了这些阅读技能后就等于有了得力助手,就可在浩如烟海的专业英语文献中,随心畅游,迅速查到自己所需要的资料。

19.2 专业英语阅读

19.2.1 The Nature of Light(光的本性)

Since the beginning of the present century, an increasing degree of uncertainty about the wave theory of light has been developing due to the discovery of the photoelectric effect, the quantum theory, and a number of allied phenomena, and due also to a detailed investigation into the nature and physical properties of the ether. Also, during this period a great deal has been learned about the structures of the atom which is not altogether limited to a wave picture of light. Hence the photon, or a light corpuscle, has been postulated.

Maxwell discovered that light is a wavelike disturbance in the electromagnetic field that travels at constant speed c in vacuum. Before Maxwell's insight shed light on the nature of light, philosophers debated its character, artists gloried in it, astronomers used it, and scientists were intrigued by it. Long before the idea of the electromagnetic field was born, people speculated about the nature and properties of light—the phenomenon that reveals all that we can see. The ancient Greeks thought light was fire, one of the four basic elements. Plato considered light to be streamers or filaments emitted by the eye. The Pythagoreans thought that light traveled from an object to one's eyes in the form of very fine particles, or corpuscles. The Greeks observed that objects cast sharp shadows and concluded that light travels in straight lines. As master geometers, the Greeks were also aware of the fact that the intensity of light decreases inversely as the square of the distance from the source. To them the particle description of light adequately fitted observations.

The particle theory remained essentially unchallenged until the seventeenth century. In 1678 Christian Huygens, a Dutch scientist, formally proposed that many of his predecessors had suspected—that light is a wavelike disturbance. Just as water waves travel form one point to another without water actually being moved from one point to the other, Huygens considered light to be waving in a medium he called the luminiferous aether. Huygens's ideas were not mere speculation. They commanded attention because he could describe many properties of light with the wave theory. The key to Huygens's theory of light was a geometric method of determining the shape of a wavefront at any time from knowledge of the wavefront at some earlier time.

A **wavefront** is the set of all points of equal phase. The phase of any wave is the argument in the functional form of the wave. For example, we learned that plane waves traveling with speed u can be described by an equation of the form $y(z, t)=f(z-ut)$.

The phase is $z-ut$, and a wavefront is the set of points in the zt plane satisfying

$z\text{-}ut$ = constant, in this case a family of straight lines. A point source of light produces spherical wavefronts, as Fig. 19-1(a). Sufficiently far from the source, the radii of the spherical wavefronts are so large that the waves appear as plane waves, as shown in Fig. 19-1(b).

The general problem of following the propagation of electromagnetic waves is extremely complicated. Some of the simpler aspects of the problem are embodied in a fundamental principle enunciated by Huygens called **Huygens's Principle**: *Every point on a wavefront can be regarded as a new source of spherical waves that spread out in all directions with the same speed and frequency as the original wave.*

Fig. 19-2 (a) and (b) illustrate this principle for spherical and plane waves, respectively. Any point on a wavefront generates new spherical waves called wavelets. During the time interval equal to one period of the wave, each wavelet travels one wavelength. To construct the new wavefront, first draw wavelets as hemispheres of radius one wavelength from each point on the original wavefront. The surface tangent to all the wavelets represents the new wavefront that has moved one wavelength. Of course, in this construction we ignored wavelets moving backward toward the source, because they are not observed. It was not until 1826 that Fresnel showed that the disturbance did not propagate backward because of interference effects. In 1883 Kirchhoff proved by mathematical analysis that the secondary waves from the individual sources destroy one another by interference except at the wavefront itself, so the wave is propagated only in the direction away from its source.

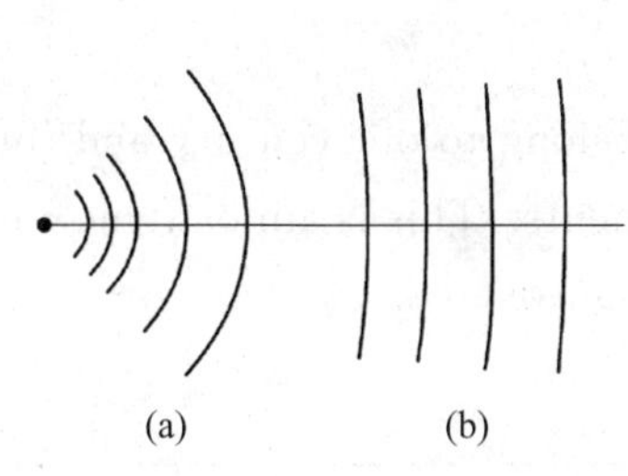

Fig. 19-1 (a) A point source of light creates spherical wavefronts. (b) Far from the source the wavefronts appears as planes

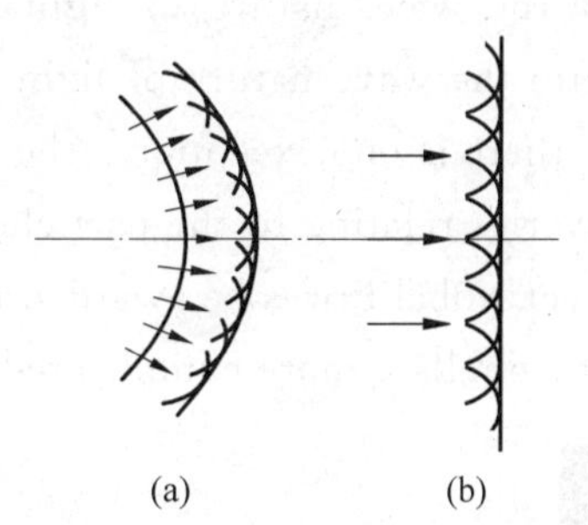

Fig. 19-2 Huygens's construction of wavefronts for (a) a spherical wave and (b) a plane wave

The ancients believed that light could be represented by straight lines called *rays*. In Huygens's construction, the rays are along the direction of propagation of the wavelets. They are lines perpendicular to the wavefronts.

19.2.2 The Wave-Particle Duality(波粒二象性)

A particle on the classical view is a concentration of energy and other properties in space and time, whereas a wave is spread out over a larger region of space and time. The question whether light are streams of particles (corpuscles) or waves is a very old one. This "either-or" formulation was classically natural and alien to the advanced "both-and" even the "neither-nor" solution of today. Early in the nineteenth century experiments were suggested and made to show that light is a wave motion. A key figure in this endeavour was Thomas Young, one of the most intelligent and clever scientists ever to live, who studied diffraction and interference of light already in 1803 with results that gave strong support to the wave theory of Christian Huygens as opposed to the particle or corpuscular theory of Isaac Newton. Further contributions were made by many other researchers, among them Augustin Jean Fresnel, who showed that light is a transverse wave.

Newton's theory of light had seemed suitable to explain the straight-line casting of sharp shadows of objects placed in a light beam. But wave theory was needed to explain interference where the light intensity can be enhanced in some places and diminished in other places behind a screen with a slit or several slits. The wave theory is also able to account for the fact that the edges of a shadow are not quite sharp.

The mathematical theory of electromagnetism by James Clerk Maxwell, set up in 1864, led to the view that light is of electromagnetic nature, propagating as a wave from the source to the receiver. Heinrich Hertz discovered experimentally the existence of electromagnetic waves at radio-frequencies in the 1880s. Maxwell died in 1879 and Hertz died only 37 years old in 1894, two years before Alfred Nobel's death.

At the end of the 19th century, which also is the time when the Nobel Prizes were instituted, the wave nature of light seemed definitely established. Thus the decisive research into the wave nature of light came too early to be considered for Nobel Prizes. However, there is one exception—the case of X-rays.

Discoveries relating to the particle nature of light belong to our century and thus one might expect Nobel Prizes be awarded for such achievements. This is almost true—but the Nobel archive tells a more complicated story as will be uncovered.

Glossary

photoelectric effect	光电效应	postulate	假定
wavelike	波状的	shed light on	阐明
intrigue	困惑	streamer	飘带
filament	细丝、长丝	emit	发射
Pythagoreans	毕达哥拉斯学派	corpuscle	微粒
geometer	几何学家	intensity of light	光的强度
luminiferous aether	传光的以太	wavefront	波前
argument	自变量	phase	相位
propagation	传播	embody	具体表达,使具体化
enunciate	系统阐明,明确表达	Huygens's Principle	惠更斯原理
wavelet	子波	hemisphere	半圆

Fresnel	菲涅尔	interference	干涉
Kirchoff	基尔霍夫	ray	(光)线
Thomas Young	托马斯·杨	diffraction	衍射
transverse wave	横波	enhance	加强
diminish	降低,减弱	slit	狭缝
Nobel Prizes	诺贝尔奖		

19.3 专业英语常用表达法-19 数字的增加与倍数

1. 在表示增加的概念时,可以有两种说法,即"A 是 B 的 N 倍"或"A 比 B 大 N－1 倍",例如"A 是 B 的 3 倍"即是"A 比 B 大 2 倍"(请注意汉语的表达不同,数字不同,但意思相同)。而英语中虽然有几种句型来表示倍数的增加,但句中的倍数却没有数字上的变化。下面是英语表达倍数的常用句型(其中 N 指倍数):

a. *N*＋times＋adj.＋than

b. *N*＋times＋as＋adj.＋as

c. *N*＋times＋noun phrase

d. verb (e.g. increase, grow, rise, be raised, multiply) (＋by)＋*N*＋times

e. verb (e.g. double, treble, quadruple)＋noun

以上句型汉译的意思为"是……的 N 倍"或"增加了 N－1 倍"。请看下面的例句:

Aluminum has nearly 5 *times* the thermal conductivity of cast iron. 铝的导热率几乎是铸铁的 5 倍。

The resistivity of aluminum is 1.6 *times* that of copper. 铝的电阻率为铜的 1.6 倍。

The mass of the proton is 1836 *times* the mass of the electron. 质子的质量为电子质量的 1836 倍。

Mercury weighs about 14 *times more than* water. 水银的重量约超过水的 14 倍。

Many planes can fly *several times faster than* sound. 许多飞机能以数倍于音速的速度飞行。

The Uranus is 14 *times as large as* the mass of the Earth. 天王星的质量是地球的 14 倍。

There is *four times as much* nitrogen *as* there is oxygen in air. 空气中的氮气比氧气多 3 倍。

The accuracy of the atomic clock is about *ten times as great as* that of the quartz clock. 原子钟的准确度约为石英钟的 10 倍。

The pinion rotates *five times as fast as* the gear. 小齿轮的转速为齿轮的 5 倍。

The resistance of the wire in circuit L_1 is *twice* the resistance of the wire in circuit L_2. L_1 电路导线的电阻为 L_2 电路导线的 2 倍。

Helium is *twice as heavy as* hydrogen. 氦有氢的 2 倍重。

If the second winding has *twice as many* turns *as* the first, the voltage will be twice as great. 如果第二个线圈的匝数比第一个线圈多 1 倍,电压也会提高 1 倍。

In a circuit of fixed resistance, if the voltage of the circuit *is doubled*, the current will also *double*. 在电阻固定的电路中，若电路电压增加1倍，则电流也会增加1倍。

2. 数字的增加还可以用分数或百分数来表示。这时的增长大多是净增长，可用汉语直接译为“增加几分之几”，“增加百分之几”。常用句型有：

a. 动词(如 increase, grow, jump, improve, exceed 等)＋(by)＋ 百分数，译为“上升了百分之……”

b. 百分数＋形容词的比较级＋than，译为“比……增百分之……”

c. 百分数＋介词(up, over) ＋名词，译为“增百分之……”

例如：

New booster can *increase* the pay load *by* 120%. 新型推动器能使有效负载增加120%。

Output of glass ceramics this year was 30 *percent higher than* last year. 今年玻璃陶瓷的产量比去年提高了30%。

They note that the amount of carbon dioxide in the air has been *up* 13% in the last one hundred years. 他们注意到过去的100年里空气中二氧化碳的含量已增加了13%。

3. 涉及 again 结构的译法

a. 在结构 as much/many...again as 与 again as much/many...as 中可译成“是……的2倍”，或“比……多一倍”。例如：

The amount left was estimated to be *again as much as* all the aluminium that has been mined. 当时估计，剩余的铝储量是已开采量的2倍。

Pulley A turns *as fast again as* pulley B. A滑轮转动比B滑轮快1倍。

b. 出现 again 的同时还出现 half，即 half as much/many...again as 或 half again as much/many...as 时，表示“是……一倍半”，“比……多一半”。

The videodisc is half *as wide again as* that one. 这个光盘比那个宽0.5倍。(或这个光盘宽度是那个的1.5倍。)

The resistance of aluminum is approximately *half again as great as* that of copper for the same dimensions. 尺寸相同时，铝的电阻约为铜的1.5倍。

LESSON 20

20.1 物理学专业英语阅读时需要注意的问题

1. 阅读材料的选择

专业阅读是为了获取信息，有时就是为了提高专业英语水平。读者多数都是成年人，理解能力强，大多数具有专业背景，但外语水平可能不尽相同，专业词汇掌握的也可能有多有少。在选择阅读材料时要特别注意应用性和原汁原味，并且要有一定的深度。但在开始学习专业英语时，难度要适合自己的水平，不要超过了自己的理解能力。随着水平的提高，阅读材料的深度可以相应加深。

应用性和原汁原味的材料是指那些未经改编、未经简写的以英语为母语的人士在实际专业交流中所使用的话语，包括日常会话、学术杂志文章、专业书籍、通知、告示等。原始材料有语境、有上下文、有完整的结构、有作者的逻辑思维和立场观点，因此，从专业的意义上讲，应用性的原始的材料可能更容易理解。

初学者可以选那些英美作者用简单词汇写成的文章，如科普文章、科学幻想、产品宣传材料等。比如下面这种程度的文章：

Most people would describe water as a colorless liquid. They would know that in very cold conditions it becomes a solid called ice, and that when heated on a fire it becomes a vapour called steam. But water, they would say, is a liquid.

We have learned that water is made up of two atoms of hydrogen and one atom of oxygen, which we describe by the formula H_2O. Chemically there is no difference between the gas, the liquid, and the solid, all of which are made of molecules with formula H_2O. And this is true of other chemical substances. Most of them can exist as gases or liquids or as solids. We may normally think of iron as a solid, but if we heat it in a furnace, it will melt and become a liquid, and at very high temperatures it will become a gas. We normally think of air as a mixture of gases, but at very low temperatures it becomes a liquid, and at lower temperatures still it becomes a white solid.

What, then, do we mean when we say that water is a liquid, air is a gas, and salt is a solid? We mean nothing more than that. This is the usual condition of things on our earth. On one of the outer planets all three substances would be solids, and on the sun all three would be gases. Most substances are only familiar to us in one state, because the temperatures required to turn them into gases are very high, or the temperatures necessary to turn them into solids are so low. Water is an exception in this respect, which is another reason why its three states have been given three different names.

对中等水平的读者，可选用科普杂志、百科全书或是用户手册上的文章，阅读面宜宽些。比如：

Chinese documents suggest that magnetism was recognized as early as about 2000 B. C. The ancient Greeks observed electric and magnetic phenomena possibly as early as 700 B. C. They found that a piece of amber, when rubbed, attracted pieces of straw or feathers. The existence of magnetic forces was known from observations that pieces of a naturally occurring stone called magnetite were attracted to iron. (The word *electric* comes from the Greek word for amber, *elektron*. The word *magnetic* comes from *Magnesia*, on the coast of Turkey where magnetite was found.)

In 1600, the Englishman William Gilbert discovered that electrification was not limited to amber but was a general phenomenon. Scientists went on to electrify a variety of objects, including chickens and people! Experiments by Charles Coulomb in 1785 confirmed the inverse-square law for the electrostatic force.

It was not until the early part of the 19th century that scientists established that electricity and magnetism are related phenomena. In 1820, Hans Oersted discovered that a compass needle, which is magnetic, is deflected when placed near an electric current. In

1831, Michael Faraday in England and, almost simultaneously, Joseph Henry in the United States showed that, when a wire loop is moved near a magnet (or, equivalently, when a magnet is moved near a wire loop), an electric current is observed in the wire. In 1873, James Clerk Maxwell used these observations and other experimental facts as a basis for formulating the laws of electromagnetism as we know them today. Shortly thereafter (around 1888), Heinrich Hertz verified Maxwell's predictions by producing electromagnetic waves in the laboratory. This achievement was followed by such practical developments as radio and television.

对英语水平较高的专业文献读者,可以选一些结合自己专业的英语原文资料、科研论文、专著等作为阅读材料。

2. 注重整体理解,避免只见树木不见森林

一开始,在进行专业英语阅读时,读者往往会过分小心仔细,总希望搞清每一个句子的意思;每遇到生词就会自觉或不自觉地去查字典,即使有时猜出了词义还觉得不放心,直到查了字典后才感觉踏实。有的读者在遇到长一些的句子就自觉不自觉地进行语法分析。还有的读者无论读什么样的文章,无论简单的还是复杂的,总是要在头脑里逐词逐句地翻译成中文后才能理解。这些习惯在专业英语阅读中应慢慢改掉。要在阅读中注重整体理解,理解掌握读物的主要内容,不需要彻底掌握读物中的每一个单词。始终记住阅读的目的是掌握读物的内容、接收读物所传递的信息,扩大词汇量和丰富语言知识是专业英语阅读的副产品。

3. 克服长难句障碍,读懂专业内容

在前面章节中,已经对专业英语中的长难句进行了介绍。应付这些长难句的有效方法之一就是我们介绍过的“**抓住主干,添枝加叶**”方法,能正确运用这种方法,一般对长句子就会做出正确分析。通过语法分析找出哪些是句子的主干,即句子的主语和谓语,哪些是修饰成分,找出各部分之间的关系。经过这样逐级逐层地分析,就能读懂长难句,最终理解文献中的专业知识内容。

4. 克服词汇障碍,提高阅读速度

扩大专业词汇量是提高专业阅读能力的一条切实可行的重要途径。要想流利地阅读专业文献资料,必须掌握大量的词汇,必须知道如何猜测生词词义。然而扩大词汇量的最佳途径就是经常地、广泛地、大量地阅读,把扩大专业词汇量变成经常的、自觉的行动。阅读和词汇既是互相牵制又是互相促进的,提高了一种能力,必然也会提高另一种能力。

尽管专业词汇量在不断扩大,但在进行专业阅读时,仍然还会遇到不少的生词,尤其是那些影响对文章理解的生词。生词不仅是一个拦路虎,而且给读者增加了压力,专业读者必须学会运用科学的猜词技巧,采用正确的方法来猜测生词的词义,这对阅读速度的提高有很大的帮助。下面就介绍一些猜测生词词义的方法。

(1) 根据释义猜测词义

在有些文章中,某些生词的词义在下文中可以得到解释或说明。这些解释有时以系表结构、定义、同位语形式出现。系表结构的句子有两种表意功能,一是表语说明主语的形状或特征,二是表语和主语是同位关系,即表语的内容等于主语的内容,后者往往是猜词的最好线索。这种结构有时用破折号、括号、冒号、引号等符号引出，还有时用 that is, that is to say, in other words 等引出。例如：

When two initially neutral objects are charged by being rubbed together, charge is not created in the process. The objects become charged because electrons are transferred from one object to the other. One object gains some amount of negative charge from the electrons transferred to it while the other loses an equal amount of negative charge and hence is left with a positive charge. That is, *charge is conserved*. The law of *conservation* of charge is a fundamental law of nature. 根据前面的解释说明，可以得出 *conserved* 和 *conservation* 具有"总量不变"的意思，实际上它们的意思就是"守恒"。

Capacitance, the ability to store electric charge, is one of the most common characteristics of electronic circuits. 由同位语我们可以很快猜出生词 *Capacitance* 的词义——电容量。

The *momentum* of an object is its velocity multiplied by its mass. 速度乘以质量就是 *momentum* 动量。

The most interesting problem centers on *inertia*—the property of matter that makes heavy things hard to get moving, but once moving, hard to stop. 破折号后面都是说明 *inertia*"惯性"的。

At the center of an atom is its *nucleus*, made up of particles called *protons and neutrons*. 位于原子中心的肯定是"原子核"*nucleus*，构成原子核的是"质子和中子"*protons and neutrons*.

(2) 根据上下文线索猜词义

某些生词的含义可以利用已知词汇的信息和未知词汇信息之间的联系以及生词前后表述的启发，弄清其含义。如：

Refraction refers to the change in direction of a wave at a boundary where it passes from one medium into another. For example, when light traveling in air is incident on a transparent material such as glass, it is partially reflected and partially transmitted. But the direction in which the transmitted light is propagated is different from the direction of the incident light, so the light is said to have been *refracted*, or bent. 根据 change in direction, the direction ...is different from... 和 bent 等词来判断，*refraction/refract* 有"改变方向、折"的意思，再根据 wave at a boundary 和 transmit 等词和专业知识，就可以判断出 *refraction/refract* 是表示"折射"。

Many quantities in physics have magnitude and direction. *Vectors* are quantities with magnitude and direction. Examples include velocity, acceleration, momentum, and force. Quantities with magnitude but no associated direction, for example, distance and speed, are called *scalars*. 从给出的例子包括"速度、加速度、动量和力"以及 quantities with magnitude and direction 就可以猜测到 *vector* 的意思应该是"矢量"，进一步也就知道了 *scalar* 的意思是"标量"。

The time it takes for a displaced object to execute a complete cycle of oscillatory motion—from one extreme to the other extreme and back—is called the *period* T. The unit of the period is second. The *reciprocal* of the period is the *frequency* ν, which is the number of cycle per second. 根据定义就会知道 *period* 是“周期”, *frequency* 是“频率”,根据频率和周期的关系就会知道 *reciprocal* 的意思是“倒数”。

(3) 利用常识和经验猜词

某些文章的生词,根据个人生活经验和基本常识也可以猜出。如:

Anyone who has observed a football moving in the air (or any object thrown in the air) has observed *projectile motion*. 根据常识可以推断: *projectile motion* 的意思是“被抛在空中的物体的运动”,也就是物理学中的抛体运动。

(4) 利用语义相反、对照信息猜词

有时文章在描述两件事物的不同特征时,使用了截然相反的词,形成鲜明的对照。这时可以给读者以暗示,从而猜出生词的意思。如:

The major points of your plan are clear to me, but the technical details are still *hazy*. 计划的要点清楚,而细节 *hazy*,显然和 clear 的意思相反,为“模糊”或“迷惑”的意思。

Suppose a person on a moving vehicle (observer A) throws a ball straight up in the air according to his frame of reference. According to observer A, the ball will move in a vertical path. On the other hand, a *stationary* observer (B), standing on the ground, will see the path of the ball as a parabola. 根据前面 on a moving vehicle 和后面的 standing on the ground 的意思相反的对比,再根据 standing on the ground 是对 *stationary* 的进一步解释,就可以猜到 *stationary* 的意思应该是“不动的、静止的”。

(5) 利用定语从句的功能猜词

定语从句是修饰和说明中心词的,根据其提供的信息来猜测词义,是一种猜词的好方法。如:

A space where there is no matter in any state is a *vacuum*. 根据定语从句的说明很容易得到 *vacuum* 的意思是“真空”。

Solids studied include crystals, solids whose molecules are in regular geometrical shapes, and *alloys*, which are combinations of different metals, such as copper and tin in bronze. 根据定语从句的说明和所举的例子青铜(包含铜与锡),就可以猜出 *alloy* 的意思是“合金”。

(6) 根据复合词的各部分猜测词义

在物理学中,有些专业复合词汇的意思可根据其各部分的意思进行猜测。例如:

wavelength 波长, microwave 微波, voltmeter 伏特计, thundercloud 雷雨云, thunderstorm 雷暴, waveform 波形。

A substance that conducts electricity is a conductor. One that does not is an insulator. One that conducts electricity without resistance is a *superconductor*, possessing the property of *superconductivity*. 此句中的 *superconductor* 一词从上下文判断、从单词组合判断可以得到“超导体”,进一步可得出 *superconductivity* 是“超导性”。

(7) 根据前后缀猜测词义

可以充分利用前后缀知识来猜测专业词汇的意思。例如：

coaxial 同轴的，coexist 共存，concentric 同心的，uncharged 不带电的，redistribute 重新分配，equipotential 等电势的，nonslip 不滑的（防滑的），nondeforming 不变形的，interatomic 原子间的，intermolecular 分子间的，liquefy 液化，massless 无质量的，counterclockwise 逆时针的。

20.2 专业英语阅读

20.2.1 Reflection and Refraction(反射与折射)

The Huygens principle is not intuitively obvious, but it gained acceptance because it could be used to explain the laws of reflection and refraction, two phenomena concerning light that had been observed since antiquity. The ancient Greeks observed that when a ray of light is reflected by a mirror, the angle of incidence is equal to the angle of reflection. This is illustrated in Fig. 20-1(a), where θ_i denotes the angle of incidence and θ_r the angle of reflection, the angles between the light ray and a line perpendicular to the mirror. The law of reflection states that $\theta_i=\theta_r$.

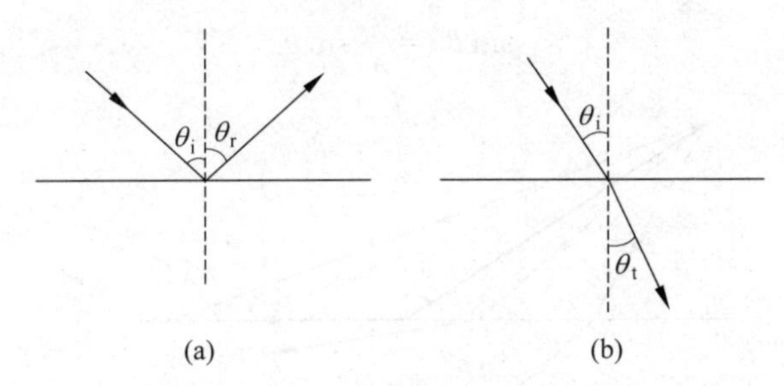

Fig. 20-1 (a) Reflection of light,(b) Refraction of light

Fig. 20-1(b) illustrates the concept of refraction. When a light ray passes from one transparent medium, such as air, into another of different density, such as water, the ray is bent or *refracted*. When refraction takes place, the angle of incidence θ_i and the angle of refraction θ_t are different. (The subscript t in θ_t suggests transmitted light.) Ptolemy had observed that for a given pair of media, such as air and water, the ratio of the angles θ_i/θ_t is nearly constant. We now know this is true only for small angles. Kepler and others made numerous measurements in an attempt to find a relation that would hold for all angles. That relation, discovered experimentally in 1621 by a Dutch mathematician Willebrord Snell, is called **Snell's law** and states that for a given pair of media the ratio $\sin\theta_i/\sin\theta_t$ is constant.

Although the Huygens principle can be used to explain both these laws, they can be explained more simply by another principle stated by Fermat in 1657. This is called **Fermat's principle of least time** and it states that a ray of light traveling in any combination of media follows the path that minimizes its total travel time. The next two examples show how Fermat's principle implies both the law of reflection and Snell's law of refraction.

A convenient number that characterizes the behavior of light in a medium is the *index of refraction* of the medium, denoted by the symbol n. by definition, the index of refraction is the ration of the speed of light in vacuum to the speed of light in the medium,

$$n = c/v \tag{20-1}$$

If light travels in a medium with index of refraction n_i and is refracted in a medium with index of refraction n_t, the general form of Snell's law of refraction is

$$n_i \sin\theta_i = n_t \sin\theta_t \tag{20-2}$$

In general, n depends on the wavelength (and therefore the color) of the light. A vivid illustration of this property is the separation of white light into a rainbow of colors by a prism. This property of waves is known as **dispersion.** For air at standard temperature and pressure the index is nearly 1. It has very nearly the same value for all visible light.

Fig. 20-2 shows a number of rays from a point source in one medium of refractive index n_1 impinging on the surface of another medium of refractive index n_2, where $n_1 > n_2$. Snell's law implies

$$\sin\theta_t = \frac{n_1}{n_2}\sin\theta_i$$

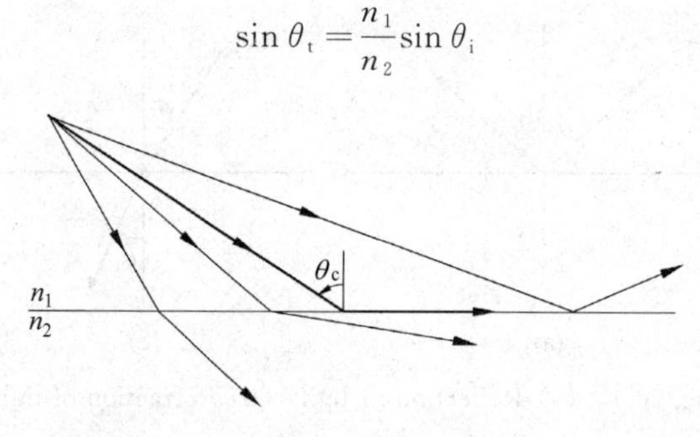

Fig. 20-2 The critical angle θ_c is the angle of incidence for which the angle of refraction is 90°. For angles greater than θ_c there is total internal reflection

where $\sin\theta_t > \sin\theta_i$ because $n_1/n_2 > 1$. For some incident angle θ_i less than 90° we have $\sin\theta_t = 1$ and hence $\theta_t = 90°$, which means the transmitted light travels parallel to the interface. This value of θ_i is called *critical angle* θ_c. When $\theta_i > \theta_c$, all the light is reflected, as shown in Fig. 20-2.

20.2.2 Interference of Light Waves(光波的干涉)

Despite Huygens's success in describing the propagation of light in terms of waves, many scientists advocated a corpuscular theory of light. Most notable among them was Isaac Newton who thought that light was best described as a stream of particles emitted by the light source. He objected to the wave theory because it did not explain why light travels in straight lines. In a lengthy dialogue with Huygens, Newton argued that if a hole is placed between a source of light and a screen, the patch of light on the screen has the same shape as the hole. One would expect this effect if light traveled in straight lines like tiny particles emitted by the source and absorbed by the screen. On the other hand, he argued, if light were truly a wave, it should bend around the hole and not produce a sharp shadow. Newton also asserted that because light travels through an evacuated glass container, light could not be a wave because waves require a medium in which to travel.

In response to Newton's objections, Huygens contended that if the wavelength of light is extremely small—much smaller than the dimensions of the object on which it shines, then bending of the waves would be too minute to observe and the object would cast sharp shadows. As far as light traveling in vacuum, Huygens replied that waves travel in the aether, a medium that permeates all of space and that is not removed when a container is evacuated.

The wave-particle controversy continued into the twentieth century when it was shown that light has the dual nature of both particles and waves. The first convincing experimental evidence for the wave nature of light appeared early in the nineteenth century in the classical investigations on interference by Thomas Young and Augustin Fresnel.

A physician by training, Young was also a self-educated linguist and natural philosopher. He made original studies of the eye while still in medical school, and his doctoral thesis dealt with the physics of sound as it relates to the human voice. In 1802, he found an ingenious way to demonstrated that light exhibits the phenomenon of interference.

When two waves encounter one another, their effects add according to the principle of superposition. If two waves are in phase at some point, they reinforce each other there, with crests adding on crests, thereby creating a stronger wave. This is called **constructive interference**. When two waves are exactly out of phase at some point, with crest arriving on top of trough, they can cancel each other and produce no light intensity at that point. This is called **destructive interference.**

Young demonstrated interference of light with two slits separated by a small distance and illuminated by light from a single source. He passed sun light through first the single slit, producing a narrow beam of light. He directed the light emerging from the single slit onto two narrow slits, and observed the result on a distant screen. The resulting interference pattern is a series of alternating bright and dark bands. Young never worked out a detailed mathematical description of interference. Instead, he drew elaborate diagrams of wave fronts to explain his observations. The task of rigorously establishing a theory was taken up later in the century by the French physicist Augustin Fresnel, who synthesized Huygens's wave description with the principle of interference. After his studies in interference, Young ventured to Egypt with Napoleon and deciphered the Rosetta Stone, which advanced the decoding of hieroglyphics.

Glossary

intuitively	直观地	antiquity	古代
angle of incidence	入射角	angle of reflection	反射角
transparent	透明的	transmitted light	透射光
Snell's law	斯涅尔定律	index of refraction	折射率
hieroglyphics	象形文字	prism	棱镜
dispersion	色散	impinge	冲击、撞击
refractive index	折射率	critical angle	临界角
angle of refraction	折射角	corpuscular theory	微粒说
lengthy	冗长的	assert	主张
evacuated	真空的	objection	反对
contend	争辩,坚持认为	vacuum	真空
permeate	弥漫,渗透	controversy	争论,争议
dual nature	二象性	interference	干涉
ingenious	巧妙的,独创的	demonstrate	演示,证明
principle of superposition	叠加原理	in phase	同相
reinforce	加强	crest	波峰
constructive interference	相长干涉	out of phase	异相
trough	波谷	cancel	抵消
light intensity	光强	destructive interference	相消干涉
interference pattern	干涉图样	wave front	波前
rigorously	严格地	synthesize	综合,合成
decipher	破译	Rosetta Stone	罗塞塔石碑
Fermat's principle of least time	费马最小时间原理		

20.3 专业英语常用表达法-20 分数 百分数 减少

1. 分数的表示法

(1) 分数可由基数词和序数词一起来表示。基数词作分子,序数词作分母,除了分子是"1"以外,其他情况下序数词都要用复数形式。

three fourths 或 three quarters 3/4 　　one third 或 a third 1/3

a half 或 one half 1/2 　　one quarter 或 a quarter 1/4

one and a half $1\frac{1}{2}$ 　　one and a quarter $1\frac{1}{4}$

three and one fourth 或 three and one quarter $3\frac{1}{4}$

$1\frac{1}{2}$hours (读作 one and a half hours,或者 an hour an a half)1.5 小时

$2\frac{3}{4}$meters （读作 two and three-fourths meters） $2\frac{3}{4}$m

4/5meter （读作 four fifths meter） $\frac{4}{5}$m

5/6inch （读作 five sixths inch） $\frac{5}{6}$in

Wire B has only *one half* as great a resistance as wire A. B 导线的电阻仅为 A 导线的 1/2。

By doubling the pressure we compressed the air to *one half of* its former volume. 通过把压强提高 1 倍，我们把空气压缩到其原体积的 1/2。

The new type of transistor operates about *one-third* faster than currently used devices. 新晶体管的操作速度比当前使用的器件约快 1/3。

If the radius of the coil were trebled, the field would be reduced to *one third* of its original strength. 假如线圈的半径增大到 3 倍，磁场强度会减少到其原强度的 1/3。

Nearly *three quarters* of the earth's surface is covered by water. 地表将近 3/4 为水所覆盖。

（2）用...times, by a factor of...，或由-fold 构成的词。

当 times, by a factor of 或-fold 与表示“减少”“降低”“小”“少”这类词连用时，汉语习惯上常化作分数表达，一般不译作“减少……倍”，而译作“减少……分之……”。例如：

Some steels can be drawn into wire *five times finer* than hair. 某些钢能被拉到头发丝的 1/5 那样细。

The moon has a mass nearly *one hundred times less* than the earth's. 月球的质量几乎是地球的 1%。

When the voltage is stepped-up by 10 times, the strength of the current is *stepped-down by 10 times*. 电压升高 10 倍，电流强度就会降低到 1/10。

During the next stroke, the air is compressed about *sixteen-fold*. 在下一冲程的过程里，空气被压缩到约为原体积的 1/16。

2. 百分数的表示法

用基数词(整数或小数)＋percent（有时写成 per cent）。

fifty percent 50%

three percent 3%

zero point one two percent 0.12%

这里的 percent 前半部 per 表示“每一”，后半部分 cent 表示“百”，所以百分数中 percent 不用复数形式。

Water covers about *seventy percent* of the Earth's surface. 水覆盖地球表面的约 70%。

The world production of aluminum is only about *one per cent* of the world production of iron. 全世界的铝产量仅约为全世界铁产量的 1%。

3. “减少”的表示方法

英语所表示的“倍数减少”,用汉语表示时,就不用“倍”来表达,而是采用分数或百分数表示,译为“减少到 $1/n$”或“减少了$(n-1)/n$”。主要句型有(以 reduce 为例,*n* 表示倍数):

reduce (by) *n* times

be reduced *n* times

reduce by a factor of *n*

reduce...*n* times

reduce to...(number)

reduce by+百分数/数字

reduce *n* times as much/many...as

n-fold reduction

n times less than

例如:

When the space between the two conductors of a capacitor is occupied by a dielectric with the relative dielectric constant ε_r, it is found that the voltage between the plate *decreases by a factor of* ε_r. 如果电容器的两个极板间充满相对介电常数为 ε_r 的电介质,极板间的电压会减少到原来的 $1/\varepsilon_r$。

This kind of film *is twice thinner than* ordinary paper, but its quality is quite good. 这种胶片的厚度只是普通纸张的一半,可是质量相当好。

The instrument *is three times as* light as that one. 这个仪器比那个轻 2/3。或:这个仪器的重量是那个的 1/3。

The advantage of new-type equipment *is a 3-fold reduction* in volume. 新型设备的优点是体积小了 2/3。

当然,也可以使用“动词(decrease, reduce, come down, drop, fall, lower, cut 等)+ to +百分数”“百分数+形容词的比较级+than”“百分数+介词+名词”的结构表示数量的减少,如:

By the year 2010, the annual rare earth output of the company is expected to *fall to* 35 *percent*. 预计到 2010 年,该公司的稀土年产量将下降 35%。

Unit Six

LESSON 21

21.1 物理学专业英语的翻译

21.1.1 翻译的总体原则

物理学专业英语是在物理学领域中进行国际学术交流的主要工具，而且其所涵盖的大量科技信息和专业术语在其他的学科中也有广泛的应用，因此在科技英语中占有重要的地位。总的来说，专业英语的翻译应力求客观、精确、通顺、简练，要充分注意其在词汇、句子、形态和文体方面的特点，以客观事物为中心，做到长而不乱，避免产生任何歧义。具体而言，翻译应力求用词上的准确明晰，论述上的逻辑严密，表述上的客观严谨，行文上的简洁通畅和修辞上的简单平实。词汇和结构是专业英语翻译的重点和难点，需要进行深入理解和反复推敲，同时注意知识的积累。只有在掌握与理解专业词汇和结构特点的基础上，恰当地使用翻译技巧，才能够达到翻译工作的基本要求。

整体上说，翻译过程包含三个阶段：理解、表述和校对。

理解包含对语法结构和逻辑关系的理解。掌握结构是基础，正确的逻辑分析才能够给出准确的翻译。由于物理学专业英语中长句较多，所以先进行结构分析会有助于理解逻辑关系。理解之后便是表述，无论采用直译还是意译，都应做到既忠实原文又表述得体。校对是必不可少的阶段，也是对前两个阶段的深化和推敲。物理学专业英语的翻译还有一个要点伴随始终，这就是专业的背景知识。语言是一种工具与桥梁，所传递的信息本身才是最重要的，优秀的译者必定要对所涉及的领域有足够的认识与理解。

翻译应首先力求正确与通达，然后才是行文的流畅、简洁和美感。因此翻译工作应遵循但不应拘泥于翻译的基本方法，应在“信”的基础上对译文进行调整与修饰，努力实现译文的“达”与“雅”。英语的各种基本翻译技巧均可以为物理学专业英语所用，例如直译、转换、省略、增补、合并、拆分等。在翻译中应将各种语法现象的基本翻译方法与翻译技巧相结合，对特殊或复杂的情况做恰当的处理，而不可一味套用固定的方法与结构。

在后面的篇幅中，我们将在专业术语、独立主格结构、名词化结构、分隔结构和长句这几个有代表性的方面，通过例子对物理学专业英语几种常见的翻译方法进行简单介绍，供读者参考。

21.1.2 物理学专业术语的翻译

词语是语言的基本要素，是信息的载体。物理学专业英语中大量使用专业术语，这些术语的意义较为固定和通用,是学术信息的载体。专业术语的翻译是准确理解专业文献的基础,更是专业写作时准确传递信息的关键。专业术语有一些基本的构成方式,了解这些方式有助于正确理解语义,同时有些词汇作为多义词在不同的语境中具有不同的含义,特别是在专业表述中有其特定的含义,需要进行认真的分析与选择。随着物理学的发展,新的专业词语也源源不断地产生,构成方式多种多样,构成的自由度也比以往更大,这就需要在平时的阅读中不断地总结和积累,并且在实际运用中以专业知识作为辅助。

1. 一般专业术语的翻译

(1) 音译

① 全部音译

物理学专业英语中很多材料的名称、物理量的计量单位(包括部分计量单位的前缀)、某些特殊方法等,通常均采用音译。例如:

ampere	安培(电流单位)
atto-	阿(托)(10^{-18})
farad	法拉(电容单位)
femto-	飞(母托)(10^{-15})
lux	勒克斯(光照度单位)
pico-	皮(可)(10^{-12})
pump	泵浦
roentgen	伦琴(辐射计量单位)
siemens	西门子(电导单位)
teflon	特氟隆

② 音译与意译结合

某些合成词包含人名、商标名或字母,一般名称和字母采取部分音译或直接保留,其余部分直译。例如:

Avogadro constant	阿伏伽德罗常数
Bose-Einstein condensate	玻色-爱因斯坦凝聚
Brinell hardness	布氏硬度
Bunsen burner	本生灯
Carnot engine	卡诺热机
Coriolis force	科里奥利力
β decay	β衰变
Fahrenheit scale	华氏温标
Fermi level	费米能级
Photovoltaic	光生伏打效应
P-N junction	PN 节
roentgen meter	伦琴计
Siemens heat	西门子热
Stokes' law	斯托克斯定律
Brookhaven National Laboratory	布鲁克海文国家实验室

这里要特别说明的是，除了为人们所熟知的著名科学家(历史人物)外，对于文献中的人名，应尽量采取保留英文拼写不变的方式，这可以提高译文的通用性，便于文献的检索和相关资料的查找。

(2) 形译与意译结合

某些物体、状态或方式是以其形状来命名的，在表示形状或状态时有些直接使用单词，有些则使用相似形象的英文字母，这类术语可以采取形译与意译相结合的方式来翻译，形译时可选择具有相应形状的汉字或物体。例如：

cross bit	十字钻头，X形钻头
delta connection	三角形连接/△接法
D-iron	圆角槽铁
H-beam/T-beam	工字梁/丁字梁
I-shaped/I-section	工字形/工形截面
L-square/T-square	直角尺/丁字尺
thumbscrew	翼形螺钉
twist drill	螺丝钻，麻花钻
U-shaped magnet/U-trap	马蹄形磁铁(U形磁铁)/虹吸管
Y-shaped network/Y-connection	星形电路/星形连接

(3) 组合与直译结合

很多专业术语是将几个单词直接组合在一起构成的，这类术语很多，构成方式也很多，例如：名词＋名词，形容词＋名词，分词＋名词，名词＋分词等。这些术语一般可采用直译的方式，直接将几个词的词义组合在一起，但为了做到翻译的准确与专业，应分析各单词之间的逻辑关系，选择最为恰当的表述。例如：

diffraction grating	衍射光栅(名词＋名词)
condensation point	凝点(名词＋名词)
beam current density	束流密度(名词＋名词＋名词)
constructive interference	相长干涉(形容词＋名词)
computational physics	计算物理学(形容词＋名词)
simple harmonic motion	简谐运动(形容词＋形容词＋名词)
convertible unsymmetrical anastigmat	可变半对称消相差镜组(形容词＋形容词＋名词)
alternating current	交流电(现在分词＋名词)
heat resisting	耐热(的)(名词＋现在分词)
battery-powered	电池供电的(名词＋过去分词)

2. 缩略词的翻译

科技进步要求信息传递的通用、快捷，而缩略词通用性强，简洁明了，因此在专业文献中被广泛使用。缩略词的翻译一般采取音译，意译或保留原来的缩写部分不变的方式，但通用性不高的缩略形式应在文章中首次使用时予以说明。

(1) 音译

某些术语的构成和意译都比较复杂，因此其缩略形式和音译已经被广泛使用并当作固定词汇，这类缩略词可直接采用音译，而无须注明其完整形式。例如：

radar　　　雷达 radio detecting and ranging
laser　　　镭射(激光) light amplification by stimulated emission of radiation
sonar　　　声呐 sound navigation and ranging

(2) 意译或保留缩写形式

当缩略词的意译表述并不太复杂时,可以采取意译,但也可以在汉语表述中保留缩写形式不变,或仅在首次使用时给出意译进行解释,其后使用缩写形式。例如:

laser	激光,这是更为通用的译法
BCS theory	BCS 理论(Bardeen,Cooper,Schrieffer 建立的一种超导微观理论)
BSCCO	铋锶钙铜氧 bismuth strontium calcium copper oxide
CD	光盘 compact disc
CMOS	互补型金属氧化物半导体 complementary metaloxide semi-conductor
EDFA	掺铒光纤放大器 erbium-doped fibre anplifier
Fortran language	Fortran 语言 formula translation language
GPS	全球定位系统 global positioning system
IC	集成电路 integrated circuit
LCD	液晶显示器 liquid crystal display
LED	发光二极管 light emitting diode
LSI	大规模集成电路 large scale integration
NMR	核磁共振 nuclear magnetic resonance
SEM	扫描电镜 scanning electron microscope
SQUID	超导量子干涉仪 superconducting quantum interference device
STM	扫描隧道显微镜 scanning tunneling microscope
TBA	紧束缚近似 tight binding approximation
TEM	透射电镜 transmission electron microscope
WDM	波分复用 wavelength division multiplexing
XRD	X 射线衍射 X-ray diffraction
YBCO	钇钡铜氧 yttrium barium copper oxide

3. 多义词的翻译

意译是物理学专业词汇翻译的一个重要方法,而正确的意译需要专业背景知识这个至关重要的因素,对于多义词的翻译更是如此。应根据文章所涉及的领域和背景、原词所涉及的概念或事实进行仔细推敲,正确选择符合句子含义或文章内容的科学概念,这些概念可能与非专业英语中的常用含义有所不同。例如:

argument	自变量
balance	天平
bank angle	自转角
bus	母线、总线
butterfly	蝶形突变
drum	鼓轮
family	族
finger	测厚规、指针

fork 音叉
grain 晶粒
hair 游丝
host 基质
moment 矩、力矩
monkey chatter 邻道干扰
normal 法线、垂直的、正交的
phosphors 荧光粉
plane 平面
spring 弹簧
transition 跃迁
work 功

21.2 专业英语阅读：Bohr's Model（玻尔模型）

The model of atom proposed by Rutherford had several severe problems. It didn't account for the periodicity of the elements，it was unable to give any clue about the origins of the spectral lines，and it implied that atoms were extremely unstable. In 1913，the Danish theoretician Niels Bohr proposed a new model incorporating Planck's discrete energies and Einstein's photon into Rutherford's model.

Scientists are always using models when attempting to understand nature. The trick is to understand the limitations of the models. Bohr challenged the Rutherford model，stating that it was a mistake to assume the atom is just a scaled-down solar system with the same rules；electrons do not behave like miniature planets. Bohr proposed a new model based on three assumption，or postulates.

First，he proposed that the angular momentum of the electron is quantized. Only momenta L equal to whole-number multiples of a smallest angular momentum are allowed. This smallest angular momentum is equal to Planck's constant h divided by 2π. Bohr's first postulate can be written

$$L_n = n\frac{h}{2\pi} \tag{21-1}$$

where n is a positive integer (whole number) known as quantum number.

In classical physics，the angular momentum L of a particle of mass m traveling at a speed v in a circular orbit of radius r is given by $L = mvr$. Therefore，the restriction on the possible values of the angular momentum puts restriction on the possible radii and speeds. Bohr showed that the only possible orbits are those for which the radii obey the relationship

$$r_n = n^2 r_1 \tag{21-2}$$

where r_1 is the smallest radius and n is the same integer that appears in the first postulate. This means that the electrons cannot occupy orbits of arbitrary size，but only a certain discrete set of allowable orbits. The numerical value of r_1 is 5.3×10^{-11} meter.

There is also a definite speed associated with each possible orbit. This means that the kinetic energies are also quantized. Because the value of the electric potential energy depends on distance, the quantization of radii means that the potential energy also has discrete values. Therefore, there is a discrete set of allowable energies for the electron.

Bohr's second postulate states that an electron does not radiate when it is in one of the allowed orbits. This statement is contrary to the observation that accelerated charges radiate energy at a frequency equal to their frequency of vibration of revolution. Bohr challenged the assumption that this property is also true in the atomic domain. Radically breaking away from what was accepted, Bohr said that an electron has a constant energy when it is in an allowed orbit.

But light is produced by atoms. If light isn't radiated by orbiting electrons, how does an atom emit light? Bohr's third postulate answered this question: A single photon is emitted whenever an electron jumps down from one orbit to another. Energy conservation demands that the photon has energy equal to the difference in the energies of the two levels. It further demands that jumps up to higher levels can only occur when photons are absorbed. The electron is normally in the smallest orbit, the one with the lowest energy. If the atom absorbs a photon, the energy of the photon goes into raising the electron into a higher orbit. Furthermore, it is not possible for the atom to absorb part of the energy of the photon. It is all or nothing.

The electron in the higher orbit is unstable and eventually returns to the innermost orbit, or ground state. If the electron returns to the ground state in a single jump, it emits a new photon with energy equal to that of the original photon. Again, this is different from our common experience. When a ball loses its mechanical energy, its temperature rises—the energy is transferred to its internal structure. As far as we know, an electron has no internal structure. Therefore, it loses its energy by creating a photon.

Occasionally, the phrase "a quantum leap" is used to imply a big jump. But the jump needn't be big. There is nothing big about the quantum jumps we are discussing. A quantum leap simply refers to a change from one discrete value to another.

Ordinarily, an amount of energy is stated in joules, the standard energy unit. However, this unit is extremely large for work on the atomic level, and a smaller unit is customarily used. The electron volt (eV) is equal to the kinetic energy acquired by an electron falling through a potential difference of 1 volt. One electron volt is equal to 1.6×10^{-19} J. It requires 13.6 electron volts to remove an electron from the ground state of the hydrogen atom.

Bohr's explanation accounted for the existence of spectral lines, and his numbers even agreed with the wavelengths observed in the hydrogen spectrum. To illustrate Bohr's model, we draw energy level diagrams in Fig. 21-1. The ground state has the lowest energy and appears in the lowest position. Higher energy states appear above the ground state and are spaced to indicate the relative energy difference between the states.

To see how Bohr's model gives the spectral lines, consider the energy level diagram. Suppose that the electron has been excited in the $n=4$ energy level. There are several ways that it can return to the ground state. If it jumps directly to the ground state, it emits the largest energy photon. The electron could also return to the ground state by making of a set of smaller jumps. It could go from the $n=4$ to the $n=2$ level and then from the $n=2$ to the ground level. The energies of these photons are smaller, so their lines occur at the end of the spectrum.

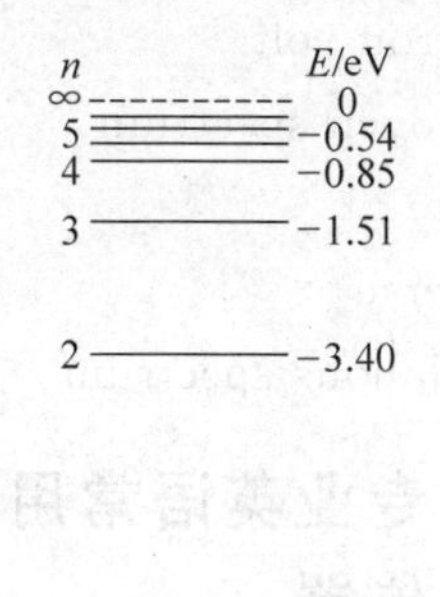

Fig. 21-1 The energy-level diagram for the hydrogen atom

Bohr's scheme also explains the absorption spectrum. The absorption spectrum is obtained when white light passed through a cool gas before it is dispersed by a prism or a diffraction grating. In this case, the photons from the beam of white light are absorbed and later reemitted. Because the reemitted photons are distributed in all directions, the intensities of these photons are reduced in the original direction of the beam. This removal of the photons from the beam leaves dark lines in the continuous spectrum.

The solution to the mystery of the missing lines in the absorption spectrum is easy within Bohr's model. For an absorption line to occur, there must be electrons in the lower level. Then they can be kicked up one or more levels by absorbing photons of the correct energy. Because almost all the electrons in the atoms of the gas occupy the ground state, only the lines that correspond to jumps from this state show up in the spectrum. Jumps up from higher levels are extremely unlikely because electrons excited to a higher level typically remain there for less than a millionth of a second. Thus, there are fewer lines in the absorption spectrum than in the emission spectrum, in agreement with the observations.

Glossary

propose	提出	periodicity	周期性
element	元素	spectral lines	光谱线
theoretician	理论家	incorporate	包含、加入
discrete	不连续的	Einstein's photon	爱因斯坦的光子
Rutherford's model	卢瑟福模型	trick	技巧
scaled-down solar system	微缩的太阳系	miniature	小型的、微小的
postulate	假定	quantize	*v*. 量子化
whole-number	整数	Planck's constant	普朗克常量
integer	整数	quantum number	量子数
restriction	限制、约束	radiate	辐射

ground state	基态	quantum leap	量子跃迁
electron volt	电子伏特	hydrogen	氢
hydrogen spectrum	氢光谱	energy level diagram	能级图
excite	激发	absorption spectrum	吸收光谱
disperse	分散	diffraction grating	衍射光栅
continuous spectrum	连续光谱	emission spectrum	发射光谱

21.3 专业英语常用表达-21 光学、近代物理及综合设计常用实验仪器

Abbe refractometer 阿贝折射计
autocollimator 自准直仪
battery of lens 透镜组
biprism 双棱镜
blazed grating 闪耀光栅
concave lens 凹透镜
convex lens, positive lens 凸透镜
diaphragm 光阑
diffractometer 衍射仪
dispersion prism 色散棱镜
double-slit interferometer 双缝干涉仪
eye lens 接目镜
eyepiece 目镜
Fabry-Perot interferometer 法布里-珀罗干涉仪
Fourier grating 傅里叶光栅
Fresnel's biprism 菲涅耳双棱镜
grating 光栅
half wave plate 半波片
holographic grating 全息光栅
interferometer 干涉仪
laser 激光
lens 透镜
Michelson interferometer 迈克耳孙干涉仪
micrometer 显微镜
monochrometer 单色仪
multi-beam interferometer 多光束干涉仪
objective lens 物镜
optical bench 光具座
optical filter 滤光片
optical lever 光杠杆
optical table 光学平台
photometer 光度计
polarimeter 旋光仪
polaroid 偏振片
prism 棱镜
quarter wave plate 四分之一波片
Rayleigh refractometer 瑞利折射计
reading micrometer 读数显微镜
reflection grating 反射光栅
reflectometer 反射仪
refractometer 折射仪
scatterometer 散射仪
spectrometer 光谱仪/分光计
telescope 望远镜
thin lens, shallow lens 薄透镜
transmission grating 透射光栅
triple prism 三棱镜
vertometer 屈度计
white-light interferometer 白光干涉仪
X-ray diffractometer X射线衍射仪
experiment facility of temperature characteristics of PN junction positive voltage drop PN 结正向压降温度特性实验仪
superconducting maglev demonstration unit 超导磁浮车演示装置

sensor experiment table　传感器实验台

plasma diagnosis experiment facility　等离子体诊断实验仪

electrical meter reform and calibration experiment facility　电表改装与校准实验仪

tester of charge-to-mass ratio of the electron　电子荷质比测定仪

comprehensive design-orientated dynamic experiment facility　动力学综合设计性实验装置

multi-function microwave plasma experiment facility　多功能微波等离子体实验装置

multi-function vacuum experiment facility　多功能真空实验仪

Doppler effect and sound velocity comprehensive experiment facility　多普勒效应及声速综合实验仪

Doppler effect experiment facility　多普勒效应实验仪

Franck-Hertz experiment facility　弗兰克-赫兹实验仪

solid measurement unit　固体测量装置

photosensor experiment facility　光电传感器实验仪

photoresistance characteristics experiment facility　光敏电阻特性实验仪

NMR experiment facility　核磁共振实验仪

glow plasma experiment facility　辉光等离子体实验仪

Young's modulus tester with Hall position sensor　带霍尔位置传感器的杨氏模量测定仪

AC circuit comprehensive experiment facility　交流电路综合实验装置

Planck constant tester/determinator　普朗克常数测定仪

atmosphere heat treatment program-controlled high-temperature furnace　气氛热处理程控高温炉

portable holograph experiment facility　轻便全息照相实验仪

sound velocity tester and signal source　声速测定仪及信号源

manual levitation force measurement unit　手动磁浮力测量装置

experiment facility of principle of digital electrical meter and design of universal meter　数字电表原理及万用表设计实验仪

double-grating weak vibration meter　双光栅微弱振动测量仪

temperature-sensing experiment facility　温度传感实验装置

physics design-orientated experimental modules　物理设计性实验模块

physics design-orientated experiment facilities　物理设计性实验装置

experiment facility of string vibration and sound　弦音实验仪

string vibration research experiment facility　弦振动研究实验仪

DC sputtering plasma experiment facility　直流溅射等离子实验装置

rotational inertia tester　转动惯量实验仪

LESSON 22

22.1 物理学专业英语中名词化结构的翻译

名词化现象是英语表达中的一种常见而又独特的语言现象，能够高度概括事物的本质，适合于表述现象、过程、性质、特征、原因和结果等，因此是物理学专业英语中重要的语法现象。其简洁、客观、正式、严密的表达可以充分的体现物理学科的严谨性、逻辑性和权威性，同时也可以简化复杂的句子结构。名词化结构同时具有名词和其他词性的功能，正确的翻译有利于信息的准确传递和对外文资料的正确理解。

名词化结构的翻译应考虑其不同的语法功能和结构特点，并结合专业知识的背景，使译文符合汉语的习惯与要求，在翻译中应先分析其地位、作用和与其他成分的关系。一些复杂的名词化结构实际上是浓缩的句子，翻译时要先理清主干和层次，并且应避免直接套用原文的结构，因为那样往往会造成译文的臃肿晦涩。只有将实际的内容、语境与翻译的方法、技巧进行良好的整合才能够得到恰当的译文。

22.1.1 动词名词化

在物理学专业英语中，动词的名词化结构最为常见，形式也比较丰富。这种名词化结构具有动词和名词的功能，名词特征使行文简洁，而动词含义则使事实表述准确，其中文词性和意义在翻译中可以灵活的处理。

1. 动作名词

动作名词指具有动词含义的名词。往往这类单词本身就具有动词和名词两种词性，翻译时应综合分析动作名词与其他成分之间的逻辑关系及其在句中的作用，将动作名词翻译为动词或是名词，并根据汉语表述的实际需要来选择通顺、易于理解的译文。

Electric current is the rate of charge *flow* past a given point in an electric circuit. 电流是单位时间内流经电路中某定点的电量。(*flow* 译为动词)

The *flow* of charge is from the positive to the negative plate when the capacitor is discharged. 放电时电荷的流动是从电容器的正极到负极。(*flow* 译为名词)或译为：放电时，电荷从电容器正极流向负极。

2. 动词的派生名词

动词的派生名词由动词加上-ment，-tion，-ance，-ence，-sm，-y，-ity，-ness，-ability 等后缀构成，这种派生名词大量存在，在句中可与其他成分构成各种逻辑关系。

(1) 派生名词＋介词＋名词

在这种结构中，如果介词后的名词在逻辑语义上是派生名词的主语，则可译为主谓结构；如果介词后的名词在逻辑语义上是派生名词的宾语，则可译为动宾结构。

The velocity *dependence of the air friction* may be very complicated so only some special cases can be treated analytically. 空气阻力随速度的变化非常复杂，因此只有一些特殊情况可以进行解析处理。(译为主谓结构，air friction 作 dependence 的逻辑主语)

The *formation of images* by lenses and eyes is on the base of the *refraction of light*. 镜头和眼睛通过光的折射成像。(formation of images 译为动宾结构，images 作 formation 的逻辑宾语；refraction of light 译为主谓结构，light 是 refraction 的逻辑主语)

The antirust paint has a *resistance to corrosion* by air or water. 防锈漆可以防止空气或水的腐蚀。(译为动宾结构，corrosion 作 resistance 的逻辑宾语)

The *superimposition of a small drift velocity by the applied electric field* on the free electron explains why the current is proportional to the voltage. 电流与电压成正比是由于外加电场在自由电子上叠加了一个很小的漂移速度。(句中有两层此类的名词化结构，其中 superimposition＋of a small drift velocity 译为动宾结构，而 superimposition＋by the applied electric field 则译为主谓结构)

(2) 介词＋派生名词

这种介词短语形式的名词化结构在句中可以表示时间、原因、条件、目的、方式、让步等，因此可译成相应的状语成分或状语从句。

The formation of rainbow is due *to dispersion* of sunlight by water droplets in the atmosphere. 彩虹的形成是由于大气中的水滴对阳光的色散。(原因状语)

Ballistic curve is the actual trajectory *in presence of* air resistance. 弹道曲线是空气阻力存在情况下的实际轨迹。(条件状语)

The electric potential at any point in space produced by any number of point charges can be calculated *from the point charge expression by simple addition* since voltage is a scalar quantity. 既然电势是标量，多个点电荷在空间某点产生的电势就可以通过单个点电荷表达式的简单叠加来计算。(方式状语)

(3) 派生名词＋短语或从句

此时派生名词前往往存在一个不表意的动词，与该派生名词构成动宾关系，因此这种结构中的派生名词往往译成动词。

The noise gives an *indication that* the signals are being disturbed. 噪声表明信号正受到干扰。(接从句)

The noise gives an *indication of* the disturbance to signals. 噪声表明信号正受到干扰。(接短语)

The existence of magnetic domains gives an *explanation to* the magnetic hysteresis. 磁畴结构的存在解释了磁滞现象。(注意其中 existence of magnetic domains 也是名词化结构，属于"派生名词＋介词＋名词"的情况)

The friction with air leads to the *decrescence of* the speed. 摩擦导致速率减小。(注意其中 friction with air 也是名词化结构，属于"派生名词＋介词＋名词"的情况)

(4) 名词＋派生名词

英语中这种结构很多，应根据名词和派生名词的逻辑关系进行翻译。

① 如果前面的名词对派生名词起到修饰的作用，短语可改写为"派生名词＋介词＋名词"的形式，则可翻译成以派生名词为中心词的名词短语，前面的名词作为前置定语。

Once the cars are lifted to the top of the hill, gravity takes over and the remainder of the ride is an experience in *energy transformation*. 一旦车被开至山顶，重力就起主要的作

用，其余的路程便是能量转化的过程。（斜体部分意为 transformation of energy）

Electrostatic shielding effectively reduces the *signal disturbance*. 静电屏蔽有效地减小了信号干扰。（斜体部分意为 disturbance to signal）

The number of significant digits does not change in the decimal *unit conversion*. 十进制的单位换算中有效数字位数不变。（斜体部分意为 conversion of unit）

② 如果名词在逻辑含义上是派生名词的宾语，则可以翻译成动宾短语。

Voltage division circuits are more often used in that case. 在这种情况下更多的使用分压电路。（voltage 是 division 的逻辑宾语）

Heat pump technology is effective for *energy efficiency improvement*. 热泵技术能够有效地提高能效。（energy efficiency 是 improvement 的逻辑宾语）

(5) 动词（短语）＋派生名词

这种用法中的动词一般没有明显的实际含义，只作为形式上的谓语，因此可以直接将动词的派生名词译为相应的谓语成分。值得一提的是，有很多这种形式的名词化结构已经成为了固定搭配，如 bear relation to，make use of，find application in，come into collision (with)，give cause/rise to，give effect to，come into action，take into consideration 等。短语的含义均来自于名词化的动词，例如：

Laser spectroscopy has *led to advances* in the precision of the measurement of the spectral line frequencies. 激光光谱提高了谱线频率的测量精度。

The measurement of the sound velocity in some medium *bears close relation to* the temperature because of the changing state of the medium along with the temperature. 由于介质的状态随温度变化，因此声速的测量结果与温度密切相关。

The resultant external force *gives cause/rise to* the change of the velocity. 合外力导致速度的变化。

The hydrostatic weighing *makes use of* the Archimedean principle. 流体静力称衡法利用了阿基米德原理。

3. 动名词

动名词的动词性更强，一般可翻译成动词，但由于包含动名词的名词化结构往往可以改写为包含相应派生名词的名词化结构，因此亦可参照这些结构的翻译方法来翻译，并且根据动名词在句中所起的作用，以及与其他成分的关系来确定译文。

The function of the rheochord in the circuit is *current limiting*. 电路中滑线变阻器的作用是限流。

Heat pump technology is effective for *improving energy efficiency*. 热泵技术能够有效地提高能效。

Doing work and *transferring heat* are the two ways for *changing* the internal energy of a system. 做功与传热是改变系统内能的两种途径。

22.1.2 形容词名词化

形容词名词化结构在专业英语中使用的也较多，该结构具有名词词性，可以做主语、宾语、同位语等，既有形容词生动形象的表意功能，又满足科技文体表述严谨的要求。

1. 形容词的派生名词

物理学中有很多由形容词加上-ness,-ity,-ancy,-ency,-ce,-th,-ism 等词缀构成的派生名词，通常表示抽象概念，如状态、性质、程度、学科等，其中大部分已经被视作普通的名词，可直译为名词。

In physics, *intensity* is a measure of the time-averaged energy flux. 物理学中，强度是时均能流的量度。

In *electromagnetism*, *permeability* is the degree of magnetization of a material that responds linearly to an applied magnetic field. 在电磁学中，磁导率是材料对外磁场线性响应的磁化程度。

Electrical *conductance* is dependent on the electrical *conductivity* and the geometric dimensions of the conducting object. 电导与导体的电导率及几何尺寸有关。（句中 conductance 是动词的派生名词，conductivity 是形容词的派生名词，注意二者在词义上的区别，conductance 强调性质，而 conductivity 强调能力）

2. the＋形容词

在英语中，“the＋形容词”的结构通常用于表示具有相应特点的一类事物，而在物理学专业英语中，这种结构实际上通常是名词结构“the＋形容词＋名词”的省略形式，而且往往并列使用，其中的名词由于已经在句中被提及而省略，以避免表述上的重复，在翻译时可以酌情进行补充。

The combined uncertainty synthetically reflects the two kinds of errors, *the systematic* and *the random*. 合成不确定度综合地反映了系统误差与随机误差。（省略了 error）

In the center-of-mass system, the curvilinear motion has two velocity components, *the tangential* and *the normal*. 在质心坐标系中曲线运动有切向和法向两个速度分量。（省略了 component）

22.2 专业英语阅读：Waves and Particles（波与粒子）

Imagine the controversy de Broglie caused. Interference of light is believable because parts of the waves pass through each slit and interfere in the overlap region. But electrons don't split in half. Every experiment designed to detect electrons has found complete electrons, not half an electron. So how can electrons produce interference patterns?

The following series of thought experiments were proposed by physicist Richard Feynman to summarize the many, many experiments that have been conducted to resolve the wave-particle dilemma. Although they are idealized, this sequence of experiments gets at crucial factors in the issue.

We will imagine passing various things through two slits and discuss the patterns that are produced on a screen behind the slits. In each case, there are three pieces of equipment: a source, two slits, and a detecting screen arranged as in Fig. 22-1.

In the first situation we shoot indestructible bullets at two narrow slits in a steel plate.

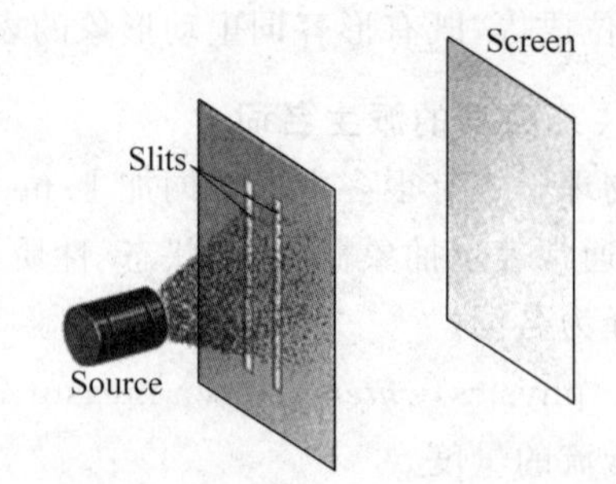

Fig. 22-1 The experimental setup for the thought experiment concerning the wave-particle dilemma

Assume that the gun wobbles, so that the bullets are fired randomly at the slits. Our detecting screen is a sand box. We simply count the bullets in certain regions in the sand box to determine the pattern.

Imagine that the experiment is conducted with the right-hand slit closed. After 1 hour we sift through the sand and make a graph of the distribution of bullets. This graph is shown in Fig. 22-2 (a). The curve is labeled N_L to indicate that the left-hand slit was open. Repeating the experiment with only the right-hand slit open yields the similar curve N_R shown in Fig. 22-2(b).

When both slits are opened, the number of bullets hitting each region during 1 hour is just the sum of the numbers in the previous experiments, that is, $N_{LR}=N_L+N_R$. This is just what we expect for bullets, or any other particles. This graph is shown in Fig. 22-2(c).

Now imagine repeating the experiment using water waves. The source is now an oscillating bar that generates straight waves. The detecting screen is a collection of devices that measure the energy (that is, the intensity) of the wave arriving at each region. Since these are waves, the detector does not detect the energy arriving in chunks, but rather in a smooth, continuous manner. The graphs in Fig. 22-3 show the average intensity of the waves at each position across the screen for the same three trials.

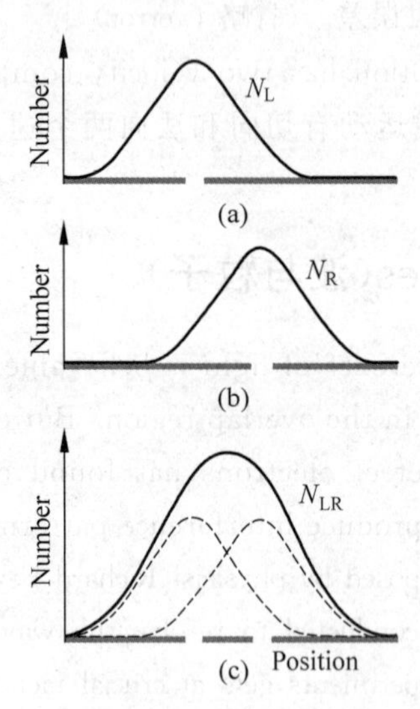

Fig. 22-2 Distribution of bullets with (a) the left-hand slit open N_L, (b) the right-hand slit open N_R, and (c) both slits open N_{LR}

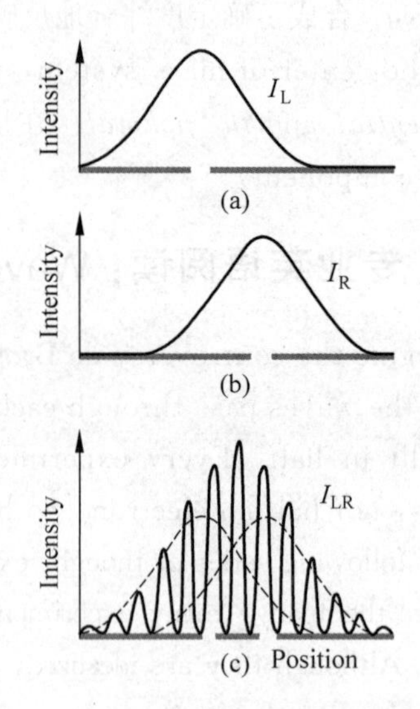

Fig. 22-3 Intensity of water waves with (a) the left-hand slit open I_L, (b) the right-hand slit open I_R, and (c) Both slits open I_{LR}

Once again, the three trials yield no surprise. We expect an interference pattern when both slits are open. We see that the intensity with two slits open is not equal to the sum of the two cases with only one slit open, $I_{LR} \neq I_L + I_R$ but this is what we expect for water waves, or any other waves.

Both sets of experiments make sense, in part, because we used materials from the macroscopic world—bullets and water waves. However, a new reality emerges when we repeat these experiments with particles from the atomic world.

This time we use electrons as our "bullets". An electron gun shoots electrons randomly toward two very narrow slits. Our screen consists of an array of devices that can detect electrons. Initially the results are similar to those obtained in the bullet experiment: Electrons arrive as whole particles. The patterns produced with either slit open are the expected ones and are the same as those in Fig. 22-2((a) and (b)). The surprise comes when we look at the pattern produced with both slits open: We get an interference pattern like the one for waves!

Note what this means: If we look at a spot on the screen that has a minimum number of electrons and close one slit, we get an increase in the number of counts at that spot. Closing one slit yields more electrons! This is not the behavior expected of particles.

One possibility that was suggested to explain these results is that the electrons are somehow affecting each other. We can test this by lowering the rate at which the source emits electrons so that only one electron passes through the setup at a time. In this case, we might expect the interference pattern to disappear. How can an electron possibly interfere with itself? Each individual electron should pass through one slit or the other. How can it even know that the other slit is open? But the interference pattern doesn't disappear. Even though there is only one electron in the apparatus at a time, interference effects are observed after a large number of electrons are measured.

The set of experiments can be repeated with photons. The results are the same. The detectors at the screen see complete photons, not half photons. But the two-slit pattern is an interference pattern. Photons behave like electrons. Photons and electrons exhibit a duality of particle and wave behavior.

Glossary

controversy	争论、争议	de Broglie	德布罗意
slit	狭缝	overlap region	叠加区
thought experiment	思想实验	dilemma	进退两难的局面
indestructible	不能破坏的	wobble	摇晃
sift	详审	chunk	大块
array	排列、编队	duality	二象性

22.3 专业英语常用表达-22 极值 约略数

1. 极值的表示

(1) 达到 as...as+number

H. K. Onnes discovered in 1911 that the resistance of the mercury suddenly disappeared when cooled to a temperature *as low as* 268.98℃ below zero which was later called superconductivity. 昂内斯在1911年发现,当温度低至零下268.98℃时,水银的电阻突然消失,这一现象称为超导。

(2) 直到 up to+number

Under different conditions, the results of the experiment varied wildly from as little as 20 percent of normal *up to* as high as 90 percent of normal. 在不同的条件下,实验结果可以从正常值的20%变化到正常值的90%。

(3) 至少 at least+number

In order to obtain reliable data the experiment should be conducted *at least* 10 times under the same conditions. 为获得可靠的数据,应在相同条件下至少测量10次。

(4) 最大 at most+number

The magnetic field should be strictly controlled so that the deviation of the exciting current is *at most* 0.5 A. 应严格控制磁场,以使励磁电流的偏差不超过0.5 A。

(5) 最大值和最小值 *maximum* and *minimum*(用作形容词或名词)

Neither the *maximum* possible temperature nor the *minimum* one should be out of the range of the thermometer.(用作形容词)

Neither the *maximum* nor the *minimum* of the possible temperatures should be out of the range of the thermometer.(用作名词)

最高和最低温度均不可超过温度计的量程。

(6) 上限与下限 *the upper limit* and *the lower limit*

This is a new method for dynamic measurement of thermocouples protected by metal tube, which can be used to raise the *upper limit* of the temperature greatly. 这是一种动态测量带保护管的热电偶温度的新方法,可大大提高测温上限。

(7) 增至最大和减至最小 *to maximize* and *to minimize*

To minimize the error of the experimental result, the precisions of all the measuring apparatus should be upgraded to the appropriate level according to the principle of equipartition for the uncertainty. 为将测量结果的误差减至最小,所有测量仪器的精度应该按照不确定度均分原理提高到适当的等级。

2. 约略数的表示

(1) 以基数词的复数形式来表示:

the early/mid-/late 70s　70年代初期/中期/晚期

the early/mid-/late 1970s　20世纪70年代初期/中期/晚期

teens/twenties/fifties　十几/二十几/五十几

dozens of/scores of　数十／几十
hundreds of　数百／几百／成百上千
thousands of　数千／几千／成千上万
tens of thousands of　数万／几万
hundreds of thousands of　数十万／几十万
millions of　数百万／几百万
tens of millions of　数千万／几千万
millions and millions of　亿万
(2) 添加表示近似的词或短语来表示：
about(around/approximately/some/nearly/roughly/almost) three meters　大约 3 m
more than/over/above 40 grams　40 多克
less than/under/below 40 grams　不到 40 g
more than a quarter/one third　四分之一强／三分之一强
upward/upwards of 60　60 以上／多于 60
40 milliliters or so　40 ml 左右
20 seconds or more/or less　20 s 以上／以下
an hour more or less　1 h 左右
of the order of 3000 volts　约为 3000 V
(3) 以表示范围的短语来表示：
greater than/less than　大于／小于
be equal or greater than/be equal or less than　大于等于／小于等于
not more than/not less than　不大于／不小于
of the order of...　约为……
range from...to.../range between...and...　范围在……之间(谓语)
with a range of...to...　范围在……之间(定语或状语)
vary from...to.../vary between...and...　从……变化至……
(4)"数量级"的表示：
order of magnitude，scale，level，order　数量级
one order/two orders of magnitude　一个／两个数量级
at the same level/at the same order (of magnitude)　在同一数量级上
the increase/decrease of three orders (of magnitude)　三个数量级的增加／减少
increase/decrease by one order of magnitude　增加／减少一个数量级
microscopic/astronomic scale　微观尺度／天文尺度的数量级
one order higher than...　比……高一个数量级
the scale of millisecond/the millisecond scale/the order of millisecond　毫秒数量级

LESSON 23

23.1 物理学专业英语中独立主格结构和分隔结构的翻译

23.1.1 独立主格结构的翻译

独立主格结构是一种很常见也很重要的语法现象，主要用于正式文体，可用于描绘事实，使行文紧凑简洁、表达清楚、形象具体，因此在专业文献中使用较多，其正确翻译在专业英语阅读与写作中都显得非常重要。

独立主格结构的实质是具有独立主语的非限定分句和无动词分句，属于从属结构，在意义上相当于一个从句，通常在句中充当状语，表示时间、条件、原因、结果、方式和伴随状况等。独立主格结构可位于句首或句末（偶见于句中），通常用逗号（或破折号）与主句分开，其主语与主句的主语一般不同。多个独立主格结构可以同时存在，通常把表示概括意义的结构置于最后。在独立结构中作主语的名词词组有时可以省去限定词。

独立主格结构可以根据不同的标准来分类，例如按照结构中是否含有动词成分可以分为两种，一种采用与动词有关的现在分词、过去分词和不定式，而另一种则采用与动词无关的形容词和介词短语等；或按照独立结构在句中作状语所表示的不同含义进行分类，如表示时间、条件、原因、结果、方式和伴随状况等。各种分类方式并不能截然分开，无论是否采用动词，均可以表示各种状语含义。因此在翻译时，应在理解逻辑关系的基础上，根据所表示的含义译为相应的状语，或直接译为句子。

1. 含有动词成分的独立主格结构

对于这种类型的独立结构，首先要分析独立成分与主句间的逻辑关系，如果独立主格结构用于表示时间、条件、原因，可以增加相应的连词，翻译成状语从句；若独立主格结构表示方式或伴随状况，则一般用来对主句加以补充，可以直接译为一个独立或并列的句子。

(1) 主语＋现在分词

在逻辑上，主语与现在分词之间为主谓关系，独立结构中的现在分词一般表示正在进行的动作，当独立结构中的动作发生在主句谓语动作之前时，应采用完成时态的现在分词。

The atoms are grouped into tiny domains, *each containing magnetically aligned atoms*. 原子被分成很多微小的磁畴，每个磁畴都包含由于磁化而整齐排列的原子。（独立结构表示状态，译为一个并列的句子）

(2) 主语＋过去分词

在逻辑上，主语与过去分词之间为动宾关系，独立结构中的过去分词一般表示被动的关系或已经发生的动作。

The two balls stuck together after collision, *a way called perfectly inelastic collision*. 两个球在碰撞后粘在一起，这种碰撞称为完全非弹性碰撞。（独立结构表示方式，过去分词表示被动，译为一个并列的句子）

（3）主语＋动词不定式

在逻辑上，主语与动词不定式之间为主谓关系，独立结构中的不定式一般表示将要发生的动作，即独立结构中的动作发生在主句谓语动作之后。

In Compton scattering, there is a perfectly elastic collision between the incident photon and an atom, *the atom's energy to increase* and *the photon's energy to decrease*. 在康普顿散射中，入射光子与电子间发生完全弹性碰撞，原子的能量增加而光子的能量减少。（不定式表示结果，译为两个并列短句）

2. 不含动词成分的独立主格结构

（1）主语＋形容词

这种结构形式上没有动词介入，实质上相当于一个"主语＋系动词＋形容词"的系表结构中省略了系动词 being，翻译时可以直接表达独立主语所处的某种状态，或增加适当的动词来表示这种状态，译为状语或并列的句子均可。

The terminal posts of an apparatus are usually in different colors, *the positive red* and *the negative black*. 仪器的接线端通常使用不同的颜色，正极为红色，负极为黑色。（译为并列短句）

（2）主语＋介词短语

这种结构形式上同样没有动词介入，实质上相当于一个"主语＋系动词＋介词短语"的系表结构中省略了系动词 being，或是"主语＋动词＋介词短语"结构中省略了动词，在翻译时可以适当添加动词，译为状语或并列的句子。

The oscilloscope can be used to show the hysteresis loop of the ferromagnetic material, *the voltage proportional to the magnetic field into the channel* 1, *the voltage proportional to the magnetic induction into channel* 2, and scanning area set at *X-Y* position. 示波器可以用于显示铁磁材料的磁滞回线，与磁场成正比的电压输入通道 1，与磁感应强度成正比的电压输入通道 2，扫描范围设置为 *X-Y* 位置。（译为并列短句，注意最后一个独立主格结构 scanning area set at *X-Y* nosition 属于"名词＋过去分词"的情况）

3. 由 with/without 引导的独立主格结构

一般来说，通常表示方式或伴随状况的独立主格结构可以转换为由 with 引导的介词词组，而表示否定意义的独立主格结构可转换为由 without 引导的介词词组。并且，在很多情况下 with 可以省略，并可同时省略 with 所引导的介词短语中的名词修饰语。

In the thermal radiation, vibrating and spinning molecules in one object give off electromagnetic radiation, *with the energy absorbed by molecules in another object*.

在热辐射过程中，物体中振动与自旋的分子发出电磁辐射，其能量被其他物体的分子所吸收。（表示结果，译为一个并列的句子，句中 with 可省略）

The two balls collide in a way of perfectly inelastic collision, *with most kinetic energy lost during the course*. 两球发生完全非弹性碰撞的过程中动能损失最多。（表示伴随的状况，原来的主干部分译为状语，句中 with 可省略）或译为：两球的碰撞方式为动能损失最多的完全非弹性碰撞。（翻译时改变主语，独立主格成分译为定语）

The two balls collide in a way of perfectly elastic collision, *without any kinetic energy loss during the course*. 两球发生完全弹性碰撞,过程中没有动能损失。(表示伴随的状况,译为一个并列的句子,原句可以改为否定形式: ..., no kinetic energy lost during the course.)或译为: 两球的碰撞方式为动能守恒的完全弹性碰撞。(翻译时改变主语,独立主格成分译为定语)

4. It 引导的独立主格结构

主语 it 通常用于表示时间、距离、天气或自然现象等,此时结构中的 being 不能够省略。

It being a very long distance between the two stars in the universe, the measurement unit should be light year. 宇宙中两恒星间的距离非常大,因此必须使用光年作为度量单位。(增加连词,译为原因状语)

5. 倒装的独立主格结构

(1) there be 句型也可以在独立主格结构中出现,此时结构中的 being 不能够省略,there be 结构一般可以意译,翻译为状语或句子。

There being a torque on the rotor, the rotor exerts an equal but opposite torque on the engine. 由于转子受到力矩的作用,转子也对引擎施加一个等大而反向的力矩。(增加连词,译为原因状语从句)

(2) such 作为表语时,独立主格结构应采取倒装形式,此时结构中的 being 一般亦不能省略。

Such being the applied condition for the Gauss's law, the system should be highly symmetric in charge distribution. 高斯定律的适用条件就是这样,系统的电荷分布必须有较高的对称性。(译为一个并列的句子)或译为: 高斯定律的适用条件是系统电荷分布的高度对称性。(主句的内容即为 such 所指的内容,因此可译为宾语)

23.1.2 分隔结构的翻译

英语是一种分析型语言,其句法关系主要通过功能词和语序 (function word and word order) 来体现,并且语序相对比较固定,即组成句子的成分在句中按其语法和语义关系都具有相对固定的位置,因而形成了五种基本句型,其语序叫作自然语序 (natural word order)。但有时为了增强语言的表现力,自然语序往往被打破,形成一些变异结构,分隔结构 (discontinuous structure) 就属于其中的一种。分隔现象在科技文体中经常出现,其形式多样,结构复杂,而物理学专业英语往往具有较多的修饰成分且表现为长句,因此分隔结构更加常见,其正确翻译有助于对原文的理解。

英语的分隔结构是指在本应紧密相连的两个成分之间插入其他成分,从而使这两部分产生分隔的现象。起分隔作用的成分主要有三种: 一是各种短语,如介词短语、不定式短语、分词短语等;二是各种从句,这也是最为常见的分隔结构;三是句中的附加成分,如插入语、同位语、独立主格结构等。分隔结构一般表现为以下几种情况: ①主谓分隔; ②动宾分隔; ③系表分隔; ④复合谓语或复合宾语的分隔; ⑤先行词与修饰成分的分隔; ⑥固定搭配的分隔。

根据分隔结构在句中所起的不同作用,分隔结构可以分为四种——结构性分隔、修饰性分隔、插入性分隔和倒装性分隔,各自具有不同的性质。结构性分隔可以起到使句子结构保持平衡匀称,避免头重脚轻的作用;修饰性分隔和插入性分隔主要起到说明、解释、补充和强调的作用;而倒装结构本身必定导致某些句子成分不在原来的位置上,从而造成成分的分隔。

在了解和掌握了分隔现象的常见形式之后,分隔结构的翻译应从句子本身入手,弄清句法结构和分隔成分的语法及语义作用,根据具体情况,灵活选取恰当的翻译方法,使译文准确表达原文的内容,准确传达原文的技术信息。以下就分隔结构的具体情况来介绍其翻译方法。

1. 主谓分隔

包含这种结构的句子其成分分析较为简单,分隔结构一般用于对主语进行说明或修饰。对于较长的分隔结构可以采用分译的方式,译为短句,而较短的分隔结构则可以采用合译的方式,译为定语,但有时也可以根据具体情况进行功能转换。例如:

Plastics, *such as nylon and artificial rubber*, are synthetic polymers. 诸如尼龙和人造橡胶这样的塑料是合成聚合物。(句中起说明作用的分隔结构比较短,因此可以采用合译,翻译为主语的定语)

Every substance in the world, *no matter how different it may seem from any other substance*, is made partly of electrons. 不管看起来与其他物质有多大的区别,世界上的各种物质都部分地由电子构成。(句中的分隔结构较长,为 no matter how 引导的让步状语从句,可采用分译)

Sensors *in a car tuned to radio signal from satellites* can locate the car precisely at any moment and warn of traffic jams. 汽车上的传感器能够随时精确地定位以及报告交通阻塞情况,这是由于可以接受卫星发出的无线电信号。(可以将句中作定语成分的分隔结构分译,并根据句意转换为原因状语从句)

Measurement, *which plays an important part in the physical experiment*, is not identical with all of the experiment. 在物理实验中起到重要作用的测量并不等同于物理实验的全部。或译为:测量并不等同于物理实验的全部,尽管它在物理实验中起到了重要的作用。(由于全句不太复杂,可以采取合译或分译)

2. 动宾分隔

动宾之间插入其他成分是比较明显的语序上的调整,因此这种分隔结构一般起到两种作用:一是由于宾语过长而将某些成分(如状语或宾语补足语)提前,此时恢复正常语序在句法上并不影响原句的意义,属于结构性分隔,因此在翻译时可进行灵活处理,此种情况较为多见;二是强调,即插入部分是原文想要强调的内容,在翻译时应保留原句的语序,即插入的成分仍紧邻动词。

Some object looks white because it sends back to *our eyes* all the colors in the light that reaches it. 某物体呈现白色是由于其将入射的各色光线均反射至我们眼中。(结构性分隔)

We call *conductors* those materials which allow electrons to pass easily. 我们把电子容易通过的材料称为导体。(结构性分隔)

Man has achieved remarkable success in the field of astronautics, the ultimate goal and purpose of which is to gain *for man himself* access to space and to other worlds. 人

们在宇航方面已取得了惊人的成就,宇航的最终目标和意图就是使人类能够亲自进入太空并到达其他星球。(强调)

3. 系表分隔

这种结构的分隔原因与动宾分隔相似,翻译中亦可采用相同的方法处理。

The compensation method is, *compared with other ones*, a more accurate and visual one for the measurement of some physical quantities. 与其他方法相比,补偿法对于某些物理量的测量更为准确与直观。

4. 复合谓语/宾语的分隔

复合谓语分隔主要是因为在并列谓语动词之间、构成时态或语态的复合动词之间、to与动词之间加入了状语成分,翻译时状语成分一般应紧随动词而不分开,但有时为了表示强调也可对语序做适当调整。复合宾语分隔在结构上与之类似,即在宾语和宾语补足语之间插入定语成分而形成分隔,翻译时也可采取相似的办法。

The dielectric can *obviously* weaken the electric field between the capacitor plates. 电介质可以明显地减弱电容器极板间的电场。

The capacitor in the circuit above is charged *from a battery* and discharged through a resistance R. 上述电路中的电容器通过电池充电,通过电阻 R 放电。

Laser *people all know* to be a very powerful device which can melt or destroy many kinds of materials known to man. 人们都知道,激光是一种非常强有力的装置,它能熔化或毁坏人们已知的很多种物质。(句中将宾语 laser 置于主谓结构之前,形成宾语和宾语补足语的分隔,是一种倒装式分隔,可用分译法进行翻译)

According to the equivalence principle, the effect produced by the gravity can *in no way* be distinguished from the effect produced by the acceleration. 根据等效原理,无法把万有引力所产生的效果与加速度所产生的效果区分开来。(介词短语 in no way 插入谓语 can be distinguished 中形成分隔,采用分译,移至谓语之前)

5. 先行词与修饰成分的分隔

这种分隔结构主要是指名词与其后置定语(或同位语)被谓语、状语或其他定语分隔,应根据句子的复杂程度选择分译或合译,若修饰成分过多,可将其中的部分修饰语转换为状语等进行翻译,以增强句子的表现力。

The beam of light is focused on the object *under the microscope* that is to be studied. 在显微镜下,光线聚焦在我们要研究的物质上。(修饰谓语动词的状语 under the microscope 在先行词和定语从句之间形成分隔,采用分译)

In experiments, another mode *is found* which is quite different from what we are familiar with. 实验中还会看到另外一种模式,它与我们所熟悉的模式有很大不同。(句中定语从句的先行词 mode 是本句的主语,且谓语 is found 较短,为平衡句子结构,避免头重脚轻,将定语从句置于谓语之后构成分隔,采用分译)

There is nothing *in the energy conservation principle or in any other law of nature* that specifies for us the direction of heat flow. 在能量守恒原理或其他任何一条自然规律中,都没有给我们规定热量的流动方向。(名词 nothing 与其后置定语 that ...被状语所分

隔，采用分译)

Scientists have found some new methods *of making electronic circuits* by which many of the electronic devices have become quite small. 科学家们已经发现了一些新的制作电子电路的方法，而这些新方法使得电子器件变得非常小。(句中中心词 methods 由两个修饰成分修饰，一是 *of making*…，一是 *by which*…，很显然必定有一个修饰成分被分隔开，此时如果将修饰成分全部翻译为定语则显得过于臃肿，因此可根据句意和语境将其中一部分转换为其他成分，使句子表达更为顺畅)

6. 固定搭配的分隔

英语中有大量的固定搭配，这些固定用法中也可以存在分隔结构，在物理学专业英语中更是如此。这是由于物理学中的定义、规律与方法等往往存在着严格的限定条件，为了表明这些限定条件，能够更完整、更准确、更严谨地进行信息传递，专业英语中经常使用分隔结构，自然也包括对固定用法的分隔。对固定用法的正确辨识有利于我们准确地把握结构和理解文意，但同时也要求我们对分隔结构做恰当的处理，使翻译准确流畅。一般应保持固定搭配的原有含义不变，将起修饰和限定作用的插入部分译为状语或定语，并根据语境调整顺序，将该状语或定语放到合适的位置。

The capacitance of the capacitor depends *to some extent* on the magnitude of the dielectric constant of the dielectric in between the plates. 电容器的电容在一定程度上取决于极板间电介质的介电常数。(固定短语 depend on 被介词短语 to some extent 分隔，其目的在于限定动词，译为状语)

We should carefully distinguish velocity *which has both magnitude and direction* from speed which has only magnitude. 我们要注意把速度和速率区分开，因为速度既有大小又有方向，而速率只有大小。(由 which 引导的限制性定语从句造成了固定搭配 distinguish …from 的分隔，翻译时可把定语从句分译)

The molecules of the gas meet with much more resistance *from the closely packed water molecules* to their free movement. 气体分子的自由运动受到了来自于紧密排列的水分子的更大的阻力。(定语分隔了名词 resistance 和介词 to 的固定搭配，插入成分译为定语)

分隔结构是极其常见的语法现象，正确翻译的要点仍是对句子主干的把握，然后将修饰成分以适当的方式添加上去，既要忠实于原文，又要避免歧义，力求表达流畅。

23.2 专业英语阅读

23.2.1 The Uncertainty Principle(测不准原理)

The interpretation that atomic particles are governed by probability left many scientists dissatisfied and hoping for some ingenious thinker to rescue them from this foolish predicament. The German physicist Werner Heisenberg showed that there was no rescue. He argued that there is a fundamental limit to our knowledge of the atomic world.

Heisenberg's idea—that there is an indeterminacy of knowledge—is often misinterpreted. The uncertainty is not due to a lack of familiarity with the topic, nor is it due to an inability to collect the required data such as the data needed to predict the outcome of a throw of dice, or the Kentucky Derby. Heisenberg was proposing a more fundamental uncertainty—one that results from the wave-particle duality.

Imagine the following thought experiment. Suppose you try to watch an electron moving through a room. The room is a complete vacuum, so there's no worry about the electron hitting gas molecules. To locate the electron you need something to carry information from the electron to your eyes. Suppose you use photons from a dim light bulb. Because this is a thought experiment, we can also assume that you have a microscope so sensitive that you will see the electron as soon as one photon bounces off the electron and enters the microscope.

You begin by using a bulb that emits low-energy photons. These have low frequencies and long wavelengths. The low energy means that the photon will not disturb the electron very much when it bounces off it. However, the long wavelength means that there will be lots of diffraction when the photon scatters from the electron. Therefore, you won't be able to determine the location of the electron very precisely.

To improve your ability to locate the electron, you now choose a bulb that emits more energetic photons. The shorter wavelength allows you to determine the electron's position relatively well. But the photon kicks the electron so hard that you don't know where the electron is going next. The smaller the wavelength, the better you can locate the electron, but the more the photon alters the electron's path.

Heisenberg argued that we cannot make any measurements on a system of atomic entities without affecting the system in this way. The more precise our measurements, the more we disturb the system. Furthermore, he argued, the measured and disturbed quantities come in pairs. The more precisely we determine one half of the pair, the more we disturb the other. In other words, the more *certain* we are about the value of one, the more *uncertain* we are about the value of the other. This is the essence of the uncertainty principle.

Two of these paired quantities are the position and momentum along a given direction. (Recall that the momentum for a particle is equal to its mass multiplied by its velocity.) As we saw in the thought experiment described above, the more certain your knowledge of the position, the more uncertain was your knowledge of the momentum. The converse is also true.

This idea is now known as Heisenberg's **uncertainty principle**. Mathematically, it says that the product of the uncertainties of these pairs has a lower limit equal to Planck's constant. For example, the uncertainty of the position along the vertical direction Δy multiplied by the uncertainty of the component of the momentum along the vertical direction Δp_y must always be greater than $\hbar/2$.

$$\Delta p_y \, \Delta y \geqslant \hbar/2$$

where $\hbar = h/2\pi$.

This principle holds for the position and component of momentum along the same direction. It does not place any restrictions on simultaneous knowledge of the vertical position and a horizontal component of momentum.

Another pair of variables that is connected by the uncertainty principle is energy and time, $\Delta E \Delta t \geqslant \hbar/2$. This mathematical statement tells us that the longer the time we take to determine the energy of a given state, the better we can know its value. If we must make a quick measurement, we cannot determine the energy with arbitrarily small uncertainty. Stated in another way, the energy of a stable state that lasts for a very long time is very well determined. However, if the state is unstable and exists for only a very short time, its energy must have some range of possible values given by the uncertainty principle.

23.2.2 Probability Waves(概率波)

Nature has an underlying rule governing this strange behavior. We need to unravel the mystery of why electrons (or photons for that matter) are always detected as single particles and yet collectively they produce wavelike distributions. Something is adding together to give the interference patterns. To see this we once again look at water waves.

There is something about water waves that does add when the two slits are open—the displacements (or heights) of the individual waves at any instant of time. Their sum gives the displacement at all points of the interference pattern; that is, $h_{12}=h_1+h_2$. The maximum displacement at each point is the amplitude of the resultant wave at that point.

With light and sound waves we observe the intensities of the waves, not their amplitudes. Intensity is proportional to the square of the amplitude. It is important to note that the resultant, or total, intensity is not the sum of the individual intensities. The total intensity is calculated by adding the individual displacements to obtain the resulting amplitude of the combined waves and then squaring this amplitude. Thus, the intensity depends on the amplitudes of each wave and their relative phase. The intensity of the combined waves is larger than the sum of the individual intensities when the displacements are in the same direction and smaller when they are in opposite directions.

There is an analogous situation with electrons. The definition for a **matter-wave amplitude** (called a wave function) is analogous to that for water waves (this is usually represented by the symbol Ψ and pronounced "psi"). The square of the matter-wave amplitude is analogous to the intensity of a wave. In this case, however, the "intensity" represents the likelihood, or probability, of finding an electron at that location and time. There is one very important difference between the two cases. We can physically measure the amplitude of an ordinary mechanical wave, but there is no way to measure the amplitude of a matter wave. We can only measure the value of this amplitude squared.

These ideas led to the development of a new view of physics known as **quantum mechanics**, which are the rules for the behavior of particles at the atomic and subatomic levels. These rules replace Newton's and Maxwell's rules. A quantum-mechanical equation, called Schrödinger's equation after the Austrian physicist Erwin Schrödinger, is a wave equation that provides all possible information about atomic particles.

Glossary

probability	概率	predicament	困境
indeterminacy	不确定性	misinterpret	曲解
thought experiment	思想实验	dim light bulb	弱光灯泡
entity	实体	uncertainty principle	测不准原理
Planck's constant	普朗克常量	probability wave	概率波
underlying rule	潜规则	unravel	阐明、解释
collectively	全体地、共同地	intensity of the wave	波的强度
wavelike distribution	波状分布	interference pattern	干涉图样
matter-wave amplitude	物质波幅	wave function	波函数
likelihood	可能性	quantum mechanics	量子力学
Schrödinger's equation	薛定谔方程		

23.3 专业英语常用表达-23 距离 间隔

1. 空间距离与间隔

(1) 表示空间上的距离、间隔、范围等的常见词汇：

distance 距离、远离

extent 广度、宽度、长度、范围

gap 裂口、缺口、间隙

interstice 空隙、间隙、裂隙

interval 空间的间隔、空隙

length 长、长度、距离

pitch (螺旋的)螺距、(齿轮的)节距

play 间隙,(例如"back play空程")

range 范围、区域、射程、距离

scope 范围、导弹的射程

separation 分隔、分开、分隔点、间隔、(导线的)间距

space 空白、距离、间隔

space length/spacing interval 间距、空间距离

spacing 间隔、(字间、行间等)空隙、空白

span 跨度、范围、全长、支点距、墩距、翼展

stretch 一段路程

way 路线、路途、路程

(2) 空间间隔的常见表达：

a distance of three kilometers/three kilometers' distance 3 km 的距离

be at a distance of three kilometers/be spaced at a distance of three kilometers/be spaced at an interval of three kilometers　在距离 3 km 处

the distance from A to B/the distance between A and B/the space interval between A and B　从 A 至 B 的距离

be 5cm (apart/away/distant/far) from point A　与 A 点相距 5 cm

be closely spaced　间距很小

be evenly spaced　等距

2. 时间距离与间隔

(1) 表示时间上的距离、间隔、范围等的常见词汇：

(time) interval　时间的间隔、间歇

distance　较长一段时间、时间上的间隔

duration　持续时间、期间

gap　时间的中断、间歇、间断

length　时间的长短、期间

space (of time)　(一段)时间、时间间隔

stretch　一段持续的时间

time lag　时间上的间隔、时滞

time limit　时限

time span/separation　时间跨度、时间间隔

cycle/period　周期

(2) 时间间隔的常见表达：

a time interval of two hours/two hours'time　2 h 的时间

two hours before/after...　在……之前/之后 2 h

every five seconds/every other four seconds　每隔 4 s/每 5 s

every forth second　每个第 4 s/每 3 s

The average *distance between* two apparatus should not be too small in case of errors caused by the mutual influence. 两仪器之间的平均距离不应过小，以免相互影响造成误差。

The screw rod goes forward *for a distance of the length of the pitch* each time it takes a turn. 螺栓每转一周可前进的距离为一个螺距。

The *time interval for* a particle to undergo a complete oscillation is called a *period*. 一个质点完成一次全振动所经历的时间为一个周期。

Half-life is *the period of time*, for a substance undergoing decay, to decrease by half. 半衰期是指某种衰变物质消耗至初始时的一半时所经历的时间。

LESSON 24

24.1　物理学专业英语中长句的翻译

由物理学科自身的特点所决定，物理学专业英语需要以准确的语言、严密的逻辑和严谨

的结构来表述物理概念、规律和方法等，因此其中经常出现结构复杂、修饰限定成分较多的长句，对这些长句的准确把握也成为专业英语阅读与翻译的重点和难点。翻译是从一种语言向另一种语言的转换，除语法上的分析之外还要正确地理解原文的含义，这就需要把握两种不同语言的特点，利用语言知识来分析，利用专业知识来理解，完成合理顺畅的转换。

24.1.1 长句翻译的基本方法

从英汉两种语言的特点来看，"英语重形合，汉语重意合"。英语句子长，动词少，语序灵活，讲究平衡，而且句中有明确的连接词作为语法标志，各种成分通过句法组合为形式上的长句整体；而汉语句子短，动词多，语序比较固定，讲究对称，句中往往并无明确的连接成分，各个短句通过意义上的联系组合在一起。

语言上的差异、物理学的深奥和长句的复杂性，使物理学专业英语长句的理解与翻译成为专业英语学习中的难点，需要仔细推敲，谨慎处理。翻译长句很难一步到位，因此不可急于求成，应遵循一定的方法，逐层分析，逐步深入，正确分析语法结构，准确判断逻辑关系，灵活运用翻译技巧。为达到这些目的，可以将长句的翻译归纳为三个基本步骤：结构分析，部分翻译，整体组合。

1. 结构分析与部分翻译

正确的结构分析是进行准确翻译的前提和基础，具体的分析方法可参见前文(Lesson 9&17)，即"抓住主干，添枝加叶"，这里不再赘述。分析的过程帮助我们弄清结构与关系，而句子的分析与理解是密不可分的，因此在结构分析的过程中，我们就已经基本完成了各个成分的翻译。

2. 确定对应关系，注意成分分隔

专业英语长句的复杂之处就在于其修饰成分繁多，关系复杂，往往表现为多个修饰成分的并列(一个中心词具有多个修饰语)或是嵌套(在一个语法单元中又出现其他的修饰成分)，同时还存在成分分隔的现象。因此在找到了句子的各个成分之后，还应明确其间的关系，例如，定语(从句)所对应的中心词是什么，状语(从句)是表示时间、原因、结果，还是表示条件等。在这一过程中应特别注意成分分隔的现象，运用专业知识来帮助分析，正确理解各成分之间的对应关系。

3. 组合成分，调整语序，正确翻译

在完成了句子主干和各个单元的分析与翻译之后，最后一步就是根据各个成分之间的逻辑关系进行组合与构建，形成完整的翻译。但直接组合后的译文很可能会晦涩难懂，这时可以在忠实原文含义的基础上根据汉语的语言习惯进行重新调整，使译文流畅并易于理解，在这一过程中可以使用一些翻译的技巧，例如顺译，倒译，转换，合并，拆解等。下面以实例进行分析。

(1) 简单句

In solid form, <u>the noble gases are held together</u> with Van der Waals forces resulting from the polarization of the electronic charge cloud on each atom.

分析：划线部分为句子主干，除 in solid form 这一状语之外，该简单句的谓语还具有一个由 with 引导的状语，在状语中，中心词 forces 又带有自己的修饰成分 resulting from...,

其中又包含了两层分别由 of 和 on 引导的嵌套修饰成分。在翻译时，如果将 forces 的定语前置，则句子的主干显得不够明确，因此可将 forces 的定语以短句的形式来翻译。

译文：处于固态时，惰性气体通过范德瓦耳斯力结合在一起，而范德瓦耳斯力来源于每个原子的电子云极化。

The value of electrical resistance associated with a circuit element or appliance can be determined by measuring the voltage across it with a voltmeter and the current through it with an ammeter and then dividing the measured voltage by the current.

分析：划线部分为句子主干，associated with...appliance 为后置定语修饰主语，by 所引导的状语成分包含两个并列成分 measuring 和 dividing，而第一个成分中 measuring 的宾语是并列的 voltage 和 current（各自带有定语），方式状语是由 with 引导的短语。

译文：电路元件或电器的电阻值可以这样测量，用电压表测量两端的电压，用电流表测量其中的电流，然后用电压值除以电流值。

（2）并列句

Most crystalline materials encountered in everyday life are polycrystalline, with the individual crystals being microscopic in scale, but macroscopic single crystals can be produced either naturally (e. g. diamonds) or artificially.

分析：划线部分为两个转折关系并列句的主干。第一个子句中，encountered in everyday life 是主语 materials 的后置定语，而 with...in scale 是表语 polycrystalline 的定语，其中又包含了分词结构 being microscopic in scale 来修饰 individual crystals。第二个子句中包含了并列状语 either...or...。

译文：大多数日常生活中所见的晶体材料是多晶体，其单晶体尺寸很微小，但宏观的单晶既可以天然形成（比如钻石）也可以人工合成。（将定语译为短句）

或者直接译为：大多数日常生活中所见的晶体材料是包含微小尺寸单晶体的多晶体，但宏观的单晶既可以天然形成（比如钻石）也可以人工合成。

The ballistic pendulum is a classic example of a dissipative collision in which conservation of momentum can be used for analysis, but conservation of kinetic energy during the collision cannot be invoked because the energy goes into inaccessible forms such as internal energy.

分析：宾语 example 后带有一个修饰语 of a dissipative collision，而 dissipative collision 又带有一个 in which 引导的定语从句，由 but 引导的第二个子句本身又是一个完整的因果关系从句。

译文：冲击摆是耗散碰撞的典型例子，可以用动量守恒进行分析，但在碰撞过程中不能应用动能守恒，这是由于某些动能转变为不可见的形式，例如内能。

从译文可以看出，尽管第一个子句中宾语的定语并不算长，但是翻译为一个独立的句子能够更好的表明“动量守恒”与“动能守恒”之间的对照关系，并且为了反映这种对照关系，将动词 invoke 译成了“应用”。可见，在翻译的过程中既应考虑到整体的布局，又要注意细节，才能够更好地表达原文的含义。

（3）复合句

The bulk of solid-state physics theory and research is focused on crystals, largely because the periodicity of atoms in a crystal—its defining characteristic—facilitates mathematical modeling, and also because crystalline materials often have electrical, magnetic, optical, or mechanical properties that can be exploited for engineering purposes.

分析：该句是一个主从复合句，划双下划线的部分为主句的主干，of solid-state physics theory and research 用于限定主语 bulk。其从句为两个并列的原因状语从句，划单下划线的部分是第一个从句的主干，of atoms in a crystal 是 periodicity 的限定成分，its defining characteristic 作为插入语，可以理解为 the periodicity of atoms in a crystal 的同位语。划波浪下划线的部分是第二个从句的主干，that can be exploited for engineering purposes 是个定语从句，用于修饰 properties，同时 electrical, magnetic, optical, or mechanical 也是用于修饰 properties 的。

译文：固体物理学的理论和研究工作的主体都集中在晶体上，这主要是由于作为晶体基本特征的原子周期性便于进行数学模拟，同时也是由于晶体材料通常具有可供工程上开发利用的电学、磁学、光学或力学性质。

The net magnetization of ferromagnetic materials in response to an external magnetic field may actually occur more by the growth of the domains parallel to the applied field at the expense of other domains rather than the reorientation of the domains themselves as implied in the sketch, which is indicated by the microscopic evidence about magnetization.

分析：该句带有一个 which 引导的非限制性定语从句，划线部分为主句的主干。主语 the net magnetization 后带有由介词短语构成的较长的修饰语，而谓语 occur 后面又带有较长的状语。但相比之下，主语的修饰语稍短，因此直接译为前置定语，而将谓语的方式状语译为句子，并且为了将意思表达清楚，可拆开译为短句。注意句中的"evidence"一词译成了"观察结果"，这是根据语境(as implied in the sketch)及物理知识分析而得到的。

译文：磁化过程的微观观察结果表明，如图所示，铁磁材料响应外磁场造成的净磁化的产生，更多地是通过平行于外磁场的磁畴破坏其他磁畴而长大，而不是磁畴自身的重新取向。

24.1.2 长句的翻译技巧

长句的翻译过程可以概括为"解构"与"重构"两个基本过程。长句的分析是一个解构过程，将长句化整为零，逐层分析，帮助我们正确地理解句子内容，得到一个基本框架并了解各成分的含义。而将各种成分进行重新组合则是翻译的重构过程，通过重构最终得到贴切、通顺的译文。解构的过程需要语法知识和专业背景的辅助，而重构的过程则需要仔细推敲并运用一些翻译技巧，从而做到译文的流畅与通达，不仅使语序与表达方式符合汉语的语言习惯，而且使句意易于被理解。

长句的翻译的基本技巧包括：顺译，倒译，拆分和转换，但在实际的翻译中应综合考虑，灵活使用。

1. 顺译

有些长句的逻辑关系表述语序大致符合汉语的语言习惯，此时可按原文语序进行翻译，无须做太大的调整，这种译法为顺译。

Stephen later discerned several new characteristics of black holes and demonstrated that the amazing forces of the Big Bang would have created mini-black holes, each with a mass about that of a terrestrial mountain, but no larger than the subatomic proton.

本句为主从复合句，包含一个主语和两个谓语 discerned 及 demonstrated，按照原句的顺序进行翻译较为合理，可将独立主格成分 each...proton 译为句子。

全句可译为：后来，Stephen 又认识到黑洞一些新的特征，并证明大爆炸中令人震惊的巨大力量可能会产生很多微小的黑洞，每一个微小黑洞的质量相当于地球上一座大山的质量，但体积不大于亚原子内的质子。

2. 倒译

英语中有大量的修饰成分后置现象，其表述的逻辑顺序可能与汉语的习惯不同，此时可改变原文的语序，进行部分或者全句的倒译。例如可以将后置的定语从句置于被修饰的中心词之前，也可以将后置的状语从句前置，按照前因后果、先提出条件后交代结果的汉语表达习惯进行翻译。在包含主语从句、宾语从句和同位语从句等名词性从句的长句中，也可根据具体情况采用倒译的方式。

By assuming that the material contains immobile positive ions and an "electron gas" of classical, non-interacting electrons, the Drude model was able to explain electrical and thermal conductivity and the Hall Effect in metals, although it greatly overestimated the electronic heat capacity.

本句为主从复合句，包含一个 although 引导的让步状语从句和一个由 by 引导的方式状语。将 although 引导的让步状语从句提前更为符合汉语的语序，此外，对 classical, non-interacting electrons 的翻译可采用倒译，改变两个定语的语序，使翻译更为符合物理专业的表述习惯。注意在 electrical and thermal conductivity 中 electrical 后省略了 conductivity 一词。全句可译为：

尽管过分地估计了电子热容，通过假定材料包含固定不动的阳离子和无相互作用的经典"电子气"，Drude 模型还是能够成功解释金属的电导性、热导性及霍尔效应的。

3. 拆分

由于英汉语言存在差异，英语多用长句，而汉语多用短句，因此拆分的方法在长句翻译中极其常见。结构复杂的英语长句直接翻译难于理解，并且繁多的修饰成分会使译文的句子冗长，采用拆分的方式可将多个修饰成分译为汉语的短句，使得句子结构简单，容易理解，表达通顺。拆分往往就意味着语序的改变，因此说翻译的方法并不唯一，应整合使用。

<u>The mirages</u> seen on hot days <u>can be explained by the optic effect</u> that the changing index of refraction of the air causes the light rays to bend and creates the appearance of specular reflections in the distance as directly from the surface of a pool of water.

分析：句子的主干(划线部分)很简单，但 optic effect 的同位语从句却比较繁琐，整个从句解释了主语 mirages 的成因，因此将同位语从句拆分为一个独立的句子进行翻译，同时将 as directly from the surface of a pool of water 也翻译为一个句子。

译文：炎热的天气中出现的海市蜃楼现象可以用光学效应来解释。空气折射率的改变引起了光线的偏折,从而导致了远处景物的镜面反射像的出现,这些反射像就好像从水面被直接反射的一样。

为了使句子更易于理解,添加了"远处景物"这一成分,此处只有了解物理光学的专业知识才能进行准确的增词翻译。

4. 转换

在翻译的过程中,由于采用倒译或拆分的方法,或是汉语表达的需要,往往会造成句子中某些成分的语法地位发生了改变,但从含义上来讲仍然是一致的。例如由宾语变为主语,由定语从句变为一个完整的短句等,这就是转换,在翻译中被大量的使用,在前面的例句中几乎都有体现。

Real crystals feature defects or irregularities in the ideal arrangements, and it is these defects that critically determine many of the electrical and mechanical properties of real materials.

译文:理想排布中的缺陷或不规则性是实际晶体的特点,而恰恰是这些缺陷决定了实际材料的电学和力学特性。

分析:由于采用了倒译,因此第一个子句中的 defects or irregularities 从宾语转换为主语,同样第二个子句中的 defects 从表语转换为主语。注意在翻译中省略了 determine 前的状语 critically 一词,这是由于通过"恰恰是……"的强调方式,使动词"决定"本身就可以表述出原文的含义。

长句的翻译方法是多种多样的,不必拘泥于某种特定的模式,在忠实于原文含义的前提下,可以根据具体的情况灵活地综合使用,同一个句子,也不局限于一种处理方法。严复先生所提出的"信、达、雅"三字翻译标准在专业英语中同样应该提倡,使专业语言的翻译在准确传递信息的基础上同样做到优美传神。

24.2 专业英语阅读:Relativity(相对论)

When observers in different inertial reference systems describe the same events, their reports don't match. In the framework of classical relativity they disagree in their descriptions of the paths and on the values of an object's velocity, momentum, and kinetic energy. On the other hand, they agree on relative positions, lengths, time intervals, accelerations, masses, and forces. Even the laws of motion and the conservation laws are the same.

We never asked, or even thought to ask, whether some of these were actually the same for all reference systems or whether we had just assumed them to be the same. In classical relativity we assumed that the concepts of length, time, and mass were the same. But are they really the same?

Albert Einstein asked this question. He reexamined the process of describing events from different reference systems with an emphasis on the concepts of space and time. This led to the development of the special theory of relativity.

Einstein arrived at the special theory of relativity by setting forth two postulates, or conditions, that are assumed to be true. He then examined the effects of these postulates on our basic concepts of space and time. The predictions of special relativity were then compared with actual experimental measurements. The theory had to agree with nature to have any validity.

The First Postulate(第一假设)

The first postulate is related to the question of whether there exists an absolute space—some signpost in the Universe from which all motion can be regarded as absolute. This postulate says that there is no absolute space; any inertial reference system is just as good as any other. Einstein's first postulate is a reaffirmation of the Galilean principle of relativity.

The laws of physics are the same in all inertial reference systems.

As we discussed in the previous chapter, Galileo argued that a traveler in the hold of a ship moving with a constant velocity could not conduct experiments that would determine whether the ship was moving or at rest. However, the Galilean principle of relativity came into question near the end of the 19th century. A theory by the Scottish physicist James Clerk Maxwell describing the behavior of electromagnetic waves, such as light and radio, yielded unexpected results.

In Newton's laws, reference systems moving at constant velocities are equivalent to each other. If, however, one system accelerates relative to another, the systems are not equivalent. Because Newton's laws depend on acceleration and not on the velocity, acceleration of a reference system can be detected, but its velocity cannot.

In Maxwell's theory, however, the *velocity* of the electromagnetic waves appears in the equations rather than their acceleration. According to the classical ideas, the appearance of a velocity indicated that inertial systems were not equivalent. In principle, you could merely turn on a flashlight and measure the speed of light to determine how your reference system was moving.

Maxwell's equations and the Galilean principle of relativity were apparently in conflict. It seemed that the physics world view could not accommodate both. During his studies, Einstein had developed a firm belief that the principle of relativity must be a fundamental part of any physical theory. At the same time, he wasn't ready to abandon Maxwell's new ideas about light. He felt that the conflict could be resolved and that both the principle of relativity and Maxwell's equations could be retained.

Meanwhile, others were pursuing different options. If there were an absolute reference system in the Universe, it should be possible to find it. The key seemed to lie in the behavior of light. It is impossible, using experiments performed within inertial systems, to distinguish any one such system from any other; therefore, no experiment can establish whether a particular inertial system is uniformly moving or at "absolute rest". The concept of absolute rest thereby loses all significance; if it can never be determined, motion is relative, not absolute. The Principle of Relativity logically abolishes the concept of absolute rest and, along with it, the concept of absolute motion—motion is relative.

The Second Postulate(第二假设)

It is difficult (if not impossible) to re-create a creative process. Although Einstein mentioned the failure to find the ether in his 1905 paper, years late he indicated that his primary motivation in formulating the *Second Postulate* was his deep belief in the principle of relativity.

The Constancy of the Speed of Light: *Light propagates in free space with a speed c that is independent of the motion of the source.*

Now this much in itself is both orthodox and reasonable; after all, the speed of sound is independent of the motion of the source. A sound wave is launched into a medium, and the speed of the disturbance is only determined by the physical characteristics of the medium. The speed of the source is irrelevant. If light is a wave in the aether, this statement makes obvious sense.

Remember that the equations of Electromagnetic Theory—Maxwell's Equations—led to a wave equation that provided the speed of light in vacuum. Assuming Maxwell's Equation are right and given the First Postulate, it must be that this same wave equation is applicable in all inertial systems—the vacuum speed of light measured on Earth or inside a rocket ship must be the same, independent of any uniform relative motion. In Maxwell's theory, the speed of light is a constant, not a motion-dependent variable. In other words, *the speed of light measured with respect to an inertial system must be the same for all such systems.*

The fact that the speed of light is independent of the motion of the source is not at all troublesome, but is it also independent of the motion of the detector (i. e. the observer)? Certainly, the speed of sound is not; if the detector rushes toward the source, mobbing with respect to the air, the measured speed of sound increases. Just imaging two identical ships headed toward a motionless sound-emitting buoy, one steaming along at full speed and the other dead in the water. On both ships, the time it takes as blast of sound to sweep from bow to stern is measured, and the speed of the wave is computed in that inertial system. Clearly, for the ship moving toward the buoy, that time will be shorter (during the interval it takes the sound to traverse the ship, the stern will advance somewhat toward the pulse, shortening the effective length), and the wave speed will be determined to be faster.

But for the light wave, the conclusion is an astonishing one: *no matter how fast a light source moves toward or away from an inertial observer, and no matter how fast the observer moves toward or away from the source, the speed of the light passing from one to the other in vacuum will always be c—the speed of light is constant*; it is absolute.

The two postulates separately are innocent enough. It's when we mix them together, when we demand that they both apply at once, that things seem to become fantastic. The paradox of Einstein's youth is now no longer a paradox—one simply cannot travel next to a light beam and catch up to it. The spaceship could be rushing either toward or away from the source at any speed you like, say, 99% of c, and still the inertial observer aboard it will measure the speed of the beam to be c! This is one of the premier conclusions of the analysis; its disturbingly "illogical" nature suggests that our familiar, comfortable understanding of space and time requires revision.

Time Dilation(时间膨胀)

Time on a clock that is moving, at speed v, with respect to an inertial observer is seen to run slower than time on a clock that is stationary with respect to that observer.

$$\Delta t_M = \frac{\Delta t_S}{\sqrt{1 - v^2/c^2}}$$

This slowing down of time is known as **time dilation**, and it's usually a very small effect. A clock aboard a commercial plane flying at top speed for≈70 000 years would lose about 1s compared to a clock on the ground.

The duration of an event Δt_S, as measured by an inertial observer who sees the event begin and end in one place is called the **proper time**; it's always shorter than the corresponding interval Δt_M, as measured by an observer who sees the event occur in a moving system.

Time dilation has been measured. Certain nuclei vibrate and emit gamma-rays with very precise frequencies. When a sample of such a substance is heated, the gamma-ray frequency is reduced. The atoms move around more rapidly, and with respect to an observer at rest in the laboratory, their nuclear clocks run more slowly.

Length Contraction(长度收缩)

A rod has one **proper length** measured by any observer at rest with respect to it, but it can also have different shorter lengths measured by people who are in uniform motion with respect to it.

$$L_M = L_S \sqrt{1 - v^2/c^2}$$

A moving observer measures an object to have a length (along the direction of motion) that is shorter than the length measured by an observer at rest with respect to the object (i. e., shorter than the proper length). This is the **length contraction**, and it applies only to the direction of motion; transverse distances are unaltered.

There is no absolute distance between Beijing and Shanghai; an atlas provides the proper distance as measure by people a rest with respect to the planet. But every traveler moving with regard to the surface sees his or her own version of that spatial interval, depending on his or her relative speed. Go fast enough, and Beijing and Shanghai can be a meter or two apart. As for the reality of all of this, the business is similar to what happens with the Doppler Effect. Run toward a source and the sound pitch increases. The wave itself doesn't physically change, but the perception of it certainly does change. What you hear and measure—the reality of the experience—most assuredly depends on how you move with respect to the source.

Glossary

classical relativity	经典相对论	perception	感知、感觉
postulate	假定、基本原理	signpost	路标
reaffirmation	再肯定	yield	产生
assuredly	确实地、确信地	flashlight	闪光灯
accommodate	容纳	retain	保持
pursue	从事	abolish	废止
aether/ether	以太	formulate	阐明
constancy	恒久不变	orthodox	正统的、传统的
irrelevant	无关的	inertial systems	惯性系
motion-dependent	与运动有关的	mob	拥挤前进
buoy	浮筒	blast of sound	一阵声波
from bow to stern	从船首到船尾	paradox	佯谬、似非而是的论点
premier	第一的、首要的	disturbingly	令人不安地、动摇地
revision	修改、修正	time dilation	时间膨胀
proper time	固有时间	corresponding	对应的
length contraction	长度收缩	proper length	固有长度
atlas	地图、地图集	spatial	空间的
pitch	音调、音高标准	special theory of relativity	狭义相对论
Galilean principle of relativity	伽利略相对性原理		

24.3 专业英语常用表达-24 位置 方位

1. 通过参照物表示物体相对位置

(1) 以 A 为参照物:

B is at the top of A.

C is in the middle of A.

D is at the bottom of A.

E is on the left of A and F is on the right of A.

B, C, D, E, and F are all inside of A.

G, H, M, and N are all outside of A.

M is over A and N is below A.

G and H are on either side/hand of A.

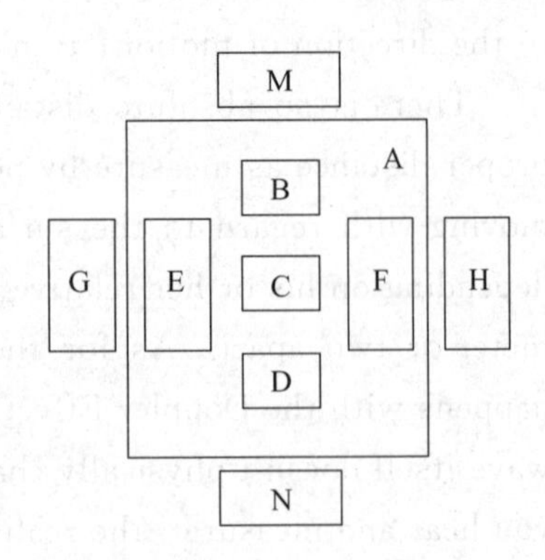

(2) 互为参照物:

B is above/over C.

B is diagonal above E/F.

D is below/under C.

E is to the left of C and F is to the right of C.

B, D, E, and F are all beside C.

G, E, C, F, and H are in the same line.

M, B, C, D, and N are in the same column.

C is between E and F/B and D.

2. 可用于方位表述的词语

above, beyond, on, on top, onto, up, upon, upwards, zenithward　在上、向上

below, beneath, down, downward, under, underneath　在下、向下

leftward, on the left (of), to the left (of), leftmost, on/to the left side (of)　在左、向左

rightward, on the right (of), to the right (of), rightmost, on/to the right side (of)　在右、向右

on the east of, to the east of, eastward　在东边、向东

on the south of, to the south of, southward, southerly　在南边、向南

on the west, to the west, westward　在西边、向西

on the north of, to the north of, southward, northerly　在北边、向北

southeast/southwest　东南/西南

northeast/northwest　东北/西北

ahead, before, farther on, fore, forth, forward, further on, in front of, in the front of, straight ahead　在前、向前

at the rear of, backward, behind, hind, in back of, in the back (of), rearwards　在后、向后

exterior, external, out, out of, outer, outside, outward, without　在外、向外

in, inner, inside, interior, internal, into, inward, within　在内、向内

among, amid, at the center of, between, centered, in the center of, in the middle (of), inbetween, midst, toward the middle　居中、在中间

aside, alongside of, at the side of, beside, on both sides (of), on the side (of), to the side (of)　在旁边、在侧面

about, ambient, circumambient, circumjacent, environmental, around, all round, surrounding, enwrap, surround, encompass, encircle　在周围、围绕、周围的

atop, at the top of, on top, on (the) top of, top half, upper part　(在)顶部

at the bottom (of), bottom side, bottom half, lower part　(在)底部

along, alongside, following　沿着

across, athwart, thwart, passing, through, throughout, crossing, cross, traverse, span, intersect, penetrate, run through, cut through, bridging over　穿过、贯穿、跨越

abreast, side by side, parallel, paratactic　平行、并列

across from, opposite to, in opposite direction, face to face　相对、相反

adjoin, be adjacent to, be close to, be contiguous to, be near, be next to, border on, bordering to, neighboring on, neighborhood, neighborhood, vicinal, vicinity　接近、邻近

far, far beyond, far off, far from, away from, in the distance, at a distance, a long way off　远离、在远处

horizontal, level, standard　水平的

vertical, normal, perpendicular, upright, erect, orthogonal　垂直的

oblique, lean, inclined, slant　倾斜的

upside down, bottom up, head over heel　颠倒的

(1) 注意：

above,over 都可表示"在……之上"。above 一般表示"在上面,高于"但不一定垂直；over 表示"在……正上方",强调垂直的关系。below,under,beneath 皆可表示"在……之下"。below 表示非垂直并且不与表面接触的下方;under 表示垂直并且不与表面接触的下方;beneath 则表示在某物之下,几乎接触或接近。above 与 below 在意义上是相对的,表示物体位置不垂直的上下关系;over 与 under 在意义上是相对的,表示物体位置垂直的上下关系。

介词 beside 和 by 均表示"在……旁边",可以替换,相当于 next to。而 near 表示的距离与前两者稍远些。介词 beyond 在表示空间位置时,指"在/向……那边","在另一边"。

(2) 特定位置的表述

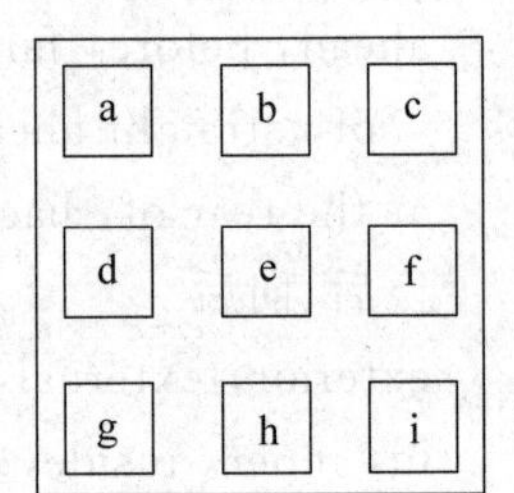

a: top left corner, upper-left

b: at the top, halfway across

c: top right corner, upper-right

d: halfway down left

e: the center, the middle

f: halfway down right

g: bottom left corner

h: at the bottom, halfway across

i: bottom right corner

3. 坐标的表示

(1) 地理坐标——经纬度

Beijing is located at 39.9°N and 116.4°E.

北京位于东经 116.4°,北纬 39.9°。

(2) 直角坐标

Taking point O as the origin, the coordinate of point P is (36.9, 45.5) in unit of centimeter.

以 O 点为原点,P 点的坐标为(36.9, 45.5)(单位: cm)。

翻译练习

Unit One

1. Although a point is a geometrical concept quite different from everyday objects such as footballs and automobiles, we shall see that the actual motion of many objects is most easily described as the motion of a single point (the "center of mass"), plus the rotation of the object about that point.

2. **An interval of time.** The fundamental unit is the **second** (s), which by international agreement is defined as the duration of 9 192 631 770 periods of radiation corresponding to the transition between the two lowest energy levels in the atomic isotope cesium 133.

3. **An interval of length.** The fundamental unit is the **meter** (m), which is defined independently of the time interval. Before 1983, by international agreement the meter was defined as exactly 1 650 763. 73 wavelengths of the orange light emitted from the isotope krypton 86. In November 1983, the length standard was defined as the distance that light travels in a vacuum in 1/299 792 458 second.

4. The cross product of two vectors $\boldsymbol{A}$ and $\boldsymbol{B}$ is defined to be a vector $\boldsymbol{C}=\boldsymbol{A}\times\boldsymbol{B}$ whose magnitude equals the area of the parallelogram formed by the two vectors. The vector $\boldsymbol{C}$ is perpendicular to the plane containing $\boldsymbol{A}$ and $\boldsymbol{B}$ in the direction given by the right-hand rule, that is, as your right hand fingers curl from the direction of $\boldsymbol{A}$ toward the direction of $\boldsymbol{B}$, the direction of $\boldsymbol{A}\times\boldsymbol{B}$ is given by your thumb. (注意定语从句、分词结构作定语的翻译)

5. For motion along a straight line, we choose a rectangular coordinate system that is oriented so that one of the axes (for example, the x axis) is along the line.

6. It is well known that all objects, when dropped, will fall toward the earth with nearly constant acceleration. In the idealized case, where air resistance is neglected, such motion is referred to as **free fall.**

7. In this section, we describe how observations made by different observers in different frames of reference are related to each other. We shall find that observers in different frames of reference may measure different displacements, velocities, and accelerations for a particle in motion. That is, two observers moving with respect to each other will generally not agree on the outcome of a measurement.

Unit Two

1. An object at rest will remain at rest and an object in motion will continue in motion

with a constant velocity(that is, constant speed in a straight line) unless it experiences a net external force(or resultant force).

2. The time rate of change of momentum of an object is equal to the resultant external force acting on the object.

3. If two bodies interact, the force exerted on body 1 by body 2 is equal to and opposite the force exerted on body 2 by body 1.

4. A force is conservative if the total work it does on a particle is zero when the particle moves around any closed path, returning to its initial position. From Fig. 6-5 we see that this definition implies that: The work done by a conservation force on a particle is independent of the path taken as the particle moves from one point to another.

5. In the past we saw that the net force acting on a system of particles is equal to the net force acting on the system because the internal forces(those exerted by the particles within the system on one another) cancel in pairs. The treatment of internal torques exerted by the particles within a system on one another leads to a similar result, that is, the net torque acting on a system equals the net external torque acting on the system.

6. Left to itself, a spring or a pendulum eventually stops oscillating because the mechanical energy is dissipated by frictional forces. Such motion is said to be damped. If the damping is large enough, as, for example, a pendulum submerged in molasses, the oscillator fails to complete even one cycle of oscillation. Instead it just moves toward the equilibrium position with a speed that approaches zero as the object approaches the equilibrium position. This type of motion is referred to as overdamped. If the damping is small enough that the system oscillates with an amplitude that decreases slowly with time-like a child on a playground swing when Mum stops providing a push each cycle-the motion is said to be underdamped. Motion with the minimum damping for nonoscillatory motion is said to be critically damped.

Unit Three

1. Suppose instead that we place the warm copper bar in a cool running stream. The bar cools until it stops contracting, at the point at which the bar and the water are in thermal equilibrium.

2. Suppose that we have two containers with identical volumes, each holding the same amount of the same kind of gas at the same temperature and pressure. If we consider the two containers as one system, we have twice the amount of gas at twice the volume, but at the same temperature and pressure.

3. Eq. (9-8), which relates the variables P, V, and T, is known as the **ideal-gas law**, and is an example of an **equation of state**. It describes the properties of real gases with low densities (and therefore low pressures). At higher densities, corrections must be made to this equation.

4. In some special cases the internal energy of a system depends only on its

temperature, not on its pressure or volume. The most familiar system having this special property is an ideal gas. For such system, if the temperature is constant, the internal energy is also constant; $\Delta E=0$, and $Q=W$, that is, any energy entering the system as heat Q must leave it again as work W done by the system. For most systems other than ideal gases the internal energy depends on pressure as well as temperature, so E may vary even when T is constant.

5. **Kelvin statement of the second law of thermodynamics**: *No system can take energy as heat from a single reservoir and convert it entirely into work without additional net changes in the system or its surroundings.*

6. **Clausius statement of the second law of thermodynamics**: *A process whose only net result is to transfer energy as heat from a cooler object to a hotter one is impossible.*

7. In the introduction, we alluded the essence of wave motion—the transfer of a disturbance through space without the accompanying transfer of matter. The propagation of the disturbance also represents a transfer of energy—thus, we can view waves as means of energy transfer. Two entries that depend on waves: mechanical waves and electromagnetic radiation. These are to be contrasted with another entry—matter transfer—in which the energy transfer is accompanied by a movement of matter through space.

Unit Four

1. The laws of electricity and magnetism play a central role in the operation of devices such as radios, televisions, electric motors, computers, high-energy particle accelerators, and a host of electronic devices used in medicine. More fundamental, however, is the fact that the interatomic and intermolecular forces responsible for the formation of solids and liquids are electric in origin. Furthermore, such forces as the pushes and pulls between objects in contact and the elastic force in a spring arise from electric forces at the atomic level.

2. Maxwell's contributions to the science of electromagnetism were especially significant because the laws he formulated are basic to all forms of electromagnetic phenomena. His work is comparable in importance to Newton's discovery of the laws of motion and the theory of gravitation.

3. Gauss's law, which relates the electric field on a closed surface to the net charge within the surface, is of fundamental importance in the study of electrostatic fields. In words, *Gauss's law states that in the electrostatic field in vacuum the net electric flux through any closed surface is equal to the net charge inside the surface divided by* ε_0.

4. In principle, Gauss's law is valid for all surfaces and all charge distributions. In practice, however, it can be used to calculate the electric field only for charge distributions that have high degrees of symmetry. For static charge distributions, Gauss's law and Coulomb's law are equivalent. However, Gauss's law is more general in that it is always

valid whether or not the charges are static.

5. The earliest known sources of magnetism were permanent magnets. One month after Oersted announced his discovery that a compass needle is deflected by an electric current, Jean Baptiste Biot and Felix Savart announced the results of their measurements of the force on a magnet near a long, current-carrying wire and analyzed these results in terms of the magnetic field produced by each element of the current. Andre-Marie Ampere extended these experiments and showed that current elements also experience a force in the presence of a magnetic field and that two currents exert forces on each other.

Unit Five

1. Experiments by Faraday, Henry, and others showed that if the magnetic flux through an area bounded by a circuit is changed by any means, an emf equal in magnitude to the rate of change of the flux is induced in the circuit.

2. This result is known as Faraday's law of induction which states the emf induced in a circuit is directly proportional to the time rate of change of magnetic flux through the circuit.

3. The negative sign in Faraday's law has to do with the direction of the induced emf, which can be found from a general physical principle known as Lenz's law.

4. The polarity of the induced emf is such that it tends to produce a current that will create a magnetic flux to oppose the change in magnetic flux through the loop.

5. Maxwell's equations, first proposed by the great Scottish physicist James Clerk Maxwell, relate the electric and magnetic field vectors $\boldsymbol{E}$ and $\boldsymbol{B}$ to their sources, which are electric charge, currents, and changing fields.

6. To study electromagnetic radiation, we must call upon our knowledge of Maxwell's equations and relate electric and magnetic fields to time-varying charge and current distributions. A primary difficulty of this task is that the charge and current distributions on antenna structures resulting from given excitations are generally unknown and very difficult to determine. In fact, the geometrically simple case of a straight conducting wire (linear antenna) excited by a voltage source in the middle has been a subject of extensive research for many years, and the exact charge and current distributions on a wire of a finite radius are extremely complicated even when the wire is assumed to be perfectly conducting.

7. Early in the nineteenth century experiments were suggested and made to show that light is a wave motion. A key figure in this endeavour was Thomas Young, one of the most intelligent and clever scientists ever to live, who studied diffraction and interference of light already in 1803 with results that gave strong support to the wave theory of Christian Huygens as opposed to the particle or corpuscular theory of Isaac Newton.

Unit Six

1. Scientists are always using models when attempting to understand nature. Bohr challenged the Rutherford model, stating that it was a mistake to assume the atom is just a scaled-down solar system with the same rules; electrons do not behave like miniature planets.

2. Jumps up from higher levels are extremely unlikely because electrons excited to a higher level typically remain there for less than a millionth of a second. Thus, there are fewer lines in the absorption spectrum than in the emission spectrum, in agreement with the observations.

3. One possibility that was suggested to explain these results is that the electrons are somehow affecting each other. We can test this by lowering the rate at which the source emits electrons so that only one electron passes through the setup at a time.

4. The set of experiments can be repeated with photons. The results are the same. The detectors at the screen see complete photons, not half photons. But the two-slit pattern is an interference pattern. Photons behave like electrons. Photons and electrons exhibit a duality of particle and wave behavior.

5. This idea is now known as Heisenberg's **uncertainty principle**. Mathematically, it says that the product of the uncertainties of these pairs has a lower limit equal to Planck's constant.

6. These ideas led to the development of a new view of physics known as **quantum mechanics**, which are the rules for the behavior of particles at the atomic and subatomic levels. These rules replace Newton's and Maxwell's rules.

7. But for the light wave, the conclusion is an astonishing one: *no matter how fast a light source moves toward or away from an inertial observer, and no matter how fast the observer moves toward or away from the source, the speed of the light passing from one to the other in vacuum will always be c—the speed of light is constant*; it is absolute.

8. Remember that the equations of Electromagnetic Theory—Maxwell's Equations—led to a wave equation that provided the speed of light in vacuum. Assuming Maxwell's Equation are right and given the First Postulate, it must be that this same wave equation is applicable in all inertial systems—the vacuum speed of light measured on Earth or inside a rocket ship must be the same, independent of any uniform relative motion.

Appendixes

附录 1　物理专业所开设的课程名称汉英对照

BASIC 语言及应用　BASIC Language & Application
C 语言　C Language
C 程序设计　Program Design in C(Language)
FORTRAN 语言　FORTRAN Language
X 射线与电镜　X-ray & Electron Microscope
半导体材料　Semiconductor Materials
半导体物理　Semiconductor Physics
毕业设计　Graduation Project
毕业实习　Graduation Practice
变分法与张量　Calculus of Variations & Tensors
材料力学　Material Mechanics
常微分方程　Ordinary Differential Equations
场论　Field Theory
超导磁体及应用　Superconductive Magnet & Application
超导及应用　Superconductivity & Application
成像原理与技术　Principles & Technique of Imaging
传感器原理及应用　Fundamentals & Application of Sensors
磁性物理　Magnetophysics
大学物理　College Physics
大学物理实验　Experiment of College Physics
大学英语　College English
大学语文　College Chinese
单片机原理　Fundamentals of Single-Chip Computers
单片机原理及应用　Fundamentals & Applications of Single-Chip Computers
弹性力学　Elasticity
导波光学　Wave guiding Optics
德育　Moral Education
等离子体工程　Plasma Engineering
邓小平理论　Theory of Deng Xiaoping
低温固体物理　Low-Temperature Solid Physics
低温物理导论　Introduction to Low-Temperature Physics
低温原理　Low-Temperature Fundamentals
电磁场　Electromagnetic Fields
电磁场与电磁波　Electromagnetic Fields & Magnetic Waves
电磁场与微波技术　Electromagnetic Fields & Microwave Technology
电磁场中的数值方法　Numerical Methods in Electromagnetic Fields
电磁场中的数值计算　Numerical Calculation in Electromagnetic Fields
电磁学　Electromagnetics
电动力学　Electrodynamics
电工与电子技术　Electrotechnics & Electronic Technology
电工电子学　Electronics in Electrical Engineering
电工学　Electrotechnics
电路　Circuits
电路分析基础　Basis of Circuit Analysis
电路分析基础实验　Basic Experiment on Circuit Analysis
电路分析实验　Experiment on Circuit Analysis
电路和电子技术　Circuit and Electronic Technique
电路理论　Theory of Circuits
电路理论实验　Experiments in Theory of Circuits
电视接收技术　Television Reception Technology
电视原理　Principles of Television
电子材料与元件测试技术　Test Technology of

Electronic Materials and Elements
电子技术 Technology of Electronics
电子技术基础 Basic Electronic Technology
电子技术基础与实验 Basis & Experiment of Electronic Technology
电子线路 Electronic Circuits
电子线路实验 Experiment in Electronic Circuits
电子学 Electronics
法律基础 Fundamentals of Law
复变函数 Complex Variables Functions
复变函数与积分变换 Functions of Complex Variables & Integral Transforms
傅里叶光学 Fourier Optics
概率论 Probability Theory
概率论与数理统计 Probability Theory & Mathematical Statistics
概率论与随机过程 Probability Theory & Stochastic Processes
高等数学 Advanced Mathematics
高分子物理 Polymer Physics
高分子物理实验 (High) Polymer Physics Experiment
高级英语听说 Advanced English Listening & Speaking
固体激光器 Solid-State Lasers
固体激光器件 Solid Laser Elements
固体激光与电源 Solid-State Laser & Power Units
固体物理 Solid State Physics
光波导理论 Optical Waveguide Theory
光电技术 Photoelectric Technology
光谱 Optical Spectrum
光谱分析 Spectral Analysis
光谱学 Spectroscopy
光纤传感器 Fiber Optical Sensors
光纤传感器基础 Fundamentals of Fiber Optical Sensors
光纤传感器及应用 Fiber Optical Sensors & Applications
光纤实验 Experiments in Optical Fiber
光纤原理与技术 Fiber Optical Operation & Technology
光学 Optics
光学信息导论 Introduction to Optical Infomation
红外物理 Infrared Physics
红外物理与技术 Infrared Physics & Technology
化学 Chemistry
画法几何及机械制图 Descriptive Geometry & Mechanical Graphing
汇编语言程序设计 Assembly Language Programming
积分变换 Integral Transforms
积分变换及数理方程 Integral Transforms & Mathematical Equations
激光基础 Basis of Laser
激光技术 Laser Technology
激光实验 Laser Experiments
激光原理 Principles of Laser
激光原理与技术 Laser Principles & Technology
计算方法 Computing Methods
计算机基础 Basis of Computer Engineering
计算机原理及应用 Principles & Applications of Computers
计算力学 Computational Mechanics
渐近方法 Asymptotical Methods
近代光学测试技术 Modern Optical Testing Technology
近代光学计量技术 Modern Optical Measuring Technology
近代物理实验 Experiment of Modern Physics
科技翻译 Scientific English Translation
科技史 History of Science & Technology
科技英语 Scientific English
科技英语基础 Elementary Scientific English
科技英语阅读 Readings of Scientific English
科学方法论 Scientific Methodology
科学技术史 History of Science & Technology
课程设计 Course Exercise
理论力学 Theoretical Mechanics
力学 Mechanics
力学实验 Experiment of Mechanics
量子力学 Quantum Mechanics
流体力学 Fluid Mechanics
模拟电路 Analog Circuitry
模拟电子技术 Analog Electronics Technique
模拟电子技术基础 Basis of Analogue Electronic Technique

模拟集成电路　Analog Integrated Circuitry
普通物理　General Physics
普通物理实验　Experiments of General Physics
热学　Thermology
热力学　Thermodynamics
热力学与统计物理　Thermodynamics & Statistical Physics
摄影技术　Techniques for Photography
生产实习　Production Practice
实验分析　Experimental Analysis
数据库原理及应用　Principles & Applications of Databases
数理方程　Mathematical Equations
数理方程积分变换　Integral Transforms of Mathematical Equations
数理方程与特殊函数　Equations of Mathmatical Physics & Special Functions
数理统计　Mathematical Statistics
数学分析　Mathematical Analysis
数学模型　Mathematical Models
数学物理方法　Methods of Mathematical Physics
数值分析　Numerical Analysis
数字电路　Digital Circuits
数字电路及微机原理　Digital Circuits & Microcomputer Principles
数字电子技术　Digital Electronic Technology
数字技术　Digital Techniques
随机过程与时序分析　Stochastic Processes & Time Series Analysis
体育　Physical Education
天线原理　Principle of Antennas
铁磁学　Ferromagnetics
铁电半导体　Ferro-Electric Semiconductor
统计物理　Statistical Physics
微波技术与天线　Microwave Techniques and Antennas
微机原理及应用　Principles & Applications of Microcomputers
物理学　Physics
物理光学　Physical Optics
物理光学实验　Optical-Physics Experiment
物理化学　Physical Chemistry
物理化学实验　Physical Chemistry Experiment
物理实验　Physical Experiment
误差理论　Theory of Error
显示技术　Display Technique
显示技术及装置　Display Technique & Equipment
线性代数　Linear Algebra
心理学　Psychology
信号变换与处理　Signal Conversion & Processing
信息光学　Information Optics
应用光学　Applied Optics
应用光学实验　Experiment of Applied Optics
应用物理实验　Experiment of Applied Physics
英语　English
英语泛读　Extensive English Reading
英语精读　Intensive English Reading
英语口语　Oral English Practice
英语听力　English Listening & Comprehension
英语写作　English Writing
英语语法　English Grammar
有限元法　Finite Element Methods
原子物理学　Atomic Physics
哲学　Philosophy
政治经济学　Political Economics
中国革命史　History of Chinese Revolutionary
专业实习　Specialty Practice
专业英语　Specialty English
专业英语阅读　Specialty English Reading
自然科学概论　Introduction to Natural Science
最优化方法　Optimigation Methods

附录 2　教学教务管理常用词汇

在职进修班　in-service training course
进修班　class for advanced studies
短训班　short-term training course
专修科　special (training) course
自学考试　self-taught examination
教职员　teaching and administrative staff

教学人员　the faculty, teaching staff
大专院校校长　president, chancellor
教务长　dean of studies
(系)主任　chairman, chairperson, department head
副主任　vice-chairman, vice department head
教授　professor
副教授　associate professor
客座教授　visiting professor, guest professor
讲师　lecturer, instructor
助教　assistant, TA
专职教师　full-time teacher
兼职教师　part-time teacher
辅导员　assistant for political and ideological work
教研室/组　teaching and research section/group
教学组　teaching group
教员休息室　staff room, common room
母校　Alma Mater
校友　alumnus, alumna
学生会　students' union/association
高年级学生　upper/higher-grade student, student in senior grades
低年级学生　lower-grade student, student in junior grades
走读生　day student, non-resident student
住宿生　boarder
旁听生　auditor
研究生　graduate student, postgraduate (student)
应届毕业生　graduating student/pupil, this year's graduate
听课　to visit a class, to sit in on a class, to attend a lecture
公开课　open class
课程　course, curriculum
必修课　required/compulsory course
选修课　elective/optional course
基础课　basic course
专业课　specialized course
课程表　school timetable
课外活动　extracurricular activity
课外辅导　instruction after class
课外阅读　outside reading, after-class reading
课堂讨论　class discussion
(大学的)研究小组,讨论会　seminar
教学大纲　teaching program, syllabus
教学内容　content of a course
学习年限　period of schooling
学历　record of formal schooling
学年　school/academic year
学期　(school)term, semester
学分　credit
分数　mark, grade
五分制　the 5-grade marking system
百分制　100-mark system
学习成绩　academic record, school record
成绩单　school report, report card, transcript
基本框架　basic framework
办学效益　efficiency in school management
基础科学　the fundamentals, basic science
学分制　credit system
三学期制　trimester system
双学士制　double bachelor's degree system
主副修制　system of a major field of specialization plus a minor field
教学、科研、生产的"三结合"　"3-in-1" combining teaching, research, and production
定向招生　oriented student recruitment
包分配　guarantee job assignment
示范试点　demonstration pilot project
动员　mobilize
多学科的　multidisciplinary
重点大学　key university
授予(学位)　confer
被授权　be authorized to do
博士后科研流动站　center for postdoctoral studies
专业　speciality
国家发明奖　National Invention Prize
国家自然科学奖　National Prize for Natural Sciences
国家科技进步奖　National Prize for Progress in Science and Technology
学术报告会,专题讨论会　symposium
毕业论文　thesis, dissertation
毕业实习　graduation field work/practice
毕业设计　graduation design/project
毕业典礼　graduation ceremony, commencement
毕业证书　diploma, graduation certificate
毕业鉴定　graduation appraisal

博士生导师　Ph. D advisor，doctoral advisor (supervisor)
授予某人学位　to confer a degree on sb.
补考　make-up examination
奖学金　scholarship
(国家)助学金　(state) stipend/subsidy
领取助学金的学生　a grant-aided student
学费　tuition (fee)
资料中心　data center
国际文化交流　international cultural communication
CET-4 College English Test(Band 4)
CET-6 College English Test(Band 6)

附录 3　物理学一、二级学科名称英文词汇

1. 一级学科
物理学　Physics
2. 二级学科
理论物理　Theoretical Physics
粒子物理与原子核物理　Particle Physics and Nuclear Physics
原子与分子物理　Atomic and Molecular Physics
等离子体物理　Plasma Physics
凝聚态物理　Condensed Matter Physics
声学　Acoustics
光学　Optics
无线电物理　Radio Physics

附录 4　数学符号及英语表达

$\frac{1}{2}$	a half *or* one half
$\frac{1}{3}$	a third *or* one third
$\frac{2}{3}$	two thirds
$\frac{1}{4}$	a quarter *or* one quarter; a fourth *or* one fourth
$\frac{1}{10}$	a tenth *or* one tenth
$\frac{1}{100}$	a [one] hundredth
$\frac{1}{1000}$	a [one] thousandth
$\frac{1}{1234}$	one over a thousand two hundred and thirty-fourth
$\frac{3}{4}$	three fourths *or* three quarters
$\frac{4}{5}$	four fifths *or* four over five
$\frac{113}{300}$	one hundred and thirteen over three hundred

$2\frac{1}{2}$	two and a half
$2\frac{7}{8}$	two and seven over eight *or* two and seven eighths
$3\frac{1}{8}$	three and one eighth
$4\frac{1}{3}$	four and a third
$125\frac{3}{4}$	a [one] hundred twenty-five and three fourths [quarters]
0.1	O point one *or* zero point one *or* nought point one
0.01	O point O one *or* zero point zero one *or* nought point nought one
0.25	nought point two five *or* point two five
0.045	decimal [point] nought four five
2.35	two point three five
$4.\dot{9}$	four point nine recurring
$3.03\dot{2}\dot{6}$	three point nought three two six, two six recurring
45.67	four five [forty-five] point six seven
38.72	three eight point seven two *or* thirty-eight decimal seven two
0.001	O point O O one *or* nought point nought nought one *or* zero point zero zero one *or* point nought nought one
+	plus; positive
−	minus; negative
±	plus or minus
∓	minus or plus
×(·)	multiplied by; times
÷	divided by
=	is equal to; equals
≡	is identically equal to; be identical with; be equitvalent to
≈≌	is approximately equal to; approximately equals
()	round brackets; parentheses
[]	square brackets
〈〉	angular brackets
{}	braces
∩	intersection
∪	union
∈	is a member of (set)
⊂	is a subset of
~	difference
*	denotes an operation
⇔	is equivalent to
⇒	implies
{} or ø	empty set
→	maps into

$\therefore$	therefore
$\because$	because
i or j	imaginary *or* square root of -1
π	pi; the ratio of the circumference of a circle to its diameter, approx. 3.14159
e or ε	① the base of natural logarithms, approx. 2.71828 ②the eccentricity of a conic section
$x!$	factorial x
$\log_n x$	log x to the base n
$\log_{10} x$	log x to the base 10(*i.e.* common logarithm)
$\log_e x$ or $\ln x$	log x to the base e(*i.e.* natural logarithm *or* Naperian logarithm)
M	modulus of common logarithms $\log_{10} e=0.4343$, $(\log_{10} x=0.4343\times\log_e x)$
M^{-1}	$\log_e 10=2.3026$, $(\log_e x=2.3026\times\log_{10} x)$
x^n	$x\cdot x\cdot x$...to n factors; the nth power of x; x to the power n
$x^{\frac{1}{n}}$ or $\sqrt[n]{x}$	the nth root of x; x to the power one over n
$x\rightarrow a$	x approaches the limit a
θ	(the angle between the radius vector and the polar axis) the angle theta
$\sin^{-1} x$	(the principal value of the angle whose sine is x)arc sine of x
sinh	sinus hyperbolicus; the hyperbolic sine
$\sum$	the sum of the terms indicated; summation of; sigma
$\sum_{n=1}^{10} x_n$	sum over n from 1 to 10 of x_n
$\prod$	the product of the terms indicated
$\vert x\vert$	the absolute value of x
$\bar{x}$	the mean value of x; x bar
b'	b prime
b''	b double prime; b second prime
b'''	b triple prime
b_1	b sub one
b_2	b sub two
b_m	b sub m
$\dot{x}$	x dot
$\ddot{x}$	x two dots
b''_m	b double prime sub m
f or F	function
$f(x)$; $F(x)$	function f(*or* F)of x
$y=f(x)$	y is a function of x
Δ	(finite difference *or* increment) delta
Δx or δx	(the increment of x) delta x
$\mathrm{d}x$	(an increment of x considered as tending to zero) dee of x; dee x; differential x
$\frac{\mathrm{d}y}{\mathrm{d}x}$ or $D_x y$	the differential coefficient of y with respect to x; the first derivative of y with respect to x
$\frac{\mathrm{d}^2 y}{\mathrm{d}x^2}$	the second derivative of y with respect to x

$\frac{d^n y}{dx^n}$	the nth derivative of y with respect to x
$\frac{\partial y}{\partial u}$	the partial derivative of y with respect to u, where y is a function of u and another variable (or variables)
$F'(x)$	the first derivative of function F of x with respect to x
∇	$\boldsymbol{i}\frac{\partial}{\partial x}+\boldsymbol{j}\frac{\partial}{\partial y}+\boldsymbol{k}\frac{\partial}{\partial z}$; del; nabla; vector differential operator
∇^n	nth del (nabla)
$\int$	integral
$\int_a^b$	integral between limits a and b; the integral from a to b
∞	infinity
$\boldsymbol{F}$	vector $\boldsymbol{F}$
$x+y$	x plus y
$(a+b)$	bracket a plus b bracket closed
$a=b$	a equals b; a is equal to b; a is b
$a\neq b$	a is not equal to b; a is not b
$a\pm b$	a plus or minus b
$a\approx b$	a is approximately equal to b
$a>b$	a is greater than b
$a\gg b$	a is much [far] greater than b
$a\geqslant b$	a is greater than or equal to b
$a<b$	a is less than b
$a\ll b$	a is much less than b
$a\leqslant b$	a is less than or equal to b
$a\perp b$	a is perpendicular to b
$x\to\infty$	x approaches infinity
$a\equiv b$	a is identically equal to b; a is of identity to b
$\angle a$	angle a
$a//b$	a is parallel to b
$a\sim b$	the difference between a and b
$a\propto b$	a varies directly as b; a is proportional to b
$a\cap b$	a cap b; a meet b; a intersection b
$a\cup b$	a cup b; a join b; a union b
$1\times1=1$	once one is one
$2\times2=4$	twice two is four
$6\times5=30$	six times [multiplied by] five equals [is equal to; are; makes; make] thirty
$30=6\times5$	thirty is five times as large as six
$s=vt$	s equals [is equal to] v multiplied by t; s equals v times t
$1:2$	the ratio of one to two
$12\div3=4$	12 divided by 3 equals [is] 4
$20:5=16:4$	the ratio of 20 to 5 equals the ratio of 16 to 4; 20 is to 5 as 16 is to 4

$a:b=c:d$	*a* is to *b* as *c* is to *d*
$a+b=c$	*a* plus *b* is[are; equals; is equal to]*c*
$c-b=a$	*c* minus *b* is [equals; is equal to] *a*; *b* from *c* leaves *a*
$v=\frac{s}{t}$	*v* equals *s* divided by *t*; *v* is *s* over *t*
$7+3<12$	7 plus 3 is less than 12
$12>7+3$	12 is greater than 7 plus 3
$72-16=56$	72 minus 16 is [equals; is equal to] 56 ;16 from 72 leaves 56
x^2	*x* square;*x* squared; the square of *x*; the second power of *x*;*x* to the second power
$5^2=25$	the second power of 5 is 25 ; 5 square is 25 ; the square of 5 is 25; 5 to the second power is equal to 25
y^3	*y* cube; *y* cubed; the cube of *y*; *y* to the third power; *y* to the third
y^{-10}	*y* to the minus tenth (power); *y* to the power (of) minus ten
$\sqrt{4}=\pm 2$	the square root of 4 is [equals] plus or minus 2
$\sqrt[3]{a}$	the cube root of *a*
$\sqrt[5]{x^2}$	the fifth root of *x* square
$\sqrt{518}$	the square root of five hundred and eighteen
$\sqrt[3]{930}$	the cubic root of nine hundred and thirty
$3x=5$	three times *x* equals 5
$\frac{x^3}{5}=y^2$	*x* raised to the third power divided by five equals *y* squared
$x^2+y^2=10$	*x* squared with *y* squared equals 10
$a=\frac{V_t-V}{t}$	*a* equals *V* sub *t* minus *V* all over (divided by)*t*
$(a+b-c\times d)\div e=f$	*a* plus *b* minus *c* multiplied by *d*, all divided by *e* equals *f*
$\left(8+6\frac{5}{8}-3.88\times 4\right)\div 2\frac{1}{2}$	eight plus six and five-eighths minus three decimal [point] eight eight multiplied by four, all divided by two and a half
$4567\div 23=198$ 余 13	23 into 4567 goes 198 times, and 13 remainder
$45+70+152=267$	45, 70 and 152 added together are 267
%	per cent
2%	two per cent
‰	per mille
5‰	five per mille
$\frac{3}{8}$%	three eighths (of one) per cent
0.3%	point three per cent
$\frac{1}{2}$ ton	half a ton
$\frac{2}{3}$ ton	two thirds of a ton
$\frac{3}{4}$ km	three quarters of a kilometer
1.75 km	one point seven five kilometers

60 mi/hr	sixty miles per hour
20°	twenty degrees
6′	①6 minutes ②6 feet
10″	①10 seconds ②10 inches
0℃	zero degree Centigrade [Celsius]
100℃	one [a] hundred degrees Centigrade
32℉	thirty-two degrees Fahrenheit

附录5 常用物理基本常量

物理常量 PHYSICAL CONSTANTS		符号	数　值
真空中光速	speed of light in vacuum	c	299 792 458 m/s
引力常量	gravitational constant	G	$6.672\ 59\times10^{-11}\ N\cdot m^2/kg^2$
阿伏伽德罗常量	Avogadro constant	N_A	$6.022\ 127\times10^{23}\ mol^{-1}$
普适气体常量	universal gas constant	R	8.314 510 J/mol·K
玻尔兹曼常量	Boltzmann constant	k	$1.380\ 66\times10^{-23}$ J/K $8.617\ 39\times10^{-5}$ eV/K
理想气体摩尔体积	molar volume (STP)	V_m	$22.413\ 996\times10^{-3}\ m^3/mol$
基本电荷(元电荷)	electron charge magnitude	e	$1.602\ 176\ 53(14)\times10^{-19}$ C
原子质量单位	atomic mass unit	u	$1.660\ 54\times10^{-27}$ kg
电子静止质量	electron mass	m_e	$9.109\ 39\times10^{-31}$ kg
电子荷质比	electron charge/mass ratio	e/m_e	$1.758\ 82\times10^{11}$ C/kg
质子静止质量	proton mass	m_p	$1.672\ 623\times10^{-27}$ kg 1.007 276 u
中子静止质量	neutron mass	m_n	$1.674\ 929\times10^{-27}$ kg 1.008 66 u
法拉第常量	Faraday constant	F	$9.648\ 53\times10^{4}$ C/mol
真空电容率	permittivity of vacuum	ε_0	$8.854\ 19\times10^{-12}\ C^2/N\cdot m^2$
真空磁导率	permeability of vacuum	μ_0	$1.256\ 64\times10^{-6}$ T·m/A
电子磁矩	electron magnetic moment	μ_e	$9.284\ 77\times10^{-24}$ J/T
质子磁矩	proton magnetic moment	μ_p	$1.410\ 618\times10^{-26}$ J/T
玻尔半径	Bohr radius	α_0	$5.291\ 77\times10^{-11}$ m
玻尔磁子	Bohr magneton	μ_B	$9.274\ 015\times10^{-24}$ J/T
核磁子	nuclear magneton	μ_N	$5.050\ 786\ 6\times10^{-27}$ J/T
普朗克常量	Planck constant	h	$6.626\ 068\ 76(52)\times10^{-34}$ J·s $4.135\ 667\ 27(16)\times10^{-15}$ eV·s
约化普朗克常量	reduced Planck constant	$\hbar$	$1.054\ 571\ 596(82)\times10^{-34}$ J·s $6.582\ 118\ 89(26)\times10^{-16}$ eV·s

续表

物理常量 PHYSICAL CONSTANTS		符号	数　值
精细结构常量	fine structure constant	α	$7.297\,352\,533\times10^{3}$
里德伯常量	Rydberg constant	R	$1.097\,373\times10^{7}$/m
电子康普顿波长	electron Compton wavelength	λ_{ec}	$2.426\,31\times10^{-12}$ m
质子康普顿波长	proton Compton wavelength	λ_{pc}	$1.321\,41\times10^{-15}$ m
热功当量	mechanical equivalent of heat		4.186 J/cal
标准大气压强	standard atmospheric pressure	1 atm	$1.013\,25\times10^{5}$ Pa
电子伏特	electron volt	1 eV	$1.602\,177\,33(49)\times10^{-19}$ J
重力加速度(标准值)	acceleration due to gravity (standard)	g	$9.806\,65$ m/s^2
电子静止能量	electron rest energy	$m_e c^2$	0.510 999 06(15)J
理想气体体积(0°C，1个大气压下)	volume of ideal gas (0°C and 1 atm)		22.414 10(19)liter/mol

附录 6　常用汉英物理学专业词汇表

BET 公式　BET formula
DLVO 理论　DLVO theory
HLB 法　hydrophile-lipophile balance method
pVT 性质　pVT property
ζ 电势　zeta potential
阿伏伽德罗常量　Avogadro number
阿伏伽德罗定律　Avogadro law
阿伦尼乌斯电离理论　Arrhenius ionization theory
阿伦尼乌斯方程　Arrhenius equation
阿伦尼乌斯活化能　Arrhenius activation energy
阿马加定律　Amagat law
艾林方程　Erying equation
爱因斯坦光化当量定律　Einstein's law of photochemical equivalence
爱因斯坦-斯托克斯方程　Einstein-Stokes equation
安托万常数　Antoine constant
安托万方程　Antoine equation
昂萨格电导理论　Onsager's theory of conductance
半电池　half cell
半衰期　half-life period
饱和吸附量　saturated extent of absorption
饱和液体　saturated liquid
饱和蒸气　saturated vapor
饱和蒸气压　saturated vapor pressure
爆炸界限　explosion limits
比表面功　specific surface work
比表面吉布斯函数　specific surface Gibbs function
比浓黏度　reduced viscosity
标准电动势　standard electromotive force
标准电极电势　standard electrode potential
标准摩尔反应焓　standard molar reaction enthalpy
标准摩尔反应吉布斯函数　standard Gibbs function of molar reaction
标准摩尔反应熵　standard molar reaction entropy
标准摩尔焓函数　standard molar enthalpy function
标准摩尔吉布斯自由能函数　standard molar Gibbs free energy function
标准摩尔燃烧焓　standard molar combustion enthalpy
标准摩尔熵　standard molar entropy
标准摩尔生成焓　standard molar formation enthalpy
标准摩尔生成吉布斯函数　standard molar formation Gibbs function
标准平衡常数　standard equilibrium constant
标准氢电极　standard hydrogen electrode
标准态　standard state
标准熵　standard entropy

标准压力 standard pressure
标准条件 standard condition
表观活化能 apparent activation energy
表观摩尔质量 apparent molar mass
表观迁移数 apparent transference number
表面 surfaces
表面过程控制 surface process control
表面活性剂 surfactant
表面超额 surface excess
表面张力 surface tension
表面质量作用定律 surface mass action law
玻意耳定律 Boyle law
玻意耳温度 Boyle temperature
玻意耳点 Boyle point
玻尔兹曼常量 Boltzmann constant
玻尔兹曼分布 Boltzmann distribution
玻尔兹曼公式 Boltzmann formula
玻尔兹曼熵定理 Boltzmann entropy theorem
玻色-爱因斯坦统计法 Bose-Einstein statistics
泊松 Poise
不可逆过程 irreversible process
不可逆过程热力学 thermodynamics of irreversible processes
不可逆相变化 irreversible phase change
布朗运动 Brownian movement
查理定律 Charle law
产额 yield
敞开系统 open system
超电势 overpotential
沉降 sedimentation
沉降势 sedimentation potential
沉降平衡 sedimentation equilibrium
弛豫时间 relaxation time
触变性 thixotropy
粗分散系统 coarse disperse system
催化剂 catalyst
单分子层吸附 monomolecule layer absorption
单分子反应 unimolecular reaction
弹式热量计 bomb calorimeter
道尔顿定律 Dalton law
道尔顿分压定律 Dalton partial pressure law
德拜和法尔肯哈根效应 Debye and Falkenhagen effect
德拜立方公式 Debye cubic formula
德拜-休克尔极限公式 Debye-Huckel's limiting equation
等焓过程 isenthalpic process
等焓线 isenthalpic line
等概率定理 theorem of equal probability
低共沸点 lower azeotropic point
低共熔点 eutectic point
低共熔混合物 eutectic mixture
低会溶点 lower consolute point
低熔冰盐结晶 cryohydrate
第二类永动机 perpetual motion machine of the second kind
第三定律熵 third-law entropy
第一类永动机 perpetual motion machine of the first kind
缔合化学吸附 association chemical absorption
电池常数 cell constant
电池电动势 electromotive force of cells
电池反应 cell reaction
电导 conductance
电导率 conductivity
电动势的温度系数 temperature coefficient of electromotive force
电动电势 electrokinetic potential
电功 electric work
电化学 electrochemistry
电化学极化 electrochemical polarization
电极电势 electrode potential
电极反应 reaction on the electrode
电极种类 type of electrodes
电解池 electrolytic cell
电流效率 current efficiency
电迁移 electromigration
电迁移率 electromobility
电渗 electroosmosis
电渗析 electrodialysis
电泳 electrophoresis
丁达尔效应 Dyndall effect
定常流过程 steady flow process
定容摩尔热容 molar heat capacity under constant volume

定容温度计 constant volume thermometer
定态近似法 stationary state approximation
定压摩尔热容 molar heat capacity under constant pressure
定压温度计 constant pressure thermometer
定域粒子系统 localized particle system
动力(学)方程 kinetic equations
动力学控制 kinetic control
独立粒子系统 independent particle system
对称数 symmetry number
对应状态原理 principle of corresponding state
多方过程 polytropic process
多分子层吸附理论 adsorption theory of multi-molecular layers
二级反应 second-order reaction
二级相变 second-order phase change
法拉第常数 Faraday constant
法拉第定律 Faraday's law
反电动势 back E. M. F.
反渗透 reverse osmosis
反应分子数 molecularity
反应级数 reaction order
反应进度 extent of reaction
反应热 heat of reaction
反应速率 rate of reaction
反应速率常数 reaction rate constant
范德瓦耳斯常数 van der Waals constant
范德瓦耳斯方程 van der Waals equation
范德瓦耳斯力 van der Waals force
范德瓦耳斯气体 van der Waals gases
范托夫方程 van't Hoff equation
范托夫规则 van't Hoff rule
范托夫渗透压公式 van't Hoff equation of osmotic pressure
方均根速率 root-mean-square speed
非定域离子系统 non-localized particle systems
非基元反应 nonelementary reaction
非体积功 non-volume work
非依时化学计量反应 time-independent stoichiometric reactions
菲克第一扩散定律 Fick's first law of diffusion
沸点 boiling point
沸点升高 elevation of boiling point
费米-狄拉克统计法 Fermi-Dirac statistics
分布 distribution
分布数 distribution number
分解电压 decomposition voltage
分配定律 distribution law
分散系统 disperse system
分散相 dispersion phase
分体积 partial volume
分体积定律 partial volume law
分压 partial pressure
分压定律 partial pressure law
分子反应力学 mechanics of molecular reactions
分子间力 intermolecular force
分子蒸馏 molecular distillation
封闭系统 closed system
弗罗因德利希吸附经验式 Freundlich empirical formula of adsorption
负极点 negative pole
负吸附 negative absorption
复合反应 composite reaction
盖-吕萨克定律 Gay-Lussac law
甘汞电极 calomel electrode
感胶离子序 lyotropic series
杠杆规则 lever rule
高分子溶液 macromolecular solution
高共沸点 upper azeotropic point
高会溶点 upper consolute point
隔离法 isolation method
格罗塞斯-德雷珀定律 Grotthus-Draper's law
隔离系统 isolated system
功 work
功函 work content
共轭溶液 conjugate solution
共沸温度 azeotropic temperature
构型熵 configurational entropy
孤立系统 isolated system
固溶胶 solid sol
固相线 solid phase line
光反应 photoreaction
光化学第二定律 the second law of photochemistry
光化学第一定律 the first law of photochemistry
光敏反应 photosensitized reactions
光谱熵 spectrum entropy

广度性质　extensive property
广延量　extensive quantity
广延性质　extensive property
规定熵　stipulated entropy
过饱和溶液　oversaturated solution
过饱和蒸气　oversaturated vapor
过程　process
过渡态理论　transition state theory
过冷水　supercooled water
过冷液体　overcooled liquid
过热液体　overheated liquid
亥姆霍兹函数　Helmholtz function
亥姆霍兹函数判据　Helmholtz function criterion
亥姆霍兹自由能　Helmholtz free energy
焓　enthalpy
赫斯定律　Hess law
亨利常数　Henry constant
亨利定律　Henry law
恒沸混合物　constant boiling mixture
恒容摩尔热容　molar heat capacity at constant volume
恒容热　heat at constant volume
恒外压　constant external pressure
恒温方程　equation at constant temperature
恒压摩尔热容　molar heat capacity at constant pressure
恒压热　heat at constant pressure
化学动力学　chemical kinetics
化学反应计量式　stoichiometric equation of chemical reaction
化学反应计量系数　stoichiometric coefficient of chemical reaction
化学反应进度　extent of chemical reaction
化学计量式　stoichiometric equation
化学计量系数　stoichiometric coefficient
化学亲合势　chemical affinity
化学热力学　chemical thermodynamics
化学势　chemical potential
化学势判据　chemical potential criterion
化学吸附　chemisorption
环境　environment
环境熵变　entropy change in environment
挥发度　volatility
混合熵　entropy of mixing
混合物　mixture
活度　activity
活化控制　activation control
活化络合物理论　activated complex theory
活化能　activation energy
霍根-华森图　Hougen-Watson chart
基态能级　energy level at ground state
基尔霍夫公式　Kirchhoff formula
基元反应　elementary reaction
积分溶解热　integration heat of dissolution
吉布斯-杜安姆方程　Gibbs-Duhem equation
吉布斯-亥姆霍兹方程　Gibbs-Helmhotz equation
吉布斯函数　Gibbs function
吉布斯函数判据　Gibbs function criterion
吉布斯吸附公式　Gibbs absorption formula
吉布斯自由能　Gibbs free energy
极化电极电势　polarization potential of electrode
极化曲线　polarization curve
极化　polarization
极限摩尔电导率　limiting molar conductivity
价数规则　rule of valence
简并度　degeneracy
键焓　bond enthalpy
胶冻　jelly
胶核　colloidal nucleus
胶凝作用　gelatification
胶束　micelle
胶体　colloid
胶体分散系统　dispersion system of colloid
胶体化学　collochemistry
胶体粒子　colloidal particles
焦耳　Joule
焦耳-汤姆孙实验　Joule-Thomson experiment
焦耳-汤姆孙系数　Joule-Thomson coefficient
焦耳-汤姆孙效应　Joule-Thomson effect
焦耳定律　Joule law
接触电势　contact potential
接触角　contact angle
结晶热　heat of crystallization
节流过程　throttling process
节流膨胀　throttling expansion
节流膨胀系数　coefficient of throttling expansion

结线 tie line
界面 interface
界面张力 interfacial tension
浸湿 immersion wetting
浸湿功 immersion wetting work
精馏 rectification
聚电解质 polyelectrolyte
聚沉 coagulation
聚沉值 coagulation value
绝对反应速率理论 absolute reaction rate theory
绝对熵 absolute entropy
绝对温标 absolute temperature scale
绝热过程 adiabatic process
绝热式热量计 adiabatic calorimeter
绝热指数 adiabatic index
卡诺定理 Carnot theorem
卡诺循环 Carnot cycle
开尔文公式 Kelvin formula
柯诺瓦洛夫-吉布斯定律 Konovalov-Gibbs law
科尔劳施离子独立运动定律 Kohlrausch's law of independent migration of ions
可逆电池 reversible cell
可逆反应 reversible reactions
可逆过程 reversible process
可逆过程方程 reversible process equation
可逆体积功 reversible volume work
可逆相变 reversible phase change
克拉珀龙方程 Clapeyron equation
克劳修斯不等式 Clausius inequality
克劳修斯-克拉珀龙方程 Clausius-Clapeyron equation
控制步骤 control step
库仑计 coulometer
扩散控制的 diffusion-controlled
拉普拉斯方程 Laplace's equation
拉乌尔定律 Raoult law
兰格缪尔-欣谢尔伍德机理 Langmuir-Hinshelwood mechanism
兰格缪尔吸附等温式 Langmuir absorption isotherm formula
冷冻系数 coefficient of refrigeration
冷却曲线 cooling curve
离解化学吸附 dissociation chemical adsorption
离解热 heat of dissociation
离解压力 dissociation pressure
离子的标准摩尔生成焓 standard molar enthalpy of formation of ions
离子电迁移率 mobility of ions
离子迁移数 transport number of ions
离子独立运动定律 law of the independent migration of ions
离子气氛 ionic atmosphere
离子强度 ionic strength
理想混合物 perfect mixture
理想气体 ideal gas
理想气体绝热指数 adiabatic index of ideal gases
理想气体微观模型 micromodel of ideal gas
理想气体反应的等温方程 isothermal equation of ideal gaseous reactions
理想气体绝热可逆过程方程 adiabatic reversible process equation of ideal gases
理想气体状态方程 state equation of ideal gas
理想稀溶液 ideal dilute solution
理想液态混合物 perfect liquid mixture
粒子 particle
粒子的配分函数 partition function of particles
连串反应 consecutive reaction
链载体 chain carrier
链反应 chain reaction
量热熵 calorimetric entropy
量子统计法 quantum statistics
量子产率 quantum yield
临界参数 critical parameter
临界常数 critical constant
临界点 critical point
临界胶束浓度 critical micelle concentration
临界摩尔体积 critical molar volume
临界温度 critical temperature
临界压力 critical pressure
临界状态 critical state
零级反应 zero-order reaction
流动电位 streaming potential
流动功 flow work
笼效应 cage effect
路易斯-兰德尔逸度规则 Lewis-Randall rule of fugacity
露点 dew point

露点线　dew-point line
麦克斯韦关系式　Maxwell relation
麦克斯韦速率分布　Maxwell distribution of speeds
麦克斯韦能量分布　Maxwell distribution of energy
毛细管凝结　capillary condensation
毛细现象　capillary phenomenon
米凯利斯常数　Michaelis constant
摩尔电导率　molar conductivity
摩尔反应焓　molar reaction enthalpy
摩尔混合熵　molar entropy of mixing
摩尔气体常数　molar gas constant
摩尔热容　molar heat capacity
摩尔溶解焓　molar dissolution enthalpy
摩尔稀释焓　molar dilution enthalpy
内扩散控制　internal diffusions control
内能　internal energy
内压力　internal pressure
能级　energy level
能级分布　energy-level distribution
能量均分原理　principle of the equipartition of energy
能斯特方程　Nernst equation
能斯特热定理　Nernst heat theorem
凝点　freezing point
凝点降低　lowering of freezing point
凝点曲线　freezing point curve
凝胶　gel
凝聚态　condensed state
凝聚相　condensed phase
浓差超电势　concentration overpotential
浓差极化　concentration polarization
浓差电池　concentration cells
帕斯卡　pascal
泡点　bubble point
泡点线　bubble-point line
配分函数　partition function
碰撞截面　collision cross section
碰撞数　number of collisions
偏摩尔量　partial molar quantity
平衡常数(理想气体反应)　equilibrium constant for reaction of ideal gases
平动配分函数　partition function of translation
平衡分布　equilibrium distribution
平衡态　equilibrium state
平衡态近似法　equilibrium state approximation
平衡状态图　equilibrium state diagram
平均活度　mean activity
平均活度系数　mean activity coefficient
平均摩尔热容　mean molar heat capacity
平均质量摩尔浓度　mean mass molarity
平均自由程　mean free path
平行反应　parallel reaction
破乳,脱乳,反乳化　demulsification
铺展　spreading
普遍化范德瓦耳斯方程　universal van der Waals equation
其他功　the other work
汽化热　heat of vaporization
气溶胶　aerosol
气体常数　gas constant
气体分子运动论　kinetic theory of gases
气体分子运动论的基本方程　fundamental equation of kinetic theory of gases
汽相线　vapor line
迁移数　transport number
潜热　latent heat
强度量　intensive quantity
强度性质　intensive property
亲水溶胶　hydrophilic sol
氢电极　hydrogen electrode
区域熔化　zone melting
热　heat
热爆炸　heat explosion
热泵　heat pump
热功当量　mechanical equivalent of heat
热函　heat content
热机　heat engine
热机效率　efficiency of heat engine
热力学　thermodynamics
热力学第二定律　the second law of thermodynamics
热力学第三定律　the third law of thermodynamics
热力学第一定律　the first law of thermodynamics
热力学基本方程　fundamental equation of thermodynamics
热力学概率　thermodynamic probability
热力学能　thermodynamic energy

热力学特性函数　characteristic thermodynamic function
热力学温标　thermodynamic scale of temperature
热力学温度　thermodynamic temperature
热熵　thermal entropy
热效应　heat effect
熔点曲线　melting-point curve
熔化热　heat of fusion
溶胶　sol
溶解焓　dissolution enthalpy
溶液　solution
溶胀　swelling
乳化剂　emulsifier
乳状液　emulsion
瑞利公式　Rayleigh equation
润湿　wetting
润湿角　wetting angle
萨克尔-泰特洛德方程　Sackur-Tetrode equation
三相点　triple point
三相线　triple-phase line
熵　entropy
熵判据　entropy criterion
熵增原理　principle of entropy increase
渗透压　osmotic pressure
渗析法　dialytic process
生成反应　formation reaction
升华热　heat of sublimation
势电解质　potential electrolyte
舒尔策-哈代规则　Schulze-Hardy rule
松弛力　relaxation force
速率方程　rate equation
速率控制步骤　rate determining step
塔费尔公式　Tafel equation
态-态反应　state-state reaction
唐南平衡　Donnan equilibrium
淌度　mobility
特鲁顿规则　Trouton rule
特性黏度　intrinsic viscosity
体积功　volume work
统计权重　statistical weight
统计热力学　statistic thermodynamics
统计熵　statistic entropy
途径　path
途径函数　path function
外扩散控制　external diffusion control
完美晶体　perfect crystal
完全气体　perfect gas
微观状态　microstate
韦斯顿标准电池　Weston standard battery
维恩效应　Wien effect
位力方程　virial equation
位力系数　virial coefficient
无热溶液　athermal solution
无限稀溶液　infinitely dilute dilution
物理化学　physical chemistry
物理吸附　physisorptions
吸附　absorption
吸附等量线　absorption isostere
吸附等温线　absorption isotherm
吸附等压线　absorption isobar
吸附剂　adsorbent
吸附量　extent of absorption
吸附热　heat of absorption
吸附质　adsorbate
析出电势　evolution/deposition potential
稀溶液的依数性　colligative property of dilute solution
稀释焓　dilution enthalpy
系统　system
系统点　system point
系统的环境　environment of system
相　phase
相变　phase change
相变焓　enthalpy of phase change
相变热　heat of phase change
相点　phase point
相对挥发度　relative volatility
相对黏度　relative viscosity
相律　phase rule
相平衡热容　heat capacity in phase equilibrium
相图　phase diagram
相倚粒子系统　system of phase-dependent particles
悬浮液　suspension
循环过程　cyclic process
压力商　pressure quotient
压缩因子　compressibility factor
压缩因子图　diagram of compressibility factor

亚稳态　metastable state
盐桥　salt bridge
盐析　salting out
盐效应　salt effect
阳极　anode
杨氏方程　Young's equation
液体接界电势　liquid junction potential
液相线　liquid phase lines
一级反应　first-order reaction
一级相变　first-order phase change
逸度　fugacity
逸度系数　coefficient of fugacity
阴极　cathode
荧光　fluorescence
永动机　perpetual motion machine
永久气体　permanent gas
有效能　available energy
逾压力　excess pressure
原电池　primary cell
约化摩尔体积　reduced molar volume
约化体积　reduced volume
约化温度　reduced temperature
约化压力　reduced pressure
增比黏度　specific viscosity
沾湿　adhesional wetting
沾湿功　work of adhesional wetting
真溶液　true solution
真实电解质　real electrolyte
真实气体　real gas
真实迁移数　true transference number
振动配分函数　partition function of vibration
振动特征温度　characteristic temperature of vibration
蒸气压下降　depression of vapor pressure
正常沸点　normal boiling point
正吸附　positive absorption
支链反应　branched-chain reaction
直链反应　straight-chain reaction
指前因子　pre-exponential factor
质量作用定律　mass action law
制冷系数　coefficient of refrigeration
中和热　heat of neutralization
轴功　shaft work
转动配分函数　partition function of rotation
转化率　rate of convertion
转化温度　conversion temperature
状态　state
状态方程　state equation
状态分布　state distribution
状态函数　state function
准静态过程　quasi-static process
准一级反应　pseudo-first order reaction
自催化　autocatalysis
自由度　degree of freedom
自由度数　number of degrees of freedom
自由焓　free enthalpy
自由能　free energy
自由膨胀　free expansion
组分数　component number
最概然分布　most probable distribution
最概然速率　most probable speed
最佳反应温度　optimal reaction temperature

附录7 常用英汉物理学专业词汇表

A

a. c. generator 交流发电机
aberration 光行差,像差
absolute index of refraction 绝对折射率
absolute refractive index 绝对折射率
absolute temperature scale 绝对温标
absolute zero 绝对零度
absorber 吸收体
absorbing power 吸收能力,吸收本领
absorptance 吸收比
absorption 吸收
absorption coefficient 吸收系数
absorption line 吸收线
absorption spectrum 吸收光谱,吸收谱
accelerate 加速
acceleration 加速度
acceleration due to gravity 重力加速度
acceleration-time graph 加速度-时间关系线图
accelerator 加速器
acceptor 受主,受体
acceptor doping 受主掺杂
acceptor impurity 受主杂质
accommodation 调节,调焦
accumulator 蓄电池
accuracy 准确度
acetate strip 醋酸酯条片
achromatic 消色差的
achromatic aberration 消色差
achromatic condenser 消色差聚光透镜
achromatic light 消色差光,白光
acoustic blur 声响模糊
acoustic board 吸音板,吸声板
acoustic navigation 声响导航,声学导航
acoustic pressure 声压
acoustics 声学
act on 施于
action 作用量,作用,作用力
action of point 尖端作用
activation 激活,活化
activation energy 激活能,活化能
active nucleus 活性核,放射性核
activity 活性,(放射性)活度
adapter 接合器
adder 加法器
adhesion 附着力,附着,黏附
adhesive force 附着力
adiabatic 绝热的
adiabatic expansion 绝热膨胀
adiabatic process 绝热过程
aerial 天线
aerial induction 天线调谐电感
aerial network 天线网络
aerodynamic force 气动力
aerodynamics 空气动力学
aerofoil 机翼
agent 剂
air blower 吹风器
air column 空气柱,气柱
air cushion 气垫
air damping 空气阻尼
air film 气膜
air track 气垫导轨
air wedge 气楔
alignment 对准,校整
alpha decay α衰变
alpha particle α粒子
alpha-particle scattering analogue α粒子散射模拟
alternating current 交流电
alternating voltage 交变电压,交流电压
alternator 交流发电机
altimeter 高度表,测高仪
ammeter 安培计
amorphous 非结晶的,无定形的
ampere 安培,安
ampere-hour 安培小时,安时
amplification 放大,放大率
amplifier 放大器

amplify 放大
amplitude 振幅
amplitude modulation 调幅,振幅调制
amyl acetate 乙酸戊酯,醋酸戊酯
analogue 模拟
analogue experiment 模拟实验
analogue signal 模拟信号
analogue-to-digital conversion 模数转换
analyzer 检偏器
AND gate 与门
anemometer 风速计
aneroid barometer 无液气压计,空盒气压表
angle of contact 接触角
angle of deviation 偏向角
angle of diffraction 衍射角
angle of dip 倾角
angle of elevation 仰角
angle of emergence 出射角
angle of incidence 入射角
angle of inclination 倾角
angle of minimum deviation 最小偏向角
angle of projection 投射角
angle of reflection 反射角
angle of refraction 折射角
angle of twist 扭转角
angstrom 埃
angular acceleration 角加速度
angular aperture 孔径角
angular displacement 角位移
angular frequency 角频率
angular impulse 角冲量
angular magnification 角放大率
angular momentum 角动量,动量矩
angular motion 角向运动,角运动
angular speed 角速率
angular velocity 角速度
angular width 角宽度
annealing 退火
anode 阳极
anomalous expansion 反常膨胀
antenna 天线
anticlockwise moment 逆时针力矩
antifreeze agent 防冻剂
antinodal line 腹线
antinode 波腹
antiparticle 反粒子
antiphase 反相
aperture 孔径
apparent depth 视深
apparent expansion 表观膨胀,视膨胀
apparent frequency 表观频率,视频率
apparent loss in weight 表观失重
apparent weight 表观重量
Appleton layer 阿普尔顿层,F电离层
aqueous layer 水层
Archimedes principle 阿基米德原理
area 面积
argon 氩
armature 电枢
artificial disintegration 人工蜕变
artificial radioactivity 人工放射性
astable 非稳态的
astable circuit 非稳态电路
astable multivibrator 非稳态多谐振荡器
astigmatism 像散
astronomical telescope 天文望远镜
at infinity 在无穷远处
at rest 静止
atmosphere 大气,大气圈,大气压
atmospheric pressure (大)气压
atom 原子
atomic bomb 原子弹
atomic bond 原子键
atomic density 原子密度
atomic diameter 原子直径
atomic energy 原子能
atomic mass 原子质量
atomic mass unit 原子质量单位
atomic model 原子模型
atomic nucleus 原子核
atomic number 原子序数
atomic radius 原子半径
atomic separation 原子间距
atomic spacing 原子间距
atomic structure 原子结构
atomic theory 原子论

atomizer 原子化器
attenuation 衰减
attraction 吸引
attractive force 吸力
audible frequency range 听频范围
audible signal 可听信号
audio frequency 声频
autofocus 自动聚焦,自动调焦
avalanche 电子雪崩
average acceleration 平均加速度
average power 平均功率
average speed 平均速率
average velocity 平均速度
Avogadro constant 阿伏伽德罗常数/常量
Avogadro number 阿伏伽德罗数
Avogadro's law 阿伏伽德罗定律
axial 轴向的,沿轴的
axial field 轴向场
axial search coil 轴向探察线圈
axis 轴
axis of rotation 转轴

B

back e. m. f. 反电动势
background radiation 本底辐射
Bainbridge mass spectrometer 班布里奇质谱仪
balance 天平,秤,平衡
balance arm 平衡臂
balance point 平衡点
balanced bridge 平衡电桥
balanced force 平衡力
ball bearing 球轴承
ballistic galvanometer 冲击电流计,冲击检流计
ballistic pendulum 冲击摆
Balmer series 巴耳末系
band spectrum 带状谱
bandwidth 带宽
bar code 条码
bar magnet 磁棒,条形磁铁
bare wire 裸线
barium 钡
barometer 气压计
barrier 障碍物
barrier layer 阻挡层,势垒层
barrier potential difference 阻挡层电势差,阻挡层电位差
Barton's pendulum 巴尔通摆
base 基极
base current 基极电流
battery 电池(组)
battery charger 电池充电器
battery holder 电池座
beaker 烧杯
beam splitter 分束器,射束分离器
beat 拍
beat frequency 拍频
becquerel 贝可(勒耳)(放射性活度单位)
bel 贝尔,贝
bell jar 钟形罩
bench mat 实验台垫
Bernoulli theorem 伯努利定理
beryllium 铍
beta decay β衰变
beta particle β粒子
biasing circuit 偏压电路
biasing voltage 偏压
biconcave lens 双凹透镜
biconvex lens 双凸透镜
bifilar pendulum 双线摆
bifocal lens 双焦距透镜
big-bang model 大爆炸模型
bimetallic strip 双金属片
binary adder 二进加法器
binary system 二进制
binding 结合
binding energy 结合能
binoculars 双筒望远镜
Biot-Savart law 毕奥-萨伐尔定律
bipolar 双极的
birefraction 双折射
bistable 双稳态的
bistable circuit 双稳电路
bistable multivibrator 双稳态多谐振荡器
bit 比特(二进制信息单位)
black-body radiation 黑体辐射
block and tackle 滑轮组

block diagram 方框图
blocking capacitor 隔直流电容器
blooming 敷霜,表面加膜
Bohr atom 玻尔原子
Bohr radius 玻尔半径
Bohr theory 玻尔理论
boil 沸腾,煮沸
boiler 锅炉
boiling point 沸点
bolometer 辐射热计
Boltzmann constant 玻尔兹曼常数
bombardment 轰击
bond energy 键能
bonding 成键
Bourdon gauge 布尔东气压计
bow wave 头波,艏波
Boyle law 玻意尔定律
Brackett series 布喇开系,布喇开光谱
Bragg angle 布拉格角
Bragg diffraction 布拉格衍射
Bragg plane 布拉格平面
Bragg's law 布拉格定律
brake 制动器
breakdown potential 击穿电势,击穿电位
breaking point 断点
breaking strength 抗断强度
breaking stress 致断应力
breeder reactor 增殖反应堆
bremsstrahlung 轫致辐射
Brewster's law 布儒斯特定律
bridge circuit 桥式电路
bridge rectifier 桥式整流器
bright fringe 亮纹
brightness 亮度
brittle 脆的,易碎的
bromine 溴
Brownian movement 布朗运动
brush 电刷
bubble chamber 气泡室
bubble raft model 泡筏模型
buffer 缓冲器
buffer circuit 缓冲电路
bulk modulus 体积弹性模量
Bunsen burner 本生灯
buoyancy 浮力
burette 滴定管
burette stand 滴定管架
burglar alarm 防盗警报器,防盗警钟
buzzer 蜂鸣器
bypass 旁路
byte 字节

C

cable 电缆
cadmium sulphide 硫化镉
caesium 铯
calcite 方解石
calibrate 校准,标上刻度
caliper 测径器,卡钳
calorie 卡路里,卡
calorimeter 量热器,量能器
camera 照相机
candela 坎(德拉)
cantilever 悬臂
capacitance 电容
capacitance substitution box 换值电容箱
capacitive circuit 电容电路
capacitive component 电容性分量
capacitive coupling 电容耦合
capacitive discharge 电容性放电
capacitive reactance 容抗
capacitor 电容器
capacitor-input filter 电容输入滤波器
capacity 容量,电容
capillarity 毛细现象,毛细作用
capillary depression 毛细下降
capillary rise 毛细上升
capillary tube 毛细管
capsule 囊
capture 俘获
carbon 碳
carbon granule 碳颗粒
carbon paper disc 圆形碳纸
carbon-14 dating 碳 14 年代测定法
carburettor 化油器,汽化器
carrier 载波

carry 进位
castor oil 蓖麻油
cathode 阴极,负极
cathode ray 阴极射线
cathode ray deflection tube 阴极射线偏转管
cathode-ray oscilloscope 阴极射线示波器,示波器
cathode-ray tube 阴极射线管
cavity resonator 空腔共振器
celestial telescope 天体望远镜
cell 电池,细胞
Celsius temperature scale 摄氏温标
centimeter 厘米
centralized low-voltage power supply unit 集中式低压电源箱
center of curvature 曲率中心
center of gravity 重心
center of mass 质心
center of oscillation 振荡中心
center-tapped transformer 中心抽头变压器
centrifugal force 离心力
centrifuge 离心机
centripetal acceleration 向心加速度
centripetal force 向心力
Chadwick 查德威克
chain reaction 链式反应
change of state 物态变化
change-over switch 换向开关
characteristic 特性,特征
characteristic curve 特性曲线
characteristic line 特征线
characteristic spectrum 特征光谱,特征谱
charge 电荷,充电,起电
charge carrier 载荷子,载流子
charge conservation 电荷守恒
charge density 电荷密度
charge distribution 电荷分布
charge-to-mass ratio 荷质比
charged particle 带电粒子
charging by contact 接触起电
charging by friction 摩擦生电
charging by induction 感应起电
charging by sharing 授受起电
charging time constant 充电时间常量
Charles law 查理定律
choke 扼流(圈)
choke circuit 扼流电路
chromatic aberration 色差
chromatic dispersion 色散
cinefilm soundtrack 电影胶片声迹
circle of least confusion 最少模糊圆
circuit 电路
circuit board 电路板
circuit breaker 断路器
circuit symbol 电路符号
circular coil 环形线圈
circular motion 圆周运动
circular orbit 圆周轨道
circular pulse 圆形脉冲
circular wave 圆形波
circular wavefront 圆形波阵面,圆形波前
clamp 夹钳,夹紧
clap-echo method 拍掌-回声法
cleave 裂开
clinical thermometer 体温计
clip 夹子
clockwise moment 顺时针力矩
closed circuit 闭合电路
closed pipe 封闭管道
closed tube 连通管
closed-loop control system 闭合环路控制系统
closed-loop gain 闭合环路增益
closed-loop voltage gain 闭合环路电压增益
closely packed 密堆积的
cloud chamber 云室
cloud chamber track 云室径迹
coaxial 共轴的,同轴的
coaxial cable 同轴电缆,同轴线
cobalt 钴
code 编码,(代)码
coder 编码器
coefficient of dynamic friction 动摩擦系数
coefficient of friction 摩擦系数
coefficient of restitution 恢复系数
coefficient of static friction 静摩擦系数
coefficient of viscosity 黏性系数
coherent 相干的

cohesion 内聚力,内聚
cohesive force 内聚力
coil 线圈
collector 集电极,集电器,集极
collector current 集电极电流,集极电流
collimator 准直管,准直仪
collision 碰撞
color 颜色
color code 色码,色标
combinational logic 组合逻辑
common emitter 共发射极
common-mode voltage 共模电压
commutator 对易式换向器,交换子
compass 指北针,罗盘
complete circuit 完整电路
component 组件,分量,组分
component force 分力
compound microscope 复显微镜
compound pendulum 复摆
compressed gas 压缩气体
compressibility 可压缩性,压缩率
compressible fluid 可压缩性流体
compression 压缩
compression spring balance 压缩弹簧天平
concave 凹,凹面,凹面的
concave lens 凹透镜
concave mirror 凹镜,凹面镜
concentric capacitor 同心电容器
condensation 凝结,凝聚
condensation nucleus 凝结核,凝聚核
condensation point 凝点,凝结点
condenser 聚光器,电容器,冷凝器
conductance 电导
conductivity 电导率
conductor 导体
conical pendulum 锥摆
conjugate foci 共轭焦点
conservation of angular momentum 角动量守恒
conversation of charge 电荷守恒
conservation of energy 能量守恒
conservation of mechanical energy 机械能守恒
conservation of momentum 动量守恒
conservative force 保守力
conserved 守恒的
constant acceleration 恒(定)加速度
constant angular acceleration 恒(定)角加速度
constant angular velocity 恒(定)角速度
constant force 恒力
constant motion 恒速运动
constant pressure 定压,恒压
constant speed 恒速率
constant temperature 定温,恒温
constant velocity 恒速度
constant volume 定容
constant volume gas thermometer 定容气体温度计
constantan 康铜
constriction 压缩,收缩
constructive interference 相长干涉
contact angle 接触角
contact area 接触面积
contact force 接触力
continuity equation 连续性方程
continuous spectrum 连续光谱,连续谱
continuous wave 连续波
contract 收缩
control experiment 对照实验,比对实验
control grid 控制栅极
control rod 控制棒
control system 控制系统
convection 运流,对流
converge 会聚
convergent lens 会聚透镜
convergent mirror 会聚镜
convex 凸,凸面,凸面的
convex lens 凸透镜
convex mirror 凸镜,凸面镜
coolant 冷却剂
cooling by evaporation 蒸发制冷
cooling correction 冷却修正
cooling curve 冷却曲线
cooling effect 冷却效应
cooling system 冷却系统
coplanar forces 共面力
core 芯,核心
Coriolis force 科里奥利力

corkscrew rule 螺旋法则
corona discharge 电晕放电
corpuscular theory of light 光的微粒学说
correction 校正,修正
cosmic radiation 宇宙辐射
coulomb 库仑
coulomb's law 库仑定律
count rate 计数率
counter 计数器
counterweight 平衡锤,配重
couple 力偶
coupled oscillation 耦合振荡
coupling 耦合
covalent bond 共价键
crane magnet 起重磁铁
creep 蠕变
crest 波峰
critical angle 临界角
critical damping 临界阻尼
critical mass 临界质量
critical point 临界点
critical temperature 临界温度
critical velocity 临界速度
crocodile clip 鳄鱼夹
cross-hairs 十字丝,叉丝
cross-sectional area 截面积
Crova's disc 克罗瓦盘
crystal 晶体
crystal diffraction 晶体衍射,晶体绕射
crystal lattice 晶格
crystal planes 晶面
crystalline 结晶的
crystallization 结晶
crystallography 晶体学
cubical expansivity 体积膨胀系数,体积膨胀率
curie 居里
curie point 居里点
current 电流
current amplification 电流放大
current amplification factor 电流放大因素
current amplifier 电流放大器
current balance 电流天平
current density 电流密度
current gain 电流增益
current intensity 电流强度
current pulse 电流脉冲
current sensitivity 电流灵敏度
current stabilizer 稳流器
current transfer characteristic 电流转移特性
current-carrying conductor 载(电)流导体
current-voltage characteristic curve 电流-电压特性曲线
curved mirror 曲面镜
cyclotron 回旋加速器
cylindrical concave lens 柱面凹透镜
cylindrical concave mirror 柱面凹镜
cylindrical convex lens 柱面凸透镜
cylindrical convex mirror 柱面凸镜
cylindrical lens 柱面透镜
cylindrical mirror 柱面镜

D

D. C. generator 直流发电机
D. C. motor 直流电动机
Dalton law of partial pressures 道尔顿分压定律
damped harmonic motion 阻尼谐动
damped harmonic oscillation 阻尼谐振
damped oscillation 阻尼振荡
damping 阻尼
damping force 阻尼力
dark fringe 暗纹
data 数据
dating 年代测定
daughter nucleus 子核
daughter nuclide 子核素
de Broglie relation 德布罗意关系
de Broglie wave 德布罗意波
dead time 失效时间
decade capacitance box 十进电容箱
decade resistance box 十进电阻箱
decay 衰变
decay analogue 衰变模拟
decay constant 衰变常量
decay curve 衰变曲线
decay law 衰变定律
decay product 衰变产物

decelerate 减速
deceleration 减速度
decibel 分贝
decode 译码,解码
decoder 译码器,解码器
deflection 偏转
deflecting plate 偏转板
deflection system 偏转系统
deflection tube 偏转管
deformation 形变
degradation of energy 能量退降
degree 度
degree Celsius 摄氏度
degree of freedom 自由度
demagnetization 退磁
demonstration diode 演示用二极管
demonstration meter 演示电表
density 密度
depletion layer 耗尽层
depth of field 景深
derived quantity 导出量
derived unit 导出单位
destructive interference 相消干涉
detection 探测,检测
detector 探测器
detector of radiation 辐射探测器
deuterium 氘
deuteron 氘核
deviation 偏差
device 器件,设备
diamagnetism 抗磁性
diaphragm 光阑,膜片
diatomic molecule 双原子分子
dielectric 介电体,电介质,介电的
dielectric constant 介电常量
dielectric polarization 电介极化
dielectric strength 电介强度
differential equation 微分方程
differential input voltage 差动输入电压
diffracted ray 衍射线,绕射线
diffraction 衍射,绕射
diffraction grating 衍射光栅,绕射光栅
diffuse reflection 漫反射
diffusion 扩散,漫射
diffusion cloud chamber 扩散云室
digital 数字的
digital display 数字显示
digital signal 数字信号
digital-to-analogue conversion 数模转换
dimension 量纲,因次,维,大小
dimensional analysis 量纲分析,因次分析
diminish 缩小
diode 二极管
diode probe 二极管微波探测器
diopter 屈光度,焦度
dipole 偶极子
direct current 直流(电)
direction 方向,方位
direction of propagation 传播方向
directly heated cathode 直热式阴极
discharge 放电
disintegration 蜕变
dislocation 位错
dispersion 色散
dispersive power 色散能力,色散本领,色散率
displacement 位移,排量
displacement can 排水罐
displacement-time graph 位移-时间图
dissipative force 耗散力
distance 距离
distance-time graph 距离-时间图
distortion 失真,扭曲,畸变
disturbance 扰动
diverge 发散
diverging lens 发散透镜
diverging mirror 发散镜
division of amplitude 振幅分割
division of wavefront 波阵面分割,波前分割
domain 畴,域
domestic circuit 家居电路
donor 施主,给体
doped semiconductor 掺杂半导体
doping 掺杂
Doppler broadening 多普勒(谱线)增宽
Doppler effect 多普勒效应
Doppler shift 多普勒频移

dosage 剂量
dose 剂量
dose rate 剂量率
double pulley 双滑轮
double refraction 双折射
double slit 双缝
double-pole double-throw switch 双刀双掷开关
doublet 双重线
drag force 曳力
drift velocity 漂移速度
driver circuit 驱动电路
driver frequency 驱动频率
driving cell 驱动电池
driving force 驱动力
driving mirror 行车后视镜
dry battery 干电池组
dry cell 干电池
dual-trace oscilloscope 双迹示波器
ductile 延性的
Dulong-Petit law 杜隆-珀蒂定律
dynamic equilibrium 动态平衡
dynamic friction 动摩擦
dynamic resistance 动态电阻
dynamics 动力学
dynamo 发电机

E

earth 接地,地线
earthquake wave 地震波
ebonite 硬质胶
ebonite rod 硬质胶棒
echo 回声,回波
eddy current 涡电流,涡流
eddy-current damping 涡流阻尼
eddy-current loss 涡流损耗
effective mass 有效质量
efficiency 效率
effort 施力
Einstein mass-energy relation 爱因斯坦质能关系
elastic collision 弹性碰撞
elastic constant 弹性常数,弹性常量
elastic deformation 弹性形变
elastic hysteresis 弹性滞后
elastic limit 弹性极限
elastic strain energy 弹性应变能
elasticity 弹性
electric bell 电铃
electric charge 电荷
electric current 电流
electric field 电场
electric field intensity 电场强度
electric field line 电场线
electric field pattern 电场图形
electric furnace 电炉
electric line of force 电力线
electric motor 电动机
electric potential 电势,电位
electric power 电功率
electric shock 电振,电击
electrical appliance 电器
electrical conductivity 导电率
electrical energy 电能
electrical oscillation 电振荡
electrical potential energy 电势能,电位能
electrical resonance 电共振
electricity 电,电学
electrode 电极
electrolysis 电解
electrolyte 电解质
electrolytic 电解质的,电解的
electrolytic capacitor 电解质电容器
electromagnet 电磁铁,电磁体
electromagnetic coil 电磁感应圈
electromagnetic contact 电磁触点
electromagnetic force 电磁力
electromagnetic induction 电磁感应
electromagnetic moment 电磁矩
electromagnetic oscillation 电磁振荡
electromagnetic radiation 电磁辐射
electromagnetic spectrum 电磁(频)谱
electromagnetic wave 电磁波
electromagnetism 电磁学
electrometer 静电计
electromotive force 电动势
electron 电子
electron beam 电子束

electron cloud 电子云
electron diffraction 电子衍射,电子绕射
electron drift 电子漂移
electron energy level 电子能级
electron gun 电子枪
electron transition 电子跃迁
electron tube 电子管
electron-volt 电子伏特,电子伏
electronics 电子学
electrophorus 起电盘
electroscope 验电器
electrostatic bond 静电键
electrostatic field apparatus 静电场仪器
electrostatic precipitation 静电沉降
electrostatics 静电学
element 元素,元件
elliptic orbit 椭圆轨道
elongation 伸长度
emergent ray 出射线
emission 发射
emission line 发射(谱)线
emission spectrum 发射光谱,发射谱
emissivity 发射率
emit 发射
emitter 发射极,射极
emitter current 发射极电流,射极电流
enamelled wire 漆包线
encode 编码
encoder 编码器
end-correction 端部修正,管口校正量
end-on collision 同向碰撞,尾追碰撞
energetic particle 高能粒子
energy 能量,能
energy band 能带
energy conversion 能量转换
energy conversion efficiency 能量转换效率
energy input 能量输入
energy level 能级
energy output 能量输出
energy transfer 能量传递
enriched uranium 浓缩铀
equation of continuity 连续性方程
equation of state 物态方程
equation of uniformly accelerated motion 匀加速运动方程
equatorial orbit 赤道轨道
equilibrium 平衡
equilibrium condition 平衡条件
equilibrium spacing 平衡间距
equipartition of energy 能量均分
equipotential 等势的,等位的
equipotential line 等势线,等位线
equipotential surface 等势面,等位面
equivalent capacitance 等效电容
equivalent inductance 等效电感
equivalent resistance 等效电阻
erect 正立,直立的,竖立的
escape velocity 逃逸速度
ether 以太
evacuation 抽成真空
evaporation 蒸发
excess pressure 逾电压
excitation 激发
excitation energy 激发能
excitation potential 激发电势,激发电位
excitation voltage 激励电压,励磁电压
excited 受激的
excited energy level 受激能级
excited state 激发态
expansion 膨胀
expansivity 膨胀率
exponential change 指数式改变
exponential decay 指数式衰变
exposure 曝光(量),照射量
extension 伸长
external force 外力
external work 外功
extrahigh tension 超高压
extranuclear structure 核外结构
extraordinary ray 非常光线
extrapolation 外推,外推法
extrinsic semiconductor 外赋半导体,非本征半导体
eye lens 接目镜
eyepiece 目镜

F

f-number　f 数，光圈数
f-stop　f 制光圈
far point　远点
farad　法拉
Faraday constant　法拉第常数，法拉第常量
Faraday law of electromagnetic induction　法拉第电磁感应定律
fast breeder reactor　快中子增殖反应堆
feedback　反馈
feedback amplifier　反馈放大器
feedback resistance　反馈电阻
ferromagnetic substance　铁磁性物质
ferromagnetism　铁磁性
fictitious force　虚假力
fidelity　保真性，保真度
field coil　场线圈
field-effect transistor　场效应晶体管
field intensity　场强
field line　场线
field magnet　场磁铁，场磁体
field of view　视场，视野
field strength　场强
figure of merit　优值，灵敏值，品质因数
filament　灯丝，丝极
filter capacitor　滤波电容器
filter circuit　滤波电路
filter pump　滤泵
final state　终态，末态
final velocity　末速度
fine-adjustment　微调，细调
fire alarm　火警警报器，火警钟
first law of thermodynamics　热力学第一定律
first-order spectrum　第一级光谱，第一级谱
fission　裂变
fission reactor　裂变反应堆
fixed point　定点
fixed pulley　定滑轮
fixed resistor　定值电阻器
flat coil　扁平线圈
flat solenoid　扁平螺线管
flat-bottomed flask　平底烧瓶
Fleming's left hand rule　弗林明左手定则
Fleming's right hand rule　弗林明右手定则
floating body　浮体
fluid　流体
fluid dynamics　流体动力学
fluorescence　荧光
fluorescent screen　荧光屏，荧光幕
flux　通量
flux density　通量密度
flyback　回扫
flywheel　飞轮
focal length　焦距
focal plane　焦面
focus　焦点，聚焦，调焦
focus control　聚焦控制
follower circuit　跟随电路
foot pump　脚踏泵
force　力
force constant　力常数，力常量
force polygon　力多边形
force resolution　力的分解
force triangle　力三角形
force-extension curve　施力-伸长曲线
forced oscillation　受迫振荡
former　框，线圈架，绕线模
forward bias　正向偏压
forward current　正向电流
fossil fuel　化石燃料
Foucault's rotating mirror method　傅科旋转镜法
frame of reference　参考坐标系，参考系
Franck-Hertz experiment　弗兰克-赫兹实验
Fraunhofer diffraction　夫琅禾费衍射，夫琅禾费绕射
Fraunhofer lines　夫琅禾费谱线
free electron　自由电子
free fall　自由下坠，自由下落
free falling body　自由落体
free oscillation　自由振荡
free path　自由程
free space　自由空间
freeze　凝固
freezing point　凝点
freon　氟利昂，二氯二氟甲烷

frequency 频率
frequency modulation 调频,频率调制
frequency response 频率响应
Fresnel biprism 菲涅耳双棱镜
Fresnel diffraction 菲涅耳衍射,菲涅耳绕射
friction 摩擦,摩擦力
friction-compensated 补偿摩擦的
frictionless motion 无摩擦运动
fringe 条纹
fringe pattern 条纹图,干涉图
fuel rod 燃料棒
fulcrum 支点
full adder 全加器
full-scale deflection 满标偏转
full-wave rectification 全波整流
full-wave rectifier 全波整流器
fundamental frequency 基频
fundamental mode of vibration 基振动模
fundamental note 基音
fundamental quantity 基本量
fundamental unit 基本单位
fuse 熔断器
fuse rating 熔断器额定值
fusion 熔化,聚变
fusion point 熔点
fusion reactor 聚变反应堆

G

G-clamp G-形钳
gain 增益
gain control 增益控制
galaxy 星系
Galileo telescope 伽利略望远镜
Galileo's thought experiment 伽利略假想实验
galvanometer 电流计,检流计
gamma radiation 伽马辐射
gamma ray 伽马射线
gap 隙
gas 气,气体
gas pressure 气体压强,气压
gaseous phase 气相
gaseous state 气态
gauge 规,计
Gauss theorem 高斯定理
Geiger counter 盖革计数器
Geiger-Marsden scattering experiment 盖革-马斯登散射实验
Geiger-Muller counter 盖革-米勒计数器
Geiger-Muller tube 盖革-米勒管
general gas equation 普适气体方程
general gas law 普适气体定律
generator 发电机
genetic effect 遗传效应
geometrical optics 几何光学
germanium 锗
ghost effect 叠影效应
glancing angle 掠射角
glass fiber 玻璃纤维
glycerine 甘油
gold foil 金箔
goldleaf electroscope 金箔验电器
gradient 梯度
graduated cylinder 量筒
Graham's law of effusion 格雷姆渗出定律
grain 粒,晶粒
gramophone record 唱片,唱碟
graph 图,线图,图表
graphical method 图解法
grating 光栅
grating spectrometer 光栅光谱仪,光栅分光计
gravitation 引力
gravitational constant 引力常量
gravitational field 引力场
gravitational mass 引力质量
gravitational potential 引力势
gravitational potential difference 引力势差
gravitational potential energy 引力势能
gravity 重力
grazing incidence 掠射
greenhouse effect 温室效应
grid 栅极
grid control 栅极控制
grid system 高压电网
groove 纹道,针槽,开槽
ground 接地
ground state 基态

guinea and feather experiment 硬币与羽毛实验

H

hair spring 游丝
half adder 半加器
half-life 半衰期
half-silvered mirror 半镀银镜
half-wave rectification 半波整流
half-wave rectifier 半波整流器
Hall effect 霍尔效应
Hall probe 霍尔探测器
Hall voltage 霍尔电压
hand stroboscope 手动式频闪仪
harmonic 谐音
harmonic motion 谐运动
harmonic oscillation 谐振荡
head-on collision 对头碰撞
heat 热,热量,加热,热学
heat absorbent 吸热剂
heat absorber 吸热器
heat absorption 吸热,热吸收
heat capacity 热容(量)
heat conduction 热传导
heat exchange 热交换,换热
heat flow 热流
heat gain 热增益,得热
heat insulation 热绝缘,隔热
heat loss 热损失,失热
heat-proof 耐热的,隔热的
heat pump 热泵
heat radiation 热辐射
heat reservoir 热库
heat sink 热壑,散热器
heat transfer 热传递,热转移
heater 热子,加热器
heat effect 热效应
heating element 加热元件
Helmholtz coils 亥姆霍兹线圈
henry 亨利,亨
hertz 赫兹,赫
high-dispersion prism 高色散棱镜
high tension 高电压
hole 空穴,孔
hollow plastic lens 空心塑料透镜
hollow plastic prism 空心塑料棱镜
hologram 全息图
holography 全息术,全息学
Hooke law 胡克定律
horizontal 水平的
horizontal component 水平分量
horizontal deflection 水平偏转
horseshoe magnet 蹄形磁铁
hot cathode 热阴极
hour 小时
Huygens principle 惠更斯原理
hydraulic press 水压机
hydroelectric power 水力发电
hydrogen bomb 氢弹
hygrometer 湿度计
hyperbolic orbit 双曲线轨道
hypothesis 假说,假设
hysteresis 滞后(效应)

I

ideal gas 理想气体
ideal gas equation 理想气体方程
ideal gas temperature scale 理想气体温标
ideal machine 理想机械
illuminate 照明,照亮
image 像,映像
image distance 像距
immerse 浸没
immersion heater 浸没式加热器
impact 碰撞
impedance 阻抗
impulse 冲量
impurity 杂质
in focus 焦点对准
in parallel 并联
in phase 同相
in series 串联
incident ray 入射线
incident wavefront 入射波阵面,入射波前
inclined plane 斜面
incoherent 非相干的
incompressible fluid 非压缩性流体

indicator 指示器
indirectly heated cathode 旁热式阴极
induced charge 感生电荷
induced current 感生电流
induced e. m. f. 感生电动势
induced voltage 感生电压
inducing charge 施感电荷
inducing current 施感电流
inductance 电感
inductance-capacitance-coupled circuit 感容耦合电路
inductance coil 电感线圈
induction 感应
induction heating 感应加热
inductive circuit 感性电路
inductive component 电感性分量
inductive reactance 感抗
inductor 电感器,感应器
inelastic collision 非弹性碰撞
inert gas 惰性气体
inertia 惯性
inertia balance 惯性秤
inertial frame 惯性坐标系,惯性系
inertial mass 惯性质量
infrared detector 红外探测器
infrared radiation 红外辐射
infrared ray 红外线
initial state 初态
initial velocity 初速(度)
input 输入
input bias current 输入偏压电流
input characteristic 输入特性
input current 输入电流
input offset current 输入补偿电流
input power 输入功率
input resistance 输入电阻
input voltage 输入电压
input-output voltage characteristic 输入-输出电压特性
instantaneous acceleration 瞬时加速度
instantaneous angular velocity 瞬时角速度
instantaneous current 瞬时电流
instantaneous power 瞬时功率
instantaneous speed 瞬时速率
instantaneous velocity 瞬时速度
instantaneous voltage 瞬时电压
insulation 绝缘
insulator 绝缘体
integrated circuit 集成电路
intensity 强度
intensity control 强度控制
intensity of current 电流强度
interaction (相)互作用
interatomic force 原子间力
interatomic potential 原子间势
interatomic separation 原子间距
intercept 截距,截段
interconversion 互换
interference 干涉
interference pattern 干涉图样
internal energy 内能
internal force 内力
internal resistance 内电阻
internal work 内功
international system of units 国际单位制,公制
interval 间隔
intrinsic semiconductor 本征半导体
inverse-square law 平方反比定律
inverted 倒的,反(向)的
inverter 反相器,逆变器
inverting input 倒相输入
ion 离子
ion-pair 离子偶,离子对
ionic bond 离子键
ionic structure 离子结构
ionization 电离
ionization chamber 电离室
ionization current 电离电流
ionization energy 电离能
ionization potential 电离电势,电离电位
ionization voltage 电离电压
ionize 电离
ionized atom 离子
ionized layer 电离层
ionizing power 电离能力,电离本领
ionizing radiation 电离辐射
ionosphere 电离层

iris 可变光阑,虹膜
iron core 铁芯
iron filings 铁粉
irreversible process 不可逆过程
isobar 等压线,同量异位素
isobaric expansion 等压膨胀
isobaric process 等压过程
isochronous oscillation 等时振荡
isotherm 等温线
isothermal process 等温过程
isotope 同位素
isovolumetric process 等容过程

J

Jaegers method 耶格法
jet propulsion 喷气推进
jockey 滑动触头
joule 焦耳
joulemeter 焦耳计
junction 结,接头
junction diode 结型二极管
junction transistor 结型晶体管

K

kaleidoscope 万花筒
kelvin 开尔文,开
kelvin temperature scale 开氏温标
Kepler law 开普勒定律
key 电键
kilogram 千克
kilowatt 千瓦特,千瓦
kilowatt-hour 千瓦小时,千瓦时
kilowatt-hour meter 电表,千瓦时计
kinematics 运动学
kinetic energy 动能
kinetic friction 动摩擦
kinetic theory (分子)动理论
kinetic theory model (分子)动理论模型
kinetic theory of gases 气体分子运动论
Kirchhoff's law 基尔霍夫定律
kit 套件,成套工具
knife-edge 刃形支承,刀刃,刀口
kryton 氪
Kundt tube 孔脱管

L

lagging 隔热层保温套
laminar flow 层流
laminated 分层的,叠片的
laser 激光,激光器
laser beam 激光束
laser material 激光材料
latch 闩锁
latent heat 潜热
lateral 横向的,侧向的,旁向的
lateral inversion 横向倒置
lateral magnification 横向放大,横向放大率
lateral search coil 横向探察线圈
lattice 点阵,晶格
lattice spacing 点阵间隔,点阵间距
law 定律
law of conservation of momentum 动量守恒定律
law of reflection 反射定律
law of refraction 折射定律
lead 超前,铅
lead-acid accumulator 铅酸蓄电池
leakage current 漏泄电流
least distance of distinct vision 最小明视距离
Leclanche cell 勒克朗谢电池
length 长度
lens 透镜,晶状体
lens formula 透镜公式
lens holder 透镜座
lens marker's formula 透镜制造者公式
Lenz law 楞次定律
lever 杠杆
light 光,光学
light beam 光束,光柱
light-dependent resistor 光敏电阻器
light-emitting diode 发光二极管
light guide 光导
light pipe (导)光管
light ray 光线
light-sensitive resistor 光敏电阻器
light source 光源
lightning 闪电

lightning conductor　避雷导线，避雷装置
lightning rod　避雷针
limiting angle　极限角
limiting friction　极限摩擦
line of action　作用线
line of force　力线
line spectrum　线状光谱,线状谱
linear air track　线性气垫导轨
linear electronics　线性电子学
linear expansivity　线性胀率
linear flow　线流
linear momentum　线动量
linear voltage amplification　线性电压放大率
linearly polarized wave　线偏振波
liquefaction　液化
liquid　液体
liquid crystal　液晶
liquid phase　液相
liquid pressure　液体压强
liquid-in-glass thermometer　玻璃管液体温度计
Lissajous figure　李萨如图形
live　载电,活线,火线
Lloyd mirror　劳埃德镜
load　负载,负荷
lodestone　磁石
logic gate 逻辑门
logic level　逻辑电平,逻辑级
logic-level indicator　逻辑电平指示器,逻辑级指示器
logic value　逻辑值
long sight　远视
long wave　长波
longitudinal magnification　纵向放大,纵向放大率
longitudinal wave　纵波
loop　回路,圈
Lorentz force　洛伦兹力
Lorentz rotating disc　洛伦兹旋转盘
loudness　响度
loudspeaker　扬声器
low-frequency A. C. generator　低频交流发电机
low voltage　低电压
low-voltage immersion heater　低压浸没式电热器
lower limit　下限
lycopodium powder　石松粉
Lyman series　莱曼系

M

Mach number　马赫数
machine　机械
macroscopic　宏观的
Magnadur magnet　玛格纳多尔磁铁
magnet　磁铁,磁体
magnet keeper　磁铁衔铁
magnetic domain　磁畴
magnetic effect　磁效应
magnetic field　磁场
magnetic field board　磁场板
magnetic field intensity　磁场强度
magnetic field strength　磁场强度
magnetic flux　磁通量
magnetic flux density　磁通密度
magnetic flux linkage　磁链
magnetic force　磁力
magnetic hysteresis　磁滞
magnetic induction　磁感应强度,磁感应
magnetic line of force　磁力线
magnetic material　磁性材料
magnetic meridian　磁子午线
magnetic moment　磁矩
magnetic north pole　磁北极
magnetic permeability　磁导率
magnetic pole　磁极
magnetic saturation　磁饱和
magnetic screen　磁屏
magnetic shielding　磁屏蔽
magnetic south pole　磁南极
magnetic susceptibility　磁化率
magnetic tape　磁带
magnetic tape recorder　磁带录音机
magnetic torque　磁矩
magnetic track　磁迹,磁道
magnetism　磁学,磁性
magnetization　磁化
magnetize　磁化
magnetizing current　磁化电流
magnification　放大,放大率

magnified 放大的
magnifying glass 放大镜
magnifying power 放大率
magnitude 量,量值,大小
mains frequency 市电频率
mains immersion heater 市电浸没式加热器
mains supply 市电电源
majority carriers 多数载流子
malleable 可锻的
Maltese cross tube 马尔塔十字管
manometer 流体压强计
mass 质量
mass defect 质量亏损
mass number 质量数
mass spectrometer 质谱仪
mass-energy relation 质能关系
matter wave 物质波
maximum error 最大误差
mean free path 平均自由程
measurement 测量
mechanical advantage 机械效益
mechanical efficiency 机械效率
mechanical energy 机械能
mechanical oscillation 机械振荡
mechanical wave 机械波
mechanism 机制,机理
medium 介质
medium wave 中波
Melde's experiment 迈尔德实验
melting 熔化
melting point 熔点
meniscus 弯液面,弯月面
meniscus lens 弯月形透镜
mercury 汞,水银
metal fatigue 金属疲劳
metal grid 金属栅
metallic bond 金属键
metastable 亚稳的,介稳的
method of dimensions 维数法,因次法
method of no-parallax 无视差法
meter 米
meter bridge 滑线电桥,米尺电桥
meter rule 米尺
mica 云母
mica capacitor 云母电容器
microammeter 微安培计,微安表
microelectronics 微电子学
micrometer 测微器,千分尺
micrometer caliper 螺旋测微器
microphone 传声器
microscope 显微镜
microscopic 微观的
microwave 微波
microwave apparatus 微波仪器
microwave receiver 微波接收器
microwave transmitter 微波发送器
milliammeter 毫安表,毫安计
Millikan experiment 密立根实验
millimeter 毫米
minimum deviation 最小偏向
minority carrier 少数载流子
minute 分,分钟
mirage 蜃景,海市蜃楼
mirror (反射)镜
mirror formula 球面镜公式
mobility 迁移率
mode 模(式)
model eye 眼球模型
model power line 输电线模型
moderator 慢化剂,减速剂
modulation 调制,调节
module 模块,模数
modulus of elasticity 弹性模量
modulus of rigidity 刚性模量
molar gas constant 摩尔气体常数,摩尔气体常量
molar heat capacity 摩尔热容
molar volume 摩尔体积
mole 摩(尔)
molecular bombardment 分子轰击
molecular force 分子力
molecular motion 分子运动
molecular polarization 分子极化
molecular separation 分子间距
molecular structure 分子结构
molecule 分子
moment 矩

moment arm　矩臂,力臂
moment of couple　力偶矩
moment of dipole　偶极矩
moment of force　力矩
moment of inertia　转动惯量
moment of momentum＝angular momentum　动量矩,角动量
momentum　动量
monatomic molecule　单原子分子
monochromatic light　单色光
motion　运动
motor　电动机
motor rule　电动机法则
movable pulley　动滑轮
moving-coil galvanometer　动圈式电流计,动圈式检流计
moving-coil loudspeaker　动圈式扬声器
moving-coil meter　动圈式电表
multiflash photography　多闪照相术
multimeter　万用电表,多用(电)表
multiple image　复像
multiple reflection　多次反射
multiplication process　倍增过程
multiplier　倍增器
multivibrator　多谐振荡器
musical quality　音品,音色
musical instrument　乐器
mutation　突变
multiple-slit interference　多缝干涉
mutual inductance　互感(系数)
mutual induction　互感(应)
mutually perpendicular　互相垂直的

N

n-type semiconductor　n 型半导体
NAND gate　与非门,非与门
naphthalene　萘(球),卫生球
natural frequency　固有频率
nature　本质
near point　近点
necking down　颈缩
negater　反相器,倒换器,非门
negative charge　负电荷
negative feedback　负反馈
negative ion　负离子
negative supply rail　负供电轨
negative terminal　负端钮,负极柱
net force　净力
network　网络
neutral　中性的,中线的
neutral equilibrium　中性平衡
neutral point　中性点
neutrino　中微子
neutron　中子
neutron number　中子数
newton　牛顿
Newton first law of motion　牛顿运动第一定律
Newton law of gravitation　牛顿万有引力定律
Newton ring　牛顿环
Newton second law of motion　牛顿运动第二定律
Newton third law of motion　牛顿运动第三定律
Newtonian fluid　牛顿流体
Newtonian mechanics　牛顿力学
nichrome wire　镍铬线,镍铬合金线
Nicol prism　尼科耳棱镜
nitrogen　氮
no-parallax　无视差
nodal line　节线
node　节点,结点
noise　噪声
noise level　噪声级
noise pollution　噪声污染
non-conservative force　非守恒力,非保守力
noninertial system　非惯性坐标系,非惯性系
non-inverting input　非反相输入
non-Newtonian fluid　非牛顿流体
non-ohmic conductor　非欧姆导体
non-ohmic resistor　非欧姆电阻器
NOR gate　或非门,非或门
normal　法线
normal incidence　正入射
normal reaction　法向反作用力
normal stress　法向应力
north pole　北极
NOT gate　非门
note　音,乐音,律音

nozzle 喷嘴
nuclear 核的
nuclear energy 核能
nuclear energy level 核能级
nuclear fission 核裂变
nuclear force 核力
nuclear fusion 核聚变
nuclear pile 核堆
nuclear radiation 核辐射
nuclear reaction 核反应
nuclear reactor 核反应堆
nuclear waste 核废料
nuclear weapon 核武器
nucleon 核子
nucleon number 核子数
nucleus (原子)核
nuclide 核素

O

object 物,物体
object distance 物距
objective 物镜
obstacle 障碍物
octave 八音度,倍频程
oersted 奥(斯特)
ohm 欧姆
Ohm's law 欧姆定律
ohmic conductor 欧姆导体
ohmic resistor 欧姆电阻器
ohmmeter 欧姆计
oil film 油膜
opaque 不透明的
open circuit 断开电路,开路
open pipe 开管
open tube 开管
open-loop control system 开环控制系统
open-loop voltage gain 开环电压增益
operating voltage 操作电压
operational amplifier 运算放大器
opposite phase 反相
optical center 光心
optical density 光学密度
optical fiber 光导纤维,光纤
optical flatness 光学平面度
optical illusion 光幻象,视错觉
optical instrument 光学仪器
optical path 光程
optical path difference 光程差
optical system 光学系统,光具组
optical thickness 光学厚度
optically anisotropic 光学各向异性的
optically denser medium 光密介质
optically isotropic 光学各向同性的
optically thinner medium 光疏介质
optics 光学
OR gate 或门
orbit 轨道
orbital electron 轨道电子
order of magnitude 数量级
ordinary ray 寻常光线
organic glass 有机玻璃
orientation 取向,定向
oscillation 振荡
oscillator 振荡器
oscillatory circuit 振荡电路
oscilloscope 示波器
Ostwald viscometer 奥斯特瓦尔德黏度计
out of focus 离焦
out of phase 异相(位)
output 输出
output characteristic 输出特性
output current 输出电流
output power 输出功率
output voltage 输出电压
overload 过载
overtone 泛音

P

p-type semiconductor p型半导体
pair production 偶生成
paraffin 石蜡
parallax 视差
parallel axis theorem 平行轴定理
parallel beam projector 平行光束投射器
parallel circuit 并联电路
parallel forces 平行力

parallel rays 平行光线
parallel resonance 并联共振
parallel resonance circuit 并联共振电路
parallel-plate capacitor 平行板电容器
parallelogram of forces 力平行四边形
paramagnetism 顺磁性
parameter 参量,参数
paraxial ray 傍轴光线
parent nucleus 母核
parent nuclide 母核素
parking orbit 驻留轨道
partial polarization 部分偏振
partial pressure 分压力
partially polarized wave 部分偏振波
particle 粒子
particle movement 粒子运动,质点运动
pascal 帕斯卡
Paschen series 帕邢系
path 路程
path difference 程差
peak value 峰值
peak voltage 峰值电压
peak-to-peak current 峰-峰电流
peak-to-peak voltage 峰-峰电压
pendulum 摆
penetrating power 贯穿本领,穿透本领
penetration depth 贯穿深度,穿透深度
percentage error 百分误差
perfectly elastic 完全弹性的
perfectly inelastic 完全非弹性的
perigee 近地点
perihelion 近日点
period 周期
periodic fading 周期性衰退
periodic motion 周期运动
periodic table 周期表
periscope 潜望镜
permanent magnet 永磁体
permeability 磁导率
permeability of vacuum 真空磁导率
permittivity 电容率,介电常数
permittivity of free space 真空电容率
perpendicular axis theorem 垂直轴定理,正交轴定理
perpendicular distance 垂直距离
Perrin tube 佩林管
phase 相,相位
phase angle 相角
phase constant 相位常量,初相位
phase difference 相差,相位差
phase lag 相位滞后
phase lead 相位超前
phasor 相(矢)量
phon 方(响度单位)
photo-timing gate 光控计时门
phototransistor 光电晶体管
photocell 光电池
photodiode 光电二极管
photoelasticity 光(测)弹性
photoelectric cell 光电池
photoelectric current 光电流
photoelectric effect 光电效应
photoelectric threshold 光电阈
photoelectron 光电子
photographic negative 照相底片
photographic plate 照相底板
photon 光子
photosensitive 光敏的
photosensitive surface 光敏面
physical optics 物理光学
physical property 物理性质
physical quantity 物理量
pick-up 拾音器,拾音
piston 活塞
pitch 螺距,倾斜度,音调
Pitot tube 皮托管
pivot 支点,支枢
Planck constant 普朗克常数,普朗克常量
plane mirror 平面镜
plane of polarization 偏振面
plane polarization 平面偏振
plane-polarized wave 面偏振波
plane wave 平面波
planetary motion 行星运动
plano-concave lens 平凹透镜
plano-convex lens 平凸透镜

plastic deformation 塑性形变
platinum-resistance thermometer 铂阻温度计
plug 插头,插塞
plutonium 钚
point action 尖端作用
point charge 点电荷
point mass 点质量
point of application of force 施力点
point of incidence 入射点
point source 点源
pointer 指针
Poiseuille's formula 泊肃叶公式
Poisson ratio 泊松比
polar 极(性)的
polarity 极性
polarization 偏振,极化
polarization grille 偏振栅
polarized light 偏振光
polarized wave 偏振波
polarizer 起偏器,起偏镜
polarizing angle 起偏振角,偏振角
polarizing filter 起偏振滤波器,偏振滤波器
polaroid 偏振片
pole 极(点)
polonium 钋
polyatomic molecule 多原子分子
polyethylene 聚乙烯
polygon of forces 力多边形
polystyrene bead 聚苯乙烯珠
polystyrene cup 聚苯乙烯杯
polythene strip 聚乙烯片
positive charge 正电荷
positive feedback 正反馈
positive ion 正离子
positive supply rail 正供电轨
positive terminal 正端钮,正接线柱
potential barrier 势垒,位垒
potential difference (电)势差,(电)位差
potential divider 分压器
potential energy 势能,位能
potential gradient 势梯度,位梯度
potentiometer 电势差计,电位差计
power 功率
power amplifier 功率放大器
power distribution 配电
power factor 功率因数
power generation 发电
power loss 功率损耗
power pack 电源箱
power rating 电功率额定值
power station 电力站,发电厂
power supply 电源,供电
power supply unit 供电设备,电源箱
power transistor 功率晶体管
power transmission 输电
pressure 压强,压力
pressure cooker 高压锅,加压蒸煮器
pressure gauge 压强表
pressure law 气压定律
pressurized water reactor 压水式反应堆
primary coil 原线圈
primary current 一次电流,原电流
primary voltage 一次电压,原电压
primary winding 初级绕组,原绕组
principal axis 主轴
principal focus 主焦点
principle 原理
principle of flotation 浮体原理
principle of moments 力矩原理
principle of superposition 叠加原理
prism 棱镜
prismatic periscope 棱镜潜望镜
probe 探头,探测器
progressive wave 前进波
projectile 抛体
projectile motion 抛体运动
projection 投影,投射
projector 投影仪,放映机
prong 音叉臂
proof plane 验电板
propagation 传播
propeller 推进器
property 性质
proportionality constant 比例常数,比例常量
protactinium 镤
protective screen 防护屏

proton 质子
proton number 质子数
pseudo-force 假力,赝力
pulley 滑轮
pulse 脉冲
pump 泵
pupil 光瞳,瞳孔
pure note 纯音
push-button switch 按钮开关
pyrometer 高温计

Q

Q-factor 品质因数
quality of note 音品,音质
quantity of electricity 电量
quantization 量子化
quantum 量子
quantum number 量子数
quarter-wave aerial 四分之一波天线
quartz 石英
quenching agent 猝灭剂
quiescent condition 宁静状态

R

radial acceleration 径向加速度
radial component 径向分量
radial field 辐向场,径向场
radiation 辐射
radiation hazard 辐射危害
radiation protection 辐射防护
radiator 散热器,辐射器,辐射体
radio frequency 射频,无线电广播频率
radio telescope 射电望远镜
radio wave 无线电波
radioactive 放射(性)的
radioactive decay 放射衰变
radioactive disintegration 放射性蜕变
radioactive fallout 放射性沉降物
radioactive isotope 放射性同位素
radioactive nucleus 放射性核
radioactive series 放射系
radioactive source 放射源
radioactive waste 放射性废物
radioactivity 放射性(活度)
radioisotope 放射性同位素
radiometer 辐射计
radionuclide 放射性核素
radiotherapy 放射治疗
radium 镭
radius of curvature 曲率半径
radius of gyration 回转半径,回旋半径
radon 氡
rail 路轨
random 无规的,随机的
random error 随机误差
random motion 无规运动,随机运动
random nature 无规性,随机性
random walk 无规行走,随机游动
range 范围,射程,量程
rarefaction 稀疏
rate of decay 衰变率
rate of disintegration 蜕变率
ratemeter 率表,率计
rating 额定值
ray 射线,光线
ray box 光线箱
ray diagram 光线图
Rayleigh criterion 瑞利判据
reactance 电抗
reaction 反作用,反(作用)力,反应
reactive component 无功分量
reactor 反应堆
reactor core 反应堆堆芯
reading error 读数误差
real depth 实深
real expansion 实膨胀,真膨胀
real focus 实焦点
real gas 真实气体
real image 实像
real object 实物
real-is-positive convention 实正虚负约定
recoil 反冲,回复
rectification 整流
rectifier 整流器
rectilinear motion 直线运动
rectilinear propagation 直线传播

red shift 红移
reed 舌簧
reed relay 舌簧继电器
reed switch 舌簧开关
reference frame 参考系
reference level 参考级,参考水平
reflectance 反射比
reflected ray 反射线
reflected wavefront 反射波阵面,反射波前
reflecting telescope 反射望远镜
reflection 反射,反映
reflection grating 反射光栅
refracted ray 折射线
refracted wavefront 折射波阵面,折射波前
refracting angle 折射角
refracting telescope 折射望远镜
refraction 折射
refractive index 折射率
refrigerator 冰箱
regular reflection 规则反射
relative density 相对密度
relative motion 相对运动
relative permeability 相对磁导率
relative permittivity 相对电容率
relative velocity 相对速度
relativity 相对论,相对性
relay 继电器
relay coil 继电器线圈
relay contact 继电器触点
rem 雷姆
remote control 遥控
repulsion 推斥,排斥
repulsive force 斥力
reservoir capacitor 存储电容器
reset switch 复位开关
resistance 电阻
resistance coil 电阻线圈
resistance substitution box 换值电阻箱
resistance thermometer 电阻温度计
resistive component 电阻分量,有功分量
resistivity 电阻率
resistor 电阻器
resolution 分解,分辨,分辨率
resolution of force 力的分解
resolution of vector 矢量的分解
resolving power 分辨本领
resonance 共振,共鸣
resonance tube 共鸣管
resonant frequency 共振频率
resonator 共振器
response time 响应时间
restoring couple 恢复力偶
restoring force 恢复力
restoring torque 恢复转矩
resultant 合量,合的
resultant displacement 合位移
resultant force 合力
resultant vector 合矢量
resultant velocity 合速度
retard 减速
retardation 减速度
retort stand 铁支架,铁架
reverberation 混响
reverberation time 混响时间
reverse bias 反向偏压
reverse current 反向电流
reversibility of light 光的可逆性
reversible process 可逆过程
reversing switch 换向开关
revolution 旋转,转数
Reynolds number 雷诺数
rheostat 变阻器
right-angled fork track 直角分叉径迹
right-hand grip rule 右手握拳定则
right-hand screw rule 右手螺旋定则
rigid body 刚体
rigidity 刚性,刚度,刚度系数
ring circuit 环形电路
ripple 纹波
ripple tank 水波槽,波动箱
rocket 火箭
rolling 滚动
rolling friction 滚动摩擦
root-mean-square value 方均根值
rotary thermometer 转动式温度计
rotary-type potentiometer 旋转电势差计,旋转电

位差计
rotating platform 旋转台
rotating vector 转动矢量
rotation 旋转,自转,转动
rotational energy 转动能
rotational motion 旋转运动
rotor 转子
rubber tubing 橡胶管
rule 定则,法则
ruling 刻度,划线
runway 跑道
Rutherford scattering 卢瑟福散射
Rutherford's atomic model 卢瑟福原子模型
Rydberg constant 里德伯常数,里德伯常量

S

safety device 安全装置
satellite 卫星
satellite communication 卫星通信
saturated vapour 饱和蒸气,饱和气
saturated vapour pressure 饱和蒸气压,饱和气压
saturation 饱和
saturation current 饱和电流
sawtooth voltage 锯齿波电压
scalar 标量
scalar product 标(量)积
scale 标度,比例尺,音阶
scaler 定标器
scan 扫描
scattering 散射
scattering analogue 散射模拟
scattering angle 散射角
scintillation 闪烁
screen 屏,幕
screened lead 屏蔽引线
screw 螺旋,螺钉
screwdriver 螺钉起子,螺丝刀
screw jack 螺旋起重器
sealed radioactive source 密封放射源
search coil 探察线圈
second 秒
second law of thermodynamics 热力学第二定律
second-order spectrum 第二级光谱,第二级谱
secondary coil 副线圈
secondary current 二次电流
secondary emission 次级发射
secondary voltage 二次电压
secondary winding 二次绕组,次级绕组
Seebeck effect 塞贝克效应
selective absorption 选择吸收
self-inductance 自感(系数)
self-induction 自感(应)
semicircular glass block 半圆玻璃块
semiconductor 半导体
semiconductor diode 半导体二极管
sensitivity 灵敏度
sensor 感应器,传感器
sequential logic 顺序逻辑
series circuit 串联电路
series resonance circuit 串联共振电路
series-wound motor 串绕电动机
set switch 设定开关
shaft (转)轴
shear force 切变应力
shear modulus 剪模量,剪切模量
shear strength 切变强度,剪切强度
shear stress 切应力,剪切应力
shearing force 剪切力
shock waves (冲)击波,激波
short circuit 短路
short sight 近视
shortwave 短波
shunt 分流器,分路
shunt-wound motor 并励电动机,并绕电动机
shutter 快门
sievert 希沃特
sign convention 符号约定
signal generator 信号发生器
significant figure 有效数字
silicon 硅
simple harmonic motion 简谐运动
simple pendulum 单摆
simulation 模拟
single-phase A.C. generator 单相交流发电机
single pulley 单滑轮
single slit 单缝

single trace oscilloscope　单迹示波器
single-pole double-throw switch　单刀双掷开关
sinusoidal wave　正弦波
slab-shaped magnet　平板形磁铁
slide projector　幻灯片放映机
slide-wire potentiometer　滑线电势差计,滑线电位差计
sliding contact　滑动接触
sliding friction　滑动摩擦
slinky spring　软弹簧
slip ring　集电环,滑环
slit　缝,狭缝
slope　斜率
slotted weight　有槽砝码
smoke cell　烟雾盒
smoothing　平波,平流
smoothing capacitor　平波电容器,平流电容器
smoothing choke　平波扼流圈,平流扼流圈
smoothing circuit　平波电路,平流电路
Snell law　斯涅耳定律
socket　插座
sodium lamp　钠灯
soft iron　软铁
soft iron core　软铁芯
solar cell　太阳能电池
solar energy　太阳能
soldering iron　烙铁
solenoid　螺线管
solid　固体
solid phase　固相
solid state　固态
solidification　凝固
solidifying point　凝固点
sonar　声呐
sonic boom　声爆,声震
sonometer　弦音计
sound　声音
sound intensity　声强(度)
sound intensity level　声强级
soundproofing　隔声,隔声法,消声
sound track　声迹,声道
sound wave　声波
south pole　(指)南极
spark　火花
spark counter　火花计数器
specific charge　荷质比,比电荷,比荷
specific heat (capacity)　比热(容)
specific heat (capacity) at constant pressure　定压比热(容)
specific heat (capacity) at constant volume　定容比热(容)
specific latent heat　比潜热
specific latent heat of fusion　熔解比潜热
specific latent heat of vaporization　汽化比潜热
spectacles　眼镜
spectral analysis　光谱分析
spectral line　(光)谱线
spectral order　光谱级
spectrometer　(光)谱仪,分光计
spectroscopy　光谱学,光谱术
spectrum　谱,光谱,波谱
spectrum tube　光谱管
specular reflection　镜(面)反射
speed　速率
speed of light　光速
speed of sound　声速,音速
speed-time graph　速率-时间图
spherical aberration　球(面像)差
spherical lens　球面透镜
spherical mirror　球面镜
spherical wave　球面波
spherically symmetric　球对称的
spin　自旋
spiral　螺线,螺旋形
spiral spring　螺旋弹簧
spirit level　气泡水平仪
spontaneous　自发的
spontaneous disintegration　自发蜕变
spontaneous emission　自发发射
spring　弹簧,簧片,发条
spring balance　弹簧秤
square wave　方波
squarer　方波产生器
stable equilibrium　稳定平衡
stable state　稳(定状)态
stackable plug　叠加式插头

standard atmospheric pressure 标准大气压
standard cell 标准电池
standard deviation 标准偏差,标准差
standard error 标准误差
standing wave 驻波
starter 起动器
starting resistance 起动电阻
starting voltage 起动电压
state 态
states of matter 物态
static equilibrium 静态平衡
static friction 静摩擦
static resistance 静态电阻
statics 静力学
stationary wave 定态波
stator 定子
steady flow 定常流(动)
steady state 定常态
steam 水蒸气,蒸汽
steam engine 蒸汽机
steam point 汽点
steam turbine 汽轮机
steel yoke 钢轭
step-down transformer 降压器
step-up transformer 升压器
stereophonic 立体声的
stiffness 刚度,劲度
stimulated absorption 受激吸收
stimulated emission 受激发射
Stokes' law 斯托克斯定律
stop watch 停表
stop clock 停钟
stopping potential 遏止电势,遏止电位
storage battery 蓄电池(组)
storage capacitor 存储电容器
stored energy 储能
straight pulse 直线脉冲
straight wave 直线波
straight wavefront 直线波阵面,直线波前
strain 应变
strain gauge 应变规
stray capacitance 杂散电容
streamline 流线
stress 应力
stress-strain curve 应力-应变曲线
stretch 伸长,拉紧
strobe frequency 闪频
stroboscope 频闪仪
stroboscopic photography 频闪照相术
strontium 锶
stylus 唱针
substation 分站,配电站,变电站
sublimation 升华
submerge 浸没
subsonic speed 亚声速,亚音速
summing amplifier 加法放大器
superconductivity 超导(电)性
superconductor 超导体
supercooling 过冷
superficial expansivity 面膨胀系数,面膨胀率
superposition 叠加
supersaturation 过饱和
supersonic speed 超声速,超音速
surface charge density 表面电荷密度
surface energy 表面能
surface tension 表面张力
susceptibility 磁化率
sweep 扫描,扫掠
sweep rate 扫描速率
switch 开关,电键
switch off 截断,关闭
switch on 接通,开启
symbol 符号
synchronization 同步
syringe 针筒,注射器
system 系统
systematic error 系统误差

T

tangential acceleration 切向加速度
tangential component 切向分量
tape chart 纸带图表
tapping key 簧片按键
telephone receiver 电话听筒
telescope 望远镜
telescope mount 望远镜座

television 电视
temperature 温度
temperature coefficient of resistance 电阻温度系数
temperature difference 温差
temperature gradient 温度梯度
temperature scale 温标
tempering 回火
temporary magnet 暂时磁铁,暂时磁体
tensile force 拉力,张力
tensile strain 拉(伸)应变
tensile strength 抗张强度
tensile stress 拉(伸)应力
tension 张力
terminal 端钮,接线柱
terminal speed 终极速率
terminal velocity 终极速度
terrestrial telescope 地面望远镜
tesla 特斯拉
test charge 检验电荷
test tube 试管
theory 理论
thermal capacity 热容量
thermal conductivity 热导率
thermal equilibrium 热平衡
thermal radiation 热辐射
thermal reactor 热中子反应堆
thermal transmission 传热,热传递
thermionic diode 热离子二极管
thermionic emission 热离子发射
thermistor 热敏电阻器
thermistor thermometer 热敏电阻温度计
thermo-electromotive force 温差电动势,热电动势
thermocouple 温差电偶,热电偶
thermocouple thermometer 温差电偶温度计,热电偶温度计
thermodynamic temperature 热力学温度
thermoelectric effect 温差电效应,热电效应
thermometer 温度计
thermometer bulb 温度计泡
thermometric property 测温性质
thermometry 计温学,测温法
thermonuclear reaction 热核反应
thermopile 温差电堆
thermostat 恒温器
thin film 薄膜
Thomson's atomic model 汤姆孙原子模型
thorium 钍
three-pin plug 三脚插头
threshold frequency 阈频(率)
threshold intensity 阈强度
threshold of hearing 听觉阈
threshold voltage 阈电压
thrust 推力
thyratron 闸流管
ticker tape 纸带
ticker-tape timer 纸带打点计时器
tidal energy 潮汐能
time 时间
time base 时基
time-base control 时基控制
time constant 时间常数
time interval 时距
time switch 定时开关,计时开关
timer 定时器,计时器
timer-scaler 计时-计数器
toroid 环状线圈
toroidal coil 环状线圈
toroidal rheostat 环形变阻器
torque 转矩
Torricelli's theorem 托里拆利定理
Torricellian vacuum 托里拆利真空
torsion 扭转
torsion constant 扭转常量
torsional force 扭力
torsional moment 扭矩
torsional oscillation 扭转型振荡
torsional pendulum 扭摆
torsional strength 抗扭强度
torsional stress 扭(转)应力
total internal reflection 全(内)反射
trace 迹,示踪
tracer 示踪物,示踪剂
track 径迹
train of waves 波列

trajectory 轨道,轨迹
transducer 换能器,传感器
transformation 变换
transformer 变压器
transient current 瞬态电流
transient state 暂态,瞬态
transient voltage 瞬态电压
transistor 晶体管
transition 跃迁,过渡
transition energy 跃迁能
translation 平移
translational energy 平动动能
translucent screen 半透明屏,半透明幕
transmission 透射,输电,传输
transmission cable 输电缆
transmission grating 透射光栅
transmittance 透射比
transmitted ray 透射光线
transmutation 嬗变,转变
transparent 透明的
transverse 横向的
transverse magnification 横向放大,横向放大率
transverse wave 横波
travelling microscope 移测显微镜,读数显微镜
travelling wave 行波
triangle of force 力三角形
triangular prism 三棱镜,三角棱镜
trigger 触发,触发器
trigger input 触发输入
triode 三极管
triple-beam balance 三杆式天平,三臂秤,三梁天平
triple point 三相点,三态点
tripod stand 三脚架
tritium 氚
trolley 小车
trough 波谷
truth table 真值表
tune 调谐
tungsten filament 钨丝
tuning fork 音叉
turbine 涡轮
turbulent flow 湍流
turn 匝,转,转动
turn off 断开,关闭
turn on 接通,开启
turn ratio 匝(数)比
turntable 转盘
twist 扭曲
two-way switch 双向开关

U

ultimate tensile stress 极限拉伸应力
ultrahigh frequency 超高频
ultraviolet radiation 紫外辐射
ultraviolet ray 紫外线
ultrasonic 超声,超音
ultrasonic wave 超声波,超音波
unbalanced force 不平衡力
unidirectional circuit 单向电路
unidirectional current 单向电流
uniform 均匀的
uniform acceleration 匀加速度
uniform electric field 均匀电场
uniform magnetic field 均匀磁场
uniform motion 匀速运动
uniform speed 匀速率
uniform velocity 匀速度
uniformly accelerated motion 匀加速运动
unit 单位,单元
unit mass 单位质量
unit temperature 单位温度
universal gas constant 普适气体常量
universal gravitational constant 万有引力常量
universal stand 普用支架
unpolarized light 非偏振光
unpolarized wave 非偏振波
unstable equilibrium 不稳定平衡
unsteady state 非稳态
unstretched length 原长
upper limit 上限
upthrust 浮力
uranium 铀
uranium ore 铀矿
uranyl nitrate 硝酸双氧铀

V

vacuum　真空
vacuum discharge　真空放电
vacuum pump　真空泵
vacuum tube　真空管
valence electron　价电子
valve　电子管,阀,活门
Van de Graaff generator　范德格拉夫起电机
van der Waals' force　范德瓦耳斯力
vaporization　汽化
vapour　蒸汽,汽
vapour density　蒸汽密度,汽密度
vapour pressure　蒸汽压,气压
variable gain　可变增益
variable motion　变速运动
variable resistor　变阻器,可变电阻
variable speed　可变速率
variable velocity　可变速度
variation　变化
vector　矢量
vector addition　矢量加法
vector product　矢积
velocity　速度
velocity ratio　速度比
velocity-time graph　速度-时间图
Venturi meter　文丘里流量计
vernier　游标
vernier microscope　游标显微镜
vertical　垂直的
vertical component　垂直分量
vertical deflection　垂直偏转
very high frequency　甚高频
vibration　振动
vibration generator　激振器
vibrator　振动器
virtual earth　虚接地
virtual focus　虚焦点
virtual image　虚像
virtual object　虚物,虚物体
viscosity　黏度,黏性
viscous　黏(性)的
viscous flow　黏性流(动),黏滞流
viscous force　黏性力,黏力
viscous resistance　黏性阻力
visibility　可见度
visible light　可见光
visible spectrum　可见光谱
visual angle　视角
volt　伏特,伏
voltage　电压
voltage amplification　电压放大(率)
voltage comparator　电压比较器
voltage divider　分压器
voltage drop　电压降
voltage gain　电压增益
voltage rating　电压额定值
voltage regulator　调压器
voltage stabilizer　稳压器
voltmeter　伏特计,电压表
volume　体积,容积
vortex　涡旋
vulcanization　硫化

W

water vapour　水蒸气,水汽
water wave　水波
watt　瓦特,瓦
watt-meter　瓦特计,功率表
wave　波
wave motion　波动
wave number　波数
wave pattern　波动图样
wave source　波源
wave speed　波速(率)
wave train　波列
wave velocity　波速(度)
wave-particle duality　波粒二象性
waveform　波形
wavefront　波阵面,波前
waveguide　波导
wavelength　波长
wavelet　子波
weber　韦伯
wedge　楔
weight　重量,砝码,权(重)

weightlessness 失重

Weston cadmium cell 韦斯顿镉电池

Wheatstone bridge 惠斯通电桥

wheel-and-axle 轮轴

white light 白光

wig-wag machine 摇摆机

Wilson cloud chamber 威尔逊云室

wind power 风力,风功率

wind tunnel 风洞

winding 绕组,绕

wire gauge 线规

wire gauze 铁丝网

wiring 接线,布线

work 功

work function 逸出功,功函数

X

X-ray X 射线

X-ray diffraction X 射线衍射

X-ray spectrum X 射线谱

X-shift 水平移

xenon 氙

xenon stroboscope 氙光频闪仪

XOR gate 异或门

Y

Y-shift 垂直移

yield point 屈服点

yield strength 屈服强度

Young modulus 杨氏模量

Young's double slit experiment 杨氏双缝实验

Young's interference fringes 杨氏干涉条纹

Z

Zener diode 齐纳二极管

zero correction 零点校正

zeroth law of thermodynamics 热力学第零定律

zeroth order 零级

参 考 文 献

[1] 仲海洋. University Physics[M]. 大连：大连海事大学出版社，2006.
[2] 董益坤，魏汝尧. 科技英语[M]. 济南：山东大学出版社，2001.
[3] 吴朔平. 科技英语速成读本[M]. 北京：新时代出版社，1982.
[4] 清华大学《英汉技术辞典》编写组. 英汉技术辞典[M]. 北京：国防工业出版社，1978.
[5] 科技英语学习[M]. 上海：上海交通大学出版社，1986.
[6] Serway & Jewett. Principles of Physics(物理学原理)[M]. Third Edition. 北京：清华大学出版社，2004.
[7] Richard P. Olenick，Tom M. Apostol & David L. Goodstein，Beyond the Mechanical Universe(影印版)[M]. 北京：北京大学出版社，2001.
[8] Hugh D. Young & Roger A. Freedman. 西尔斯物理学[M]. 北京：机械工业出版社，2003.
[9] http://www.ccp2009.tw/index.asp
[10] ESP 理论与实践. http://www.EnMajor.com.
[11] 叶云屏，邢清清，张剑. 科技英语读写：方法与实践[M]. 北京：高等教育出版社，2007.
[12] 魏汝尧. 致用科技英语[M]. 北京：国防工业出版社，2007.
[13] 马新英，孙学涛，林易. 科技英语教程[M]. 北京：国防工业出版社，2005.
[14] 闫文培. 实用科技英语翻译要义[M]. 北京：科学出版社，2008.